DIRECTIONS IN TROPICAL AGROFORESTRY RESEARCH

FORESTRY SCIENCES

Volume 53

The titles published in this series are listed at the end of this volume.

Directions in Tropical Agroforestry Research

Adapted from selected papers presented to a symposium on Tropical Agroforestry organized in connection with the annual meetings of the American Society of Agronomy, 5 November 1996, Indianapolis, Indiana, USA

Edited by

P. K. R. NAIR and C. R. LATT

University of Florida, Gainesville, Florida 32611, USA

In cooperation with CATIE (Centro Agronómico Tropical de Investigación y Enseñanza), Turrialba, Costa Rica

Reprinted from *Agroforestry Systems*, Volume 38, Nos. 1–3 (1997)

Springer-Science+Business Media, B.V.

Library of Congress Cataloging-in-Publication Data

A C.I.P. Catalogue record for this book is available from the Library of Congress.

ISBN 978-90-481-5025-0 ISBN 978-94-015-9008-2 (eBook)
DOI 10.1007/978-94-015-9008-2

Printed on acid-free paper

Originally published by Kluwer Academic Publishers in 1998
Softcover reprint of the hardcover 1st edition 1998

Table of Contents

Agroforestry Systems **38**: 1–2, 1998.

Preface

In any long-term activity, periodic review and evaluation of progress and setbacks is an essential part of program development and implementation. If the establishment of ICRAF in 1977 is taken as the starting point of organized research in agroforestry, we have just crossed its 20-year mark. It is appropriate at this juncture that the agroforestry research community takes stock of its accomplishments and failures. With this objective, a one-day symposium on tropical agroforestry research was organized on the occasion of the annual meetings of the tri-societies of American Society of Agronomy, Crop Science Society of America, and Soil Science Society of America, at Indianapolis, Indiana, USA, on 5 November 1996. Acknowledged world leaders in various aspects of tropical agroforestry were invited to present state-of-the-art reviews of the developments in their respective research areas, and subsequently were requested to submit the texts of their presentations for publication. This volume represents 10 of the 13 papers; all have been peer-reviewed, updated and/or otherwise improved, and edited for style and consistency.

In accordance with the objective of the symposium, each paper synthesizes the results of research, summarizes the current state of knowledge, identifies knowledge gaps, and outlines the directions that research should take in the years ahead in the respective subject areas. Not all aspects of past or current research in agroforestry are covered in the publication in deserving detail, nor are all major geographical/ecological regions of the tropics represented. In spite of these drawbacks, we believe that the volume makes a significant contribution to the growing body of knowledge in tropical agroforestry, and it complements well the several noteworthy books and publications with which agroforestry literature has lately been blessed.

While organizing the symposium and putting together this publication, it became quite clear to us that agroforestry research has been somewhat skewed in favor of biophysical investigations. The apparent "neglect" of some important areas, as discussed in Chapter 10, is a serious deficiency in the context of overall thrust and developments in agroforestry research today. We hope that the research- and donor communities will accept the challenge of addressing this glaring deficiency in the near future.

The generous financial support of CATIE (Centro Agronomico Tropical de Investigacion y Enseñanza – The Tropical Agricultural Center for Research and Education), Turrialba, Costa Rica, in this endeavor was very timely and appropriate. Without it, it would not have been possible to bring out this publication so expeditiously. We express our sincere thanks to Dr Rubén Guevara Moncada, Director-General of CATIE and the staff of its Agroforestry Division, notably Dr John Beer, Head, for their splendid support. Our special for improvement of traditional, as well as evolving, systems. Research on interactions in AF systems, which was of a relatively low-key nature until the early 1990s, is now receiving increasing attention. Anderson and Sinclair (1993) and Nair et al. (1994) examined the topic from an ecological perspective, whereas various chapters in the recent book on tree–crop interactions (Ong and Huxley, 1996) approached it from a more quantitative and physiological perspective. Additionally, other recent reviews have examined specific aspects of interactions, e.g., pest and disease relations in agroforestry (Singh Rathore, 1995). Interactions in shade-tree/perennial-crop systems and mixed-tree systems have been discussed separately in this volume by Beer et al. (1997) and Khanna (1997), respectively. Thus, considerable progress has been made lately in our understanding of biophysical interactions in AF systems. But, so far, the efforts have been confined mostly to strategic research, with little or no progress in translating this knowledge into management options. The emphasis in this paper, therefore, is to synthesize the available information and discuss the agronomic significance of biophysical interactions in the major agroforestry systems, focusing on systems not covered by other reviewers in this volume. By following this approach, the paper complements the efforts of Sanchez (1995) and stresses the importance of management implications.

In the study of biophysical interactions in AF, some efforts were made to extrapolate from research on annual intercropping, on the assumption that the principles of interactions between component species for growth resources are common in all types of plant associations. Thus, early research (during the 1980s) on resource 'capture' in agroforestry systems concentrated on the interception and use of radiation, although radiation is seldom the most limiting growth resource in the tropics (Ong and Black, 1994). Agroforestry systems are much more complex than annual intercropping systems because there are major differences in the nature and arrangement of components and, therefore, in the nature and extent of interactions. Of considerable importance is the fact that the components in AF systems are 'unequal,' the trees being dominant and perennial. Thus, interactions in AF systems are continuous, rather than seasonal as in annual systems, and the course and extent of interactions are largely determined by the system's tree component.

Interaction is defined as the effect of one component of a system on the performance of another component and/or the overall system (Nair, 1993). Furthermore, for this review, AF systems are broadly categorized into two groups: 1) simultaneous systems, in which trees and crops are grown together in different spatial arrangements; examples include trees on croplands, hedgerow intercropping (HI), intercropping in perennial-tree–crop stands and multistrata systems; and, 2) sequential systems, in which trees and crops are grown in rotation; examples include rotational bush fallow or planted tree fallows followed by crops (Sanchez, 1995). Some systems, such as taungya, rotational HI, and relay planted tree fallows in crops combine the features of both simultaneous and sequential systems.

Table 1. Major tree–soil–crop interaction processes in tropical agroforestry systems.

Nature of interaction	Process
Soil fertility: chemical	
Carbon	• Increases in active pools of SOM through litterfall, root turnover and incorporation of tree prunings and crop residues
Nitrogen	• Increased soil N supply through – Nitrogen fixation – Deep soil N capture – Reduced leaching
Phosphorus	• Transformation of less available inorganic P forms into readily plant-available forms
Cations (Ca, Mg, K)	• Relocation in soil profile
Aluminum	• Organic acids binding Al • Localized Al detoxification
Soil fertility: physical	• Improved soil aggregation, porosity and pore connectivity • Reduced soil bulk density • Break up of hardpans/compacted soil layers
Soil fertility: biological	• Build up of soil macrofauna and microbial populations • Build up of VAM and rhizobial populations • Reduced/increased soil insect pests and pathogens
Competition	• Sharing of growth resources: light, water and nutrients by trees and crops
Microclimate	• Shading: reduced soil and air temperature • Shelter: protection from wind • Rainfall interception and re-distribution
Conservation	• Reduced soil erosion • Reduced leaching
Biological:	
Weeds	• Reduced weed populations • Shifts in weed species • Decreased viability of perennial weed rhizomes • Decay of annual weed seed bank
Pests and diseases	• Reduced/increased pest–parasite/predator populations
Allelopathy	• Release of growth affecting chemicals into soil environment

The study of interactions in agroforestry requires the examination of a number of complex processes (Table 1). However, it can be simplified if the interactions are categorized according to the factors that are most affected (Akyeampong et al., 1995a). These relate to soil fertility (F) [which includes soil chemical (Sc), soil physical (Sp) and soil biological (Sb) interactions], competition (C) [which includes competitive interactions for soil water (Sw), soil nutrients (Sn) and radiation (r)], microclimate (M), pests and diseases (P) [which include interactions related to weeds (Pp), insects (Pi) and diseases (Pd)], soil conservation (L) and allelopathy (A).

In general terms, the interaction effect (I) on crop yields in the two major groups of AF systems can be expressed as follows:

simultaneous systems: I = F + C + M + P + L + A; and
sequential systems: I = F + M + P + L + A.
[F includes Sc, Sp, and Sb; C includes Sw, Sn, and r; P includes Pp, Pi and Pd].

An important difference between these two groups of systems is the absence of (C) in sequential systems, where interactions between trees and crops are indirect, rather than direct as in simultaneous systems. The resource competition that occurs among tree species and between trees and weeds during the fallow period will have no direct effect on subsequent crops. The effect of (M) in sequential systems occurs mostly in slash-and-mulch systems rather than in slash-and-burn systems. On small farms, the distinction between simultaneous and sequential systems may lose its meaning as spatial interactions and boundary effects will occur in fine-grained mosaics.

It needs to be emphasized that many of the interactions are interdependent; for example, soil chemical, physical and biological changes are closely associated with organic inputs. Therefore, interaction effects for these variables cannot be experimentally estimated independent of one another. Some factors may be interdependent over time; for example, part of the (C) effect in a season may be the basis for the (F) effect in the following season (Cannell et al., 1996). Nevertheless, studying the interactions in a factor-wise framework will help us determine the relative importance of each factor for any given system, and subsequently enable us to focus on the more important factors. Moreover, quantifying the magnitude of interactions over a range of species, soil, management, and climatic conditions will help determine the biophysical limits of the system.

With this background, we review and discuss biophysical interactions in simultaneous and sequential AF systems. Within the former category, systems with contrasting methods of tree management are considered, for example, hedgerow intercropping, which manages trees as hedges, and trees on croplands, which allows trees to grow normally. The review draws heavily on the results of research in sub-Saharan Africa because, compared with tropical Asia and Latin America, this region has contributed substantially more to agroforestry research and publication of results in international media.

Simultaneous systems

Hedgerow intercropping

In tropical agroforestry, hedgerow intercropping (HI) is synonymous with alley cropping, and involves the growing of crops between hedgerows of regularly coppiced woody species. The term HI is preferred in this paper to distinguish

this technology from the growing of arable crops in alleys formed by widely spaced rows of uncoppiced trees, a practice that is increasingly referred to as alley cropping in temperate agroforestry. HI was promoted initially (in the 1980s) for improving soil fertility and sustainability of crop production on nutrient-depleted soils in the humid and subhumid tropics, where bush fallowing had been traditionally practiced for the purpose. However, HI became attractive to researchers and development organizations in other parts of the tropics because it combined the benefits of tree fallow and continuous cropping. The technology was then extended to fodder production systems (by using fodder species as hedgerows) and to erosion control efforts on sloping lands (by using contour-aligned hedgerows as live barriers). The potential of HI for these different purposes has been investigated under several agroclimatic conditions and the results have been reviewed by many authors (Kang et al., 1990; Nair, 1990; Kang, 1993; Akyeampong et al., 1995a; Sanchez, 1995; Cooper et al., 1996).

As a simultaneous AF system, all previously mentioned categories of tree–crop interactions occur in HI (Akyeampong et al., 1995a). However, the major interactions that affect crop yields are related to soil fertility, competition, weed control, and soil conservation particularly on sloping lands (Table 2). Hedgerows might increase the yield of the closest crop rows on the sheltered side (Monteith et al., 1991; Huxley et al., 1994), and hedgerow prunings might reduce soil evaporation, if applied as mulch (Tian et al., 1993). However, microclimate changes that result from hedgerows pruned regularly at optimum height have a small overall effect on crop yield, relative to other factors (Monteith et al., 1991; Ong et al., 1991). Barrier hedgerows have been shown to control soil erosion and surface loss of nutrients (Lal, 1989b; Kiepe and Rao, 1994). Compared to traditional soil conservation technologies, HI systems have been found to increase crop yields in a shorter period

Table 2. Net effect on crop yield of tree–soil–crop interactions in hedgerow intercropping systems in different climates, assuming a moderately a fertile soil.

Process	Semiarid	Subhumid	Humid
Nutrient availability to alleycrops	positive (S → L)	positive (L)	positive (L)
Soil chemical changes	positive (S)	positive (S)	positive (L)
Soil physical changes	positive (S → L)	positive (S → L)	positive (S → L)
Soil biological changes	neutral	positive (S → L)	positive (L)
Soil conservation	positive (S → L)	positive (L)	positive (L)
Water availability to alleycrops	negative (L)	neutral/negative (S)	neutral
Shading	neutral	negative (S)	negative (L)
Microclimate changes	positive (S)/neutral	neutral	neutral
Weed suppression	positive (S)	positive (L)	positive (L)
Crop yield	negative (S → L)	positive (S → L)	positive (S → L)

S = small; L = large.

of time (Banda et al., 1994). Soil conservation effects of hedgerow systems are discussed further in this volume by Craswell (1997). The various mechanisms of tree–soil–crop interactions in HI with respect to soil fertility, competition and weed control are reviewed in the subsequent sections.

Soil fertility: chemical

Hedgerows exercise short- and long-term effects on alley-planted crops. The short-term effects, which influence current-season crops, are governed by the nutrients released through mineralization of recently added hedgerow prunings and root decay. The long-term effects, which influence the sustainable production of crops, are governed by changes in soil nutrient stocks that occur over the years from regular additions of prunings and root turnover. In any given season, only a relatively small proportion (10% to 20%) of N in prunings may be taken up by the alleycrop (Mulongoy and van der Meersch, 1988; Xu et al., 1993a; Akinnifesi et al., 1997; D.N. Mugendi, pers. comm., 1997); a substantial part of the N is leached from the soil–plant system and the rest is humified in the soil (Xu et al., 1993b). The unutilized N from prunings will contribute to the build up of soil N 'capital.'

The chemical aspects of soil fertility have received much emphasis in the research on HI. Major benefits of regular additions of hedgerow prunings to soils are increases in the dynamic pools of soil organic matter (SOM) and plant-available soil nutrients. The magnitude of changes depends on the amount and quality of prunings applied, soil type, system management, climate, and duration of the practice of the system. In general, significant improvement in soil nutrient status under HI has been observed in inherently good soils in subhumid and humid tropics, such as Oxic Paleustalfs in Nigeria (Lal, 1989c; Kang et al., 1990), and a loamy clay Tropofluvent in Peru (Salazar et al., 1993). On the other hand, soil nutrient status under HI has either declined – as occurs in annual cropping systems – in acid infertile soils of the humid tropics, for example, Typic Paleudults in Peru (Szott et al., 1991), or has remained relatively unaffected, even on good soils, in the semiarid tropics (Mathuva et al., 1997). Absence of any positive soil chemical changes in these environments is attributable to the small quantity of prunings produced by the hedgerows.

Hedgerow species have also showed significant differences in ability to influence soil chemical properties. In Nigeria, Kang (1993) observed the relative ranking of species for soil surface organic C as a function of litter quantity and quality, after five years of alleycropping, as follows: leucaena (*Leucaena leucocephala*) > gliricidia (*Gliricidia sepium*) > alchornea (*Alchornea cordifolia*) > dactyladenia (*Dactyladenia barteri*). Yamoah et al. (1986b) ranked the three species they compared for improving a number of soil chemical properties as *Senna siamea* > gliricidia > flemingia (*Flemingia macrophylla*). In Rwanda, calliandra (*Calliandra calothyrsus*) showed the maximum positive impact on soil fertility, followed by *Senna spectabilis*, *Leucaena diversifolia* and leucaena; differences were largely related to the amount of tree biomass produced (Balasubramanian and Sekayange, 1991). Species differences were due primarily to differences in their adaptation to soil and climatic conditions, rooting characteristics, biological N_2 fixation (BNF), and ability to establish mycorrhizal associations.

The major mechanisms by which hedgerows increase or maintain nutrient status in the crop rooting zone are: (1) nitrogen input into the system through BNF in the case of N_2-fixing species (Dommergues, 1995), (2) reduced soil erosion, (3) reduced leaching of nutrients (Hauser and Kang, 1993), and (4) uptake of soil nutrients from deep soil layers that are beyond the crop root zone and recycling them to the soil surface via prunings. Although hedgerow systems do not increase total soil P, they have been found to increase plant available P (Haggar et al., 1991; Hands et al., 1995). External sources of P are essential on P-deficient soils to sustain the productivity of HI (Palm, 1995). Nutrient cycling through aboveground prunings is many times higher than that which occurs through root turnover (Schroth and Zech, 1995; Govindarajan et al., 1996).

Soil fertility: physical

Hedgerow intercropping, as compared with annual crops, has considerably improved soil physical properties. These improvements include better soil aggregation (Yamoah et al., 1986b; Mapa and Gunasena, 1995), lower bulk density (Yamoah et al., 1986b; Lal, 1989d; Hulugalle and Kang, 1990; Dalland et al., 1993; Jama et al., 1995; Mapa and Gunasena, 1995; Alegre and Rao, 1996), lower resistance to penetration (Lal, 1989d; Dalland et al., 1993), reduced surface sealing (Hulugalle and Ndi, 1993) and improved soil porosity (Lal, 1989d; Hulugalle and Kang, 1990; Dalland et al., 1993). The consequences of these changes are increased water infiltration (Lal, 1989e; Halugalle and Kang, 1990; Hulugalle and Ndi, 1993; Mapa and Gunasena, 1995), higher saturated hydraulic conductivity (Jama et al., 1995; Alegre and Rao, 1996), and higher water holding capacity (Yamoah et al., 1986b; Lal, 1989e) under HI, as compared with crop-only controls, in a range of climatic and soil conditions. The improved soil physical conditions under HI are primarily due to increased SOM and root activity of perennial hedgerows, and secondarily to increased soil biological activity by soil macrofauna. It is doubtful that improved soil physical conditions will increase available soil water to alleycrops in water-limiting situations because of the dominance of hedgerows over crops.

Soil fertility: biological

Soil biological activity, mediated through soil fauna and flora, is crucial in low-input systems because, in these systems, nutrient supply to plant growth is largely due to the decomposition of newly added organic residues and mineralization of SOM. Yamoah et al. (1986b) observed 46% higher soil microbial biomass C (a measure of biological activity) under HI with gliricidia and *Senna* sp. than under sole cropping, in the 0–15 cm soil layer. More earthworm activity was reported under HI than under annual cropping in a number of trials on Alfisols in Nigeria (Kang et al., 1990; Hauser, 1993). In a seven-year trial, there were nearly five times more worm casts under leucaena hedgerows (117 Mg ha^{-1}) than in the middle of the alley (24 Mg ha^{-1}) or in sole maize (*Zea mays*) (28 Mg ha^{-1}) (Hauser, 1993). Under HI, worm casts recycled three times more N, K and Mg to the surface, as compared with sole maize. Hedgerow species (e.g., dactyladenia and gliricidia) that maintained higher shade and water near the soil surface increased earthworm activity more than other tree species (Kang et al., 1994). Similarly, higher populations of earthworms, ants and termites were noted under HI with gliricidia and erythrina (*Erythrina poeppigiana*) in Costa Rica (Hands et al., 1995). Soil macrofauna (especially earthworms) will play a significant role in improving soil structure and, in turn, soil water relations and nutrient availability to crops.

Competition for growth resources

Competition for growth resources is inevitable between hedgerows and crops, but its importance is determined by climate, management, soils, and hedgerow and crop species. In humid and subhumid environments, where adequate rainfall favors rapid tree growth, competition is primarily for light, resulting in significant negative effects (crop shading) on the intercrops (Lawson and Kang, 1990). The relative importance of various hedge/crop interactions to crop yields in humid tropics is illustrated by the results of an ongoing long-term experiment at Lampung, Indonesia (av. rainfall 2200 mm yr^{-1}) for three contrasting tree species – gliricidia, leucaena, and peltophorum (*Peltophorum dasyrrachis*). The interaction effects considered were: (1) soil fertility (F), which was partitioned into the cumulative long-term effect of soil chemical, physical and biological changes (determined by removing hedgerows eight years after their establishment) and short-term effect primarily due to current addition of mulch (determined by transferring to sole maize an equivalent amount of mulch produced by hedgerows in HI), and (2) competition (C), which was partitioned into competition for light (or shade effect) and below-ground resources (determined with/without root barriers). Effects due to mulch and root interactions were small for all three species (Figure 1). Both the fast-growing leucaena and the open-canopy gliricidia provided large positive fertility effects and negative shade effects. Peltophorum – a slow-growing, non-N_2-fixing species, with a compact canopy – also exhibited a substantial positive fertility effect, but contrary to the other species, showed only a small negative effect due to shade. The combined result of the different interactions on intercropped maize yield was negative for leucaena and gliricidia because the positive fertility effect was offset by the negative shade effect; only peltophorum, which showed the lowest shade effect, resulted in a net positive effect on maize (ICRAF, 1995). The study questions fast growth and N_2 fixation as criteria for selection of species for HI, and suggests that species

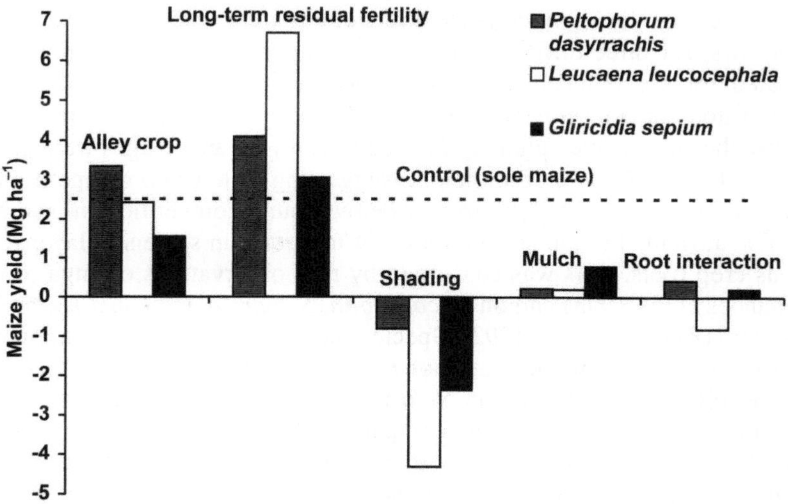

Fig. 1. Net effects, expressed in increase or decrease of maize grain yield, of long-term residual soil fertility (estimated by removing hedgerows), short-term mulch (estimated in terms of response of sole maize to equivalent amount of prunings produced by hedgerows), root interactions (estimated from yields with and without root pruning) and shading (estimated by difference) in an eight-year-old hedgerow intercropping system with three contrasting tree species in North Lampung, Indonesia. The results are based on yields from the 1994 first rainy season, with the above treatments applied eight years after the establishment of the trial. The alleycrop yield is assumed to reflect the net result of various hedge–crop interaction effects. (*Source:* ICRAF, 1995).

that supply high N per unit shade (such as peltophorum) are appropriate for the humid tropics (van Noordwijk et al., 1995).

The shade of hedgerows may have no effect or beneficial effect on shade-tolerant and short-duration crops (e.g., beans: *Phaseolus* sp.), while it may generally have negative effect on other short-statured and longer-duration crops (e.g., rice: *Oryza sativa*). The negative effect could be more pronounced if hedgerows are spaced closely and their pruning is delayed. In the humid tropics, an alley spacing wider than the 4 m that is commonly used elsewhere would be appropriate. Based on simulation studies combining the effects of light interception and N supply from erythrina prunings in Costa Rica, Nygren and Jiménez (1993) recommended for optimum alleycrop yields a between-row spacing of 6 or 8 m, a within-row spacing of 1 or 2 m, and an east-west orientation of rows.

Belowground competition for nutrients appears to be the major factor affecting the performance of HI in acid infertile soils of the humid tropics. On a Typic Paleudult in Yurimaguas, Peru, hedgerows of inga (*Inga edulis*) reduced the yield of rice in rows closest to the hedgerows by up to 50% by one year; root pruning of hedgerows at 25 cm from the hedgerow base, to 20-cm soil depth, significantly increased the rice yield by 20%, compared to yields from alleys without root pruning (Fernandes et al., 1993). On alluvial soils at the same location, crop-yield reductions were noted up to 1.5 m from hedgerows, for three different tree species. Input of additional mulch compensated for the reduced yields to a large extent (Salazar et al., 1993). Competition for nutrients was reported as the reason for yield reductions because the site received high rainfall (2200 mm yr^{-1}) and hedges were pruned periodically to 0.65 to 0.75 m height, suggesting little or no competition for water or light. Such a high degree of belowground competition in acid soils could be attributed to the concentration of tree roots in same soil layer (0–20 cm) as crop roots. This was confirmed by root observations on inga in Peru (Fernandes et al., 1993) and on *A. cordifolia*, *S. siamea* and *Gmelina arborea* in Nigeria (Ruhigwa et al., 1992). Species such as *D. barteri* and peltophorum that root deeper, below the crop root zone, appear to be better suited to acid soils (Ruhigwa et al., 1992; van Noordwijk et al., 1995).

In the semiarid tropics, the competition of hedgerows with crops is primarily for water. On Alfisols near Hyderabad, India, with 600 to 700 mm of rainfall during the main cropping season, yields of a number of crops in HI with leucaena were severely reduced (up to 90%) compared with the crop-only control. Installation of a root barrier, up to 0.5-m depth, between leucaena hedges and crop rows removed the effect of belowground competition and restored crop yields almost to the level in sole cropping (Singh et al., 1989; Rao et al., 1990; Ong et al., 1991; Rao et al., 1991a).

Detailed measurements of soil water in the 165-cm soil profile during three consecutive seasons under HI with contrasting *Senna* species (*S. siamea* and *S. spectabilis*) and a maize–cowpea (*Vigna unguiculata*) annual crop system at Machakos, Kenya (av. rainfall 760 mm yr^{-1} in two rainy seasons), highlighted the importance of competition of hedgerows for water in semiarid environments (Figure 2). Soil water under both HI systems was lower than in the annual crop system throughout the study period and the differences were greater in periods of water stress. Soil water depletion was greater under the fast-growing and high-biomass-producing *S. spectabilis* than under the slower-growing and less-biomass-producing *S. siamea*. The soil profile was never fully recharged, even when rainfall was 547 mm (50% higher than normal) during the 'short rainy' season of 1994–95, because of severe water depletion in the previous season. This points out that interseasonal water deficits in HI can contribute to water shortage for crops in any given season. Similar observations were made in other trials conducted in Machakos, Kenya (Govindarajan et al., 1996; McIntyre et al., 1997). Root studies revealed the lack of spatial complementarity between the tree and crop components in water use, as a large percentage of fine roots of many hedgerow species were in the top 0.5 m soil layer where crop roots were also concentrated (Rao et al., 1993; ICRAF, 1996).

Light competition was negligible in the above studies in semiarid India and Kenya because hedgerows were pruned to 0.5 m or 0.75 m twice per season, and yields of crop rows closest to hedges were similar to those in the

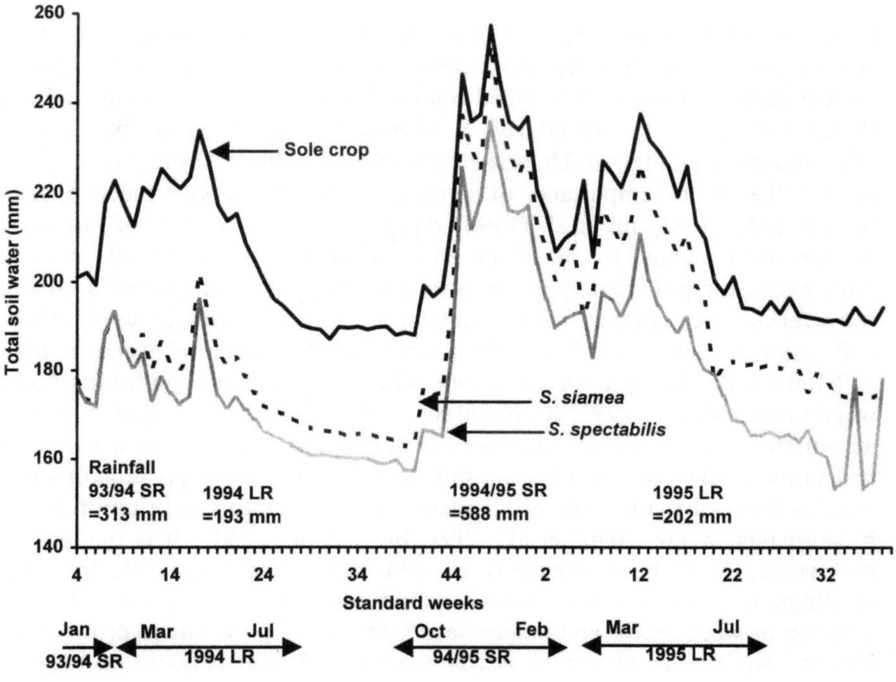

LR = long rainy season; SR = short rainy season

Fig. 2. Weekly changes in the total soil water content in 165-cm profile under hedgerow intercropping with *Senna siamea* and *S. spectabilis*, and sole annual crop (maize–cowpea (*Vigna unguiculata*) sequential) system during three cropping seasons (January 1994 to November 1995) (*Source:* H. Odhiambo and C. K. Ong, unpublished).

center of the alleys (Rao et al., 1990; Ong et al., 1991; Jama et al., 1995; Govindarajan et al., 1996). Substantial reduction of crop yields even when crops were fertilized at recommended rates confirms that the inadequacy of water to meet the demands of hedgerows and crops is the principal factor determining crop yields in HI in semiarid climates (Rao et al., 1991b; Jama et al., 1995).

Weed control

One of the major effects of trees in HI is reduction of weeds compared with annual cropping. This reduction is probably due to the complementary processes of: (1) mulch from hedgerow prunings providing a ground cover (Yamoah et al., 1986a; Kang, 1993); (2) hedgerows shading the alleys (Yamoah et al., 1986a; Jama et al., 1991) and competing with weeds for other growth resources (Rippin et al., 1994); and (3) potential allelopathy from hedgerow species (Obondo, 1987; Weston, 1996).

Striking differences were observed among hedgerow species in their ability to control weeds because species differ in biomass production and decomposition rates, and canopy spread. Species with slowly decomposing biomass, such as inga, achieved greater weed control than leucaena or erythrina in Peru (Salazar et al., 1993). Similarly, *S. siamea* controlled weeds better than gliricidia and flemingia in Nigeria (Yamoah et al., 1986a). In regard to weed control, the relative importance of competition and mulching effects depends on tree species. Rippin et al. (1994) reported that competition of gliricidia hedgerows accounted for all of the 28% reduction in weed growth in these plots because rapid decomposition prevented any effect of gliricidia mulch. In erythrina plots, competition accounted for a 13% reduction in weed growth and mulch accounted for 39%. The potential for HI to control weeds is high in humid and subhumid environments where the climatic conditions favor weed growth (Kang, 1993; Rippin et al., 1994), and when prunings are used as mulch. Another significant process that occurs in HI – as a result of hedgerow shading and mulch – is a shift over time in the composition of weed species from difficult-to-control grasses and sedges to easily controllable broad-leaved weeds (Jama et al., 1991; Siaw et al., 1991). It is likely that, over years, HI might reduce the weed seed-bank in the soil, especially if the prunings are used as mulch, which would smother weeds, prevent further addition of weed seeds, and reduce seed viability by preventing germination. HI can also control perennial weeds, which are less dependent on the soil seed-bank, as indicated by the observations in long-term trials in southwestern Nigeria, where HI with gliricidia and leucaena reduced *Imperata* infestation by shading and probably allelopathy (Kang, 1993).

Allowing hedgerows to grow between cropping seasons will increase their potential to shade and smother weeds in the alleys. Three months of uninterrupted growth of leucaena hedgerows, between the two cropping seasons, reduced weed growth by 93% at the subhumid coast of Kenya (Jama et al., 1991). However, long growth periods could lead to the potential danger of the tree species becoming a 'weed,' especially in the case of profusely seeding species such as leucaena and flemingia. The closer the hedgerows are spaced, the greater would be the potential for weed control (Jama et al., 1991), but close spacing also means loss of space for crops and greater competition of hedgerows with crops. Therefore, when determining the optimum spacing of hedgerows, the primary objective should be the improvement of crop yields. The negative effects of competition versus the positive effects of biomass production for nutrients should be the main considerations, with weed control assuming lesser importance.

Effect on crop yields

The effect of HI on crop yields could be positive or negative depending on climate and soil conditions. Most studies have evaluated the system based on short-term results (two to three years) and a high proportion of trials reported increased yields from HI (Young, 1997). However, early success or failure does not indicate the realistic long-term potential of the system, especially if the early seasons are characterized by atypical rainfall for the site. As the effects of trees are likely to be cumulative, they take a longer time to stabilize. Additionally, many experiments have suffered from the disadvantages of small plots, in which the sole crop yields could be underestimated because of insidious exploitation of water and nutrients by tree roots from HI plots. Akyeampong et al. (1995a) reviewed the results of 2 to 3-year-old experiments conducted by ICRAF in sub-Saharan Africa and observed no benefit of HI at sites (50% of total sites) where rainfall was less than 1000 mm yr^{-1}, primarily because the negative effect of competition for water exceeded the positive effect of improved soil fertility. Woomer et al. (1995) reviewed the short-term results of a number of trials conducted on diverse soils, mostly in subhumid and humid West Africa. They observed, on average (n = 44 observations), 183% yield increase for maize in HI over sole crop control, but as these results were based on small plots, tree root interference with maize controls might have exaggerated the advantage of HI.

In order to assess the long-term performance of HI, we have reviewed the results of 29 trials, mostly with small plots, conducted for four or more years over a wide range of soils and climates. Experiments on sloping lands where the primary benefit is likely to be soil conservation are not included in this analysis. Twenty-eight of these trials evaluated the potential of HI without application of N fertilizer to crops, but the majority received P fertilizer; one trial evaluated the hedgerows for fodder production, with crop fertilization at the recommended rate (Rao et al., 1991b). When the trials involved multiple hedgerow species or different alley widths, results of the best HI system were considered. The average yields of annual crops from HI relative to those of the sole system are presented in Figure 3. Yields of sequential crops in bimodal rainfall sites are presented separately if crops involved are different (hence more than 29 observations in Figure 3). Tree species used in these trials were leucaena (n = 12), *Senna siamea* (n = 3), *L. diversifolia* (n = 1), gliricidia alone or mixed with calliandra (n = 2) and *Albizia lebbeck* (n = 1) in the semiarid and subhumid climates; and calliandra (n = 3), erythrina (n = 4), inga, peltophorum and *Paraserianthes falcataria* at one site each in the humid tropics.

The results showed both positive (n = 15 for cereals, n = 8 for non-cereal crops such as beans, cowpea and cassava (*Manihot esculenta*)) and negative (n = 13 for cereal crops and n = 1 for sweet potato (*Ipomoea batatas*) and taro (*Colocasia esculenta*)) effects of HI on crop yields across the tropics, indicating that the system performance is location specific and sensitive to management, and that generalizations are difficult to make. Ignoring < 15% yield increases as unattractive to farmers, we find that only two of ten studies in semiarid sites (< 1000 mm rainfall) gave substantial yield increases. One of the sites with yield increases was a highland area in Rwanda with bimodal rainfall, where typical crop yields were very low (Balasubramanian and Sekayange, 1991); at the other site, soil and hedgerow management were very good (Bunderson, 1992). In subhumid environments (rainfall between 1000

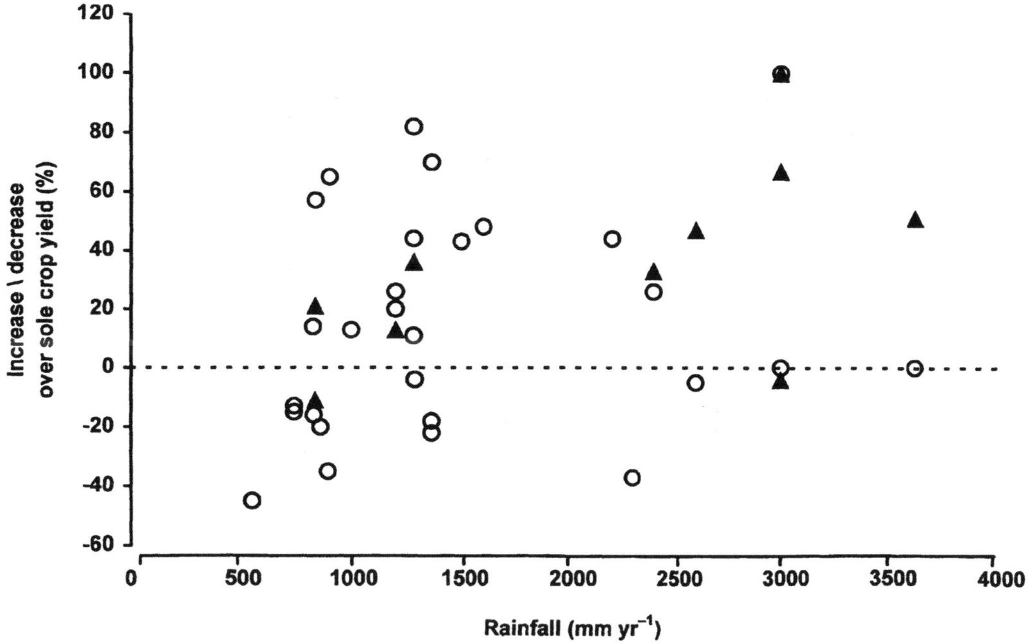

Fig. 3. Crop yields in hedgerow intercropping expressed as a percent of yields in sole annual cropping in 29 experiments conducted throughout the tropics. Open circles represent the average relative yields of cereal alley-crops (maize (*Zea mays*), sorghum (*Sorghum bicolor*), pearlmillet (*Pennisetum glauca*) or rice (*Oryza sativa*)) grown singly or in mixtures with other crops. Closed triangles represent the average relative yields of non-cereal alley-crops grown throughout the experimental period such as taro (*Colocasia esculenta*) or in the second season where two crops were grown per year such as beans (*Phaseolus vulgaris*) or cowpea (*Vigna unguiculata*). (*Sources:* Akondé et al., 1996; Akyeampong and Hitimana, 1996; Balasubramanian and Sekayange, 1991; Bundersen, 1992; Chiyenda and Materechera, 1989; Duguma et al., 1994; Evensen et al., 1995; Fernandes et al., 1993; Hauser and Kang, 1993; ICRAF, 1995; Kang, 1993; Kang et al., unpublished; Kass et al., 1995; Lal, 1989a; Matthews et al., 1992a, b; Mathuva et al., 1997; Mittal and Singh, 1989; Mugendi et al., 1997; Mureithi et al.,1994; Rao et al., 1991a; Rosecrance et al., 1992; Schroth et al., 1995b; Shannon et al., 1994; Singh et al., 1989; Tilander et al., 1995; van Noordwijk et al., 1995).

and 1600 mm), significant positive yield responses were observed in seven out of eleven studies where the soils were either inherently fertile (Kang, 1993; Hauser and Kang, 1993; Mureithi et al., 1994; Shannon et al., 1994) or the acid infertile sites received external nutrients and lime (Akyeampong and Hitimana, 1996; Matthews et al., 1992a,b). In the humid tropics (rainfall > 2000 mm), maize and taro did not benefit from HI in four out of eight trials, but interestingly bean and cowpea yields invariably increased.

Whereas low yield of hedgerow prunings (2 to 3 Mg ha^{-1} yr^{-1}) and competition of hedgerows for water were the major reasons for the negative results in water-limited areas (Ong et al., 1991; Rao et al., 1991a; Mathuva et al., 1997), low yield of prunings and competition of hedgerows for nutrients were responsible for negative results in poor soils (Matthews et al., 1992a, b; Fernandes et al., 1993). Inadequate water limited the response of crops even though HI improved soil fertility in certain sites of the semiarid tropics. In favorable environments, hedgerows produced 8 to 12 Mg ha^{-1} yr^{-1} of prunings (dry weight) which increased SOM and ensured adequate supply of nutrients, especially N, to the alleycrops. However, to realize the benefits of HI, hedgerows must be frequently pruned to avoid shade competition (Kang, 1993). The reported increases of legume yields in HI is surprising, but may have occurred because these short-duration C_3 crops do not suffer from shade competition, experienced a low incidence of pests and diseases, and benefited from improved microclimate.

In Peru, on Typic Paleudults characterized by high acidity and Al toxicity, annual crops did not produce economic yields beyond two or three years without applications of lime and fertilizer. On this site, HI maintained yields of Al-tolerant rice and cowpea crops at 1 and 0.5 Mg ha^{-1}, respectively, for many seasons (ICRAF, 1995; Palm et al., 1995). Despite declining maize yields in sole-crop systems, HI produced higher yields over a long period in acid soils of Indonesia (van Noordwijk et al., 1995; ICRAF, 1995) and higher yields in the later years after one season fallow in acid soils of Burundi (Akyeampong and Hitimana, 1996). In both sites, rainfall was adequate, phosphorus was not limiting or was added through fertilizer, and 5 to 12 Mg ha^{-1} of prunings were harvested per year. Reduced Al toxicity to maize, in addition to increased supply of nitrogen, could be responsible for the favorable response of HI in these situations. Many studies have indicated the reduction of Al toxicity with the addition of such a large amount of organic residues because their decomposition products would bind Al and reduce Al saturation (Wong et al., 1995).

Conclusions

Interactions related to soil fertility (F) and competition for growth resources (C) dominate hedge/crop interactions in HI on lands with slopes of less than 15%; the relative magnitude of interactions determines their net effect on crop yields. HI generally results in increased crop yields when (F) is sufficiently large to overcome the negative effect of (C). The advantages of HI are more common on relatively fertile, N-deficient soils in subhumid and humid environments; in areas where hedgerows produce a large quantity of prunings; and where there is adequate water for both hedge and crop growth. In semiarid environments with inadequate water (< 1000 mm rainfall yr^{-1}) and on acid infertile soils, hedgerows produce too little biomass and/or become too competitive to benefit crops. If (C) is substantially reduced, HI may increase crop yields in some of these situations, even for a relatively small (F), but the necessary management of hedges could be too intensive to be practical. On acid soils, HI may give a positive response over the long term and with external inputs (lime and P fertilizer), but many small farmers may not cope with these requirements.

In agroclimates where HI is suitable, it can provide most of the N needed for moderate yields of cereals, but it will not supply adequate N or P to realize maximum yields. HI does not seem appropriate for tuber crops and cotton. Sustainable production from HI requires P fertilization and use of deep-rooted, N_2-fixing trees. Leucaena and gliricidia are well suited for base-rich soils, whereas calliandra, *Acacia auriculiformis*, peltophorum, dactyladenia, erythrina, *L. diversifolia* and inga are suited for acid soils in specific climatic conditions.

Trees in cropland systems

Three distinct systems are considered under this category – scattered trees, boundary plantings, and intercropping with annual crops between widely spaced rows of trees. The degree of tree/crop interactions increases with increasing tree density, from low density in scattered tree systems to high density in tree/crop intercropping systems. Shade trees in perennial-crop systems, described by Beer et al. (1997) in this volume, are not considered here.

Scattered or dispersed trees in cropland, often known as 'parklands,' are a widespread traditional practice in the semiarid tropics (SAT). The best known systems are those involving faidherbia (*Faidherbia albida*), prevalent throughout the SAT of sub-Saharan Africa (CTFT, 1988; Laike, 1992; Saka et al., 1994); 'néré' (*Parkia biglobosa*), 'karité' (*Vitellaria paradoxa*) and neem (*Azadirachta indica*) in West Africa (Kater et al., 1992; Kessler, 1992; Vandenbeldt and Williams, 1992; Tilander et al., 1995); prosopis (*Prosopis cineraria*), neem and *Acacia nilotica* in India (Tejwani, 1994); and mango (*Mangifera indica*), melia (*Melia volkensii*), *Acacia* spp. and baobab (*Adansonia digitata*) in the semiarid parts of east and southern Africa. Nair (1989; 1991) has documented many other systems of this type in the tropics. Trees in these systems are rarely planted but are derived from natural regeneration and are protected by farmers. Trees generally occur in low density, for example two to three trees ha^{-1} in the case of large trees such as néré (Kater et al., 1992), five to ten trees ha^{-1} in the case of karité (Kessler, 1992), 5–50 trees ha^{-1} in the case of faidherbia (Depommier et al., 1992), and 10–45 trees

ha^{-1} in the case of prosopis (Tejwani, 1994). Farmers' objectives for maintaining trees in these systems are to provide products such as fodder, fruits, and fuelwood, and to reduce risk.

In contrast to trees in parklands, trees in boundary plantings and intercropping systems, practiced throughout the tropics, are deliberately planted and managed. Boundary plantings involve trees on farm and field boundaries, soil conservation structures and terrace risers. Intercropping systems use widely spaced rows of trees (rather than hedgerows), and include fast-growing trees such as *Populus deltoides* in northern India (Singh et al., 1993), and upperstory trees such as cedrela or cedro (*Cedrela odorata*) and grevillea (*Grevillea robusta*) in banana and bean fields in East Africa (Akyeampong et al., 1995b). The role of trees in these systems is mainly to provide poles, timber, fuelwood and/or fodder.

Three distinct tree/crop interaction zones can be distinguished in these systems: (1) a zone of light and root competition, mostly under the tree crown, (2) a zone of root competition, some area beyond the tree crown, and (3) open cropped areas that are relatively free from the interference of trees. The extent of the first two interaction zones depends on tree species, climate and soil, and spreads progressively over time with age and size of the tree. An understanding of these zones is essential for determining the optimum tree density and developing appropriate pruning regimes to maximize resource use and productivity of the systems. The major tree–soil–crop interactions that affect crop yields in these agroforestry systems are related mostly to chemical aspects of soil fertility, soil physical properties and water relations, and microclimate, especially shading (Table 3).

Table 3. Comparison of the effects of tree–soil–crop interactions under tree crowns and in open areas, in agroforestry systems with dispersed trees on cropland.

Parameter	Relative effect
Soil organic matter	under crown > open area
Soil N, P and cations	under crown > open area
Soil pH	under crown ≥ open area
Mineralizable N	under crown > open area
Soil particle size	under crown ≤ open area
Soil bulk density	under crown < open area
Water infiltration	under crown > open area
Light intensity	under crown < open area
Soil temperature	under crown < open area
Wind speed	under crown < open area
Relative humidity	under crown > open area
Soil evaporation	under crown < open area
Microbial biomass	under crown > open area
Crop yield	under crown > open area
Crop yield	under crown < open area

Soil fertility

Several studies have reported improved soil fertility in terms of SOM, extractable P, and exchangeable cations under trees, compared with treeless open areas. These studies included a wide variety of trees, both N_2-fixing species such as faidherbia (Kamara and Haque, 1992), prosopis (Aggarwal, 1980) and *Acacia* spp. (Belsky et al., 1993), and non-N_2-fixing species such as néré and karité (Kater et al., 1992; Kessler, 1992; Tomlinson et al., 1995), and a number of other species (Campbell et al., 1994). The magnitude of reported changes under trees varied considerably. In four faidherbia parklands in Burkina Faso, the soils under trees had higher nutrient status than soils in open areas: 14% to 100% more organic C, 13% to 117% more organic N, 18% to 36% more extractable P, 2% to 67% more exchangeable Ca, and 60% to 100% more exchangeable K (Depommier et al., 1992). The fertility improvement was mostly noted in the topsoil (0–20 cm). Generally, there was a fertility gradient with fertility decreasing from a tree's base to the edge of its crown or beyond. Higher SOM and organic forms of nutrients nearer the tree suggest that there could be increased mineralization and greater availability of plant available nutrients under trees than in the open areas during the cropping season (Rhoades, 1995).

Soil changes under trees may be species-specific and also dependent on size and age of the trees, and site conditions. For example, soils under leguminous trees such as *Prosopis cineraria*, *Acacia tortilis* and néré tended to be more acidic than soils under non-leguminous trees such as baobab and karité (Belsky et al., 1993; Kater et al., 1992). While both karité and néré trees significantly improved soil C and exchangeable Mg under their canopies, compared with open areas, only karité trees showed higher exchangeable Ca and K (Kater et al., 1992). The magnitude of changes in the soil properties under trees was generally greater in light soils than in heavy soils (Campbell et al., 1994) and in less fertile upland areas than in more fertile valley bottoms (Depommier et al., 1992).

Several mechanisms could be responsible for increased soil fertility under trees in parkland systems. Many leguminous trees contribute nitrogen to the system through BNF. Trees with deep roots can take up subsoil nutrients that are beyond the reach of crops and recycle them to the surface through litterfall (Buresh and Tian, 1997). Large trees, which are characteristic of parkland systems, often have extensive lateral root spread and so can take up nutrients from a large surrounding area and eventually concentrate them under their crowns. Néré trees, with crowns of 7-m radius, extended lateral roots up to 20 m from the tree base (Tomlinson et al., 1995). Lateral roots of *Acacia seyal* extended up to 26 m and those of *Sclerocarya birrea* extended up to 50 m (Groot and Soumaré, 1995). This process does not bring additional nutrients into the system but causes a redistribution of nutrients. Nevertheless, observations of improved soil fertility under trees, even after trenching has cut off surface lateral roots, prove the ability of tree processes, such as nutrient cycling and addition of litter, to improve soil fertility (Campbell et al., 1994).

The excreta of livestock resting under the shade of trees during the dry season and of birds nesting in canopies could also contribute to higher nutrients under trees (Tejwani, 1994; Tomlinson et al., 1995). In environments that are subjected to wind erosion, such as the Sahel, tree canopies can intercept the fine particles of fertile topsoil carried by winds and deposit them under the canopies via stem flow and throughfall in the rainy season (Tomlinson et al., 1995). In addition to providing nutrients, this could be the major mechanism responsible for changes in soil texture under trees. Phosphorus accumulation under some agroforestry tree species is aided by their symbiotic association with endomycorrhizae which explore a large soil volume (Mason and Wilson, 1994; Tomlinson et al., 1995).

It has also been demonstrated that trees can improve soil physical conditions under their crowns, compared with open areas. Improvements are reflected through lower bulk density (Belsky et al., 1993), lower surface resistance to penetration (Campbell et al., 1994), increased porosity and greater aggregate stability. These changes would lead to higher rain water infiltration into the soil and higher availability of water to crops. Improvement in soil physical conditions is closely associated with increased SOM. Also, as old tree roots decay, they leave channels that increase water infiltration (van Noordwijk et al., 1991). However, the beneficial effects of improved soil physical conditions might be obliterated by compaction caused by animals sheltering under the trees during dry-season grazing.

It is possible that naturally regenerating trees in parkland systems establish and grow better in pre-existing islands of higher fertility (Geiger et al., 1994) and sites characterized by subterranean termite activity (Brouwer et al., 1992). In such circumstances, it is difficult to disprove the claim that trees are indicators of soil fertility rather than agents of soil improvement. Faidherbia trees apparently regenerate on fertile sites, but it is not known to what extent this is true of other tree species.

Competition for growth resources

Competition for light is unlikely to be a major factor under faidherbia because of its 'reverse phenology' of dropping leaves at the beginning of the rainy season (Vandenbeldt and Williams, 1992), nor under prosopis because of heavy lopping of foliage during the dry season for livestock feeding (Tejwani, 1994). However, shade could be a major problem under large, unmanaged, evergreen trees such as néré, karité, *A. nilotica* and mango. Under large néré trees, as little as 20% of photosynthetically active radiation (PAR) reached the ground near the bases of trees and 91% at 14 m away from the bases; PAR transmission generally increased from 43% near the bases of trees to 90% beyond their crowns (Kessler, 1992). Jonsson et al. (1997) recorded on average only 25% of the PAR of the open area under the canopies of both néré and karité, which could cause about 50% yield reduction in pearlmillet (*Pennisetum glaucum*). While competition for light is the major limiting factor for crop growth under tree canopies, competition for water and nutrients could be the major factors beyond the crown areas.

Aboveground competition (i.e., for light) and belowground competition (i.e., for water and nutrients) may both be important under boundary plantings, as was demonstrated in a study in Uganda that used a root mesh to prevent lateral root spread (Okorio et al., 1994). When trees of four species were 3.5 years old, crop yields adjacent to trees without root mesh were 20% to 55% of yields in the treeless open area. When root mesh was installed to 0.5-m depth and 0.5 m away from trees, yields increased by 152% adjacent to *Maesopsis eminii*, 57% adjacent to *Markhamia lutea* and 16% adjacent to *Casuarina cunninghamiana*. However, the presence of root mesh did not restore crop yields to open-area levels, suggesting the existence of aboveground competition. No study, to our knowledge, has separated the trees' competition for nutrients from that for water, although competition for water is considered to be far more important than that for nutrients in semiarid and subhumid climates. Aboveground competition can be reduced by side-pruning of trees and belowground competition by periodic root pruning (e.g., by deep plowing), although the feasibility for managing root competition is questionable on many tropical farms.

Microclimate

Trees in croplands bring about microclimate changes under their canopies by reducing soil and air temperature, irradiance and wind speed. These changes will have direct influence on soil water evaporation and humidity, which in turn may significantly affect crop growth, depending on the climate. In the SAT savanna of Kenya, baobab and *Acacia tortilis* trees reduced soil temperature under their crowns by 6 °C at 5 cm and 10 cm depth, compared with open areas (Belsky et al., 1993). In the Sahel, faidherbia trees lowered soil temperature at 2-cm depth by 5° to 10 °C depending on the movement of shade (Vandenbeldt and Williams, 1992). Soil temperatures in the Sahel often reach lethal levels of 50° to 60 °C at the beginning of the rainy season, causing poor establishment of millet (Soman et al., 1987). A reduction of 10 °C in soil temperature greatly reduces the detrimental effect of high temperature and improves crop stand establishment. Recent studies in Burkina Faso showed that topsoil temperature under néré and karité trees was lower than in the open areas throughout the cropping season (Figure 4). Shade of 30- to 100-year-old trees, with 60- to 130-m^2 crown areas, resulted in lower soil temperatures than shade of smaller trees (Jonsson et al., 1997). Wind speed was also reduced under the canopies, particularly in the later part of the season, which is dominated by the desiccating harmattan winds (Jonsson et al., 1997). The reduced radiation load and windspeed under canopies reduce water demand by crops and soil evaporation, which are particularly important during drought periods.

Trees contribute to loss of rainfall through evaporation of canopy interception, and modify soil water conditions under canopies by shading and redis-

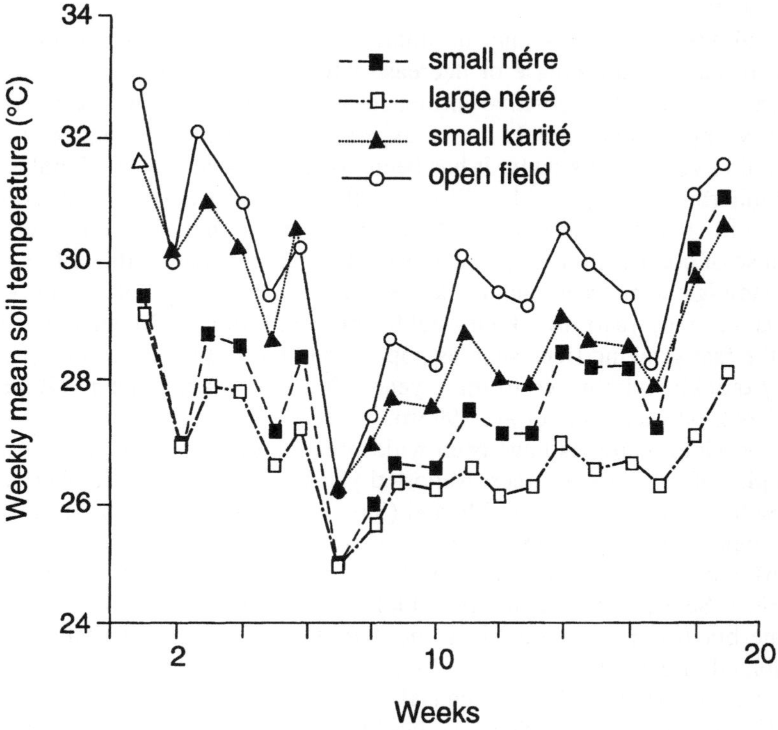

Fig. 4. Weekly mean soil temperature at 5-cm depth, measured at Saponé, Burkina Faso, from 29 June to 15 November 1994 under large and small trees of néré (*Parkia biglobosa*) and small trees of karité (*Vitellaria paradoxa*), and in the open field. Tree crown diameters ranged from an average of 8.6 m to 13.1 m. (*Source:* ICRAF, 1996).

tribution of intercepted rain through stem flow and canopy drip (Wallace, 1996). Fully foliated trees may cause up to a 50% loss of rain water for small storms (< 15 mm precipitation) but a 10 to 20% loss for larger storms (> 15 mm precipitation). Individual trees on an annual basis may account for 10% to 15% loss of rainfall in the tropics (Breman and Kessler, 1995). However, on the positive side, tree canopies will reduce evaporation of water from soil, which is particularly important in areas where rainfall frequently occurs in light showers, such as in the highlands. Wallace et al. (1997) predicted that the reduction in soil evaporation under tree canopies would be sufficient to offset the increased losses due to canopy interception at a semiarid site in Machakos, Kenya. Their model was used to demonstrate how annual total soil evaporation varied with soil type and rainfall climate. The largest amounts of soil evaporation and the greatest saving in soil evaporation due to canopy shade were found in clay and sandy loam soils, and the least were found in sandy soils. Rainfall intensity and annual total were shown to be important determinants of soil evaporation under tree canopies.

Crop yields

Crop yields under trees may be unaffected during the early years of tree growth, but could increase or decrease when trees are large, depending on the tree species. Any positive influence of trees on crop yields through soil fertility improvement would not be expected for many years, as trees modify the soil environment slowly. It has been postulated that the beneficial effects of faidherbia on crop yields will start only after the trees are from 20 to 40 years old (Poschen, 1986). On the other hand, the negative effect of trees because of competition for growth resources can be noted within a few years. The slow-growing trees such as faidherbia and acacia may not influence crop yields for many years after their establishment (Okorio and Maghembe, 1994), but the fast-growing trees such as poplars and eucalyptus reduce crop yields fairly quickly, within two to three years after their establishment (Ralhan et al., 1992; Akyeampong et al., 1995b).

Crop yield increases have been widespread under open and well-managed canopies of fully grown trees. Reported yield increases under faidherbia are: maize by more than 100% in Malawi (Saka et al., 1994), and 76% in Ethiopia (Poschen, 1986); sorghum (*Sorghum bicolor*) by 36% in Ethiopia (Poschen, 1986) and 125% in Burkina Faso (Depommier et al., 1992); and millet by 250% in Senegal (Charreau and Vidal, 1965). These yield increases under faidherbia (often referred to as the 'albida effect') are attributed to the combined effects of improved soil fertility, soil water and microclimate. The relative importance of these causal factors depends on soil and climatic conditions. Increased nutrient availability through mineralization of SOM is responsible for higher yields in nutrient-depleted soils (Rhoades, 1995). In the Sahel, about 60% of the 'albida effect' was estimated to be due to increased nitrogen availability and 40% due to increased phosphorus availability (ICRAF, 1997). Other contributory factors could be more important in drier climates. For example, reduction of soil temperature by partial shading of the soil surface is important in the Sahel to improve crop stand, particularly when early season rainfall is erratic, and to reduce the detrimental effect of high temperature on crop growth later in the season (Vandenbeldt and Williams, 1992; Jonsson et al., 1997).

Crop yield declines, relative to yields in open, treeless fields, are noted mostly under the canopies of large, evergreen, unmanaged trees. In Burkina Faso, sorghum yields under karité and néré trees were reduced on average by 50% and 70% respectively (Kessler, 1992). Sorghum yields were reduced by up to 60% under canopies of both these species in southern Mali (Kater et al., 1992). In India, wheat yields were reduced by up to 60% (Puri and Bangarwa, 1992) and mustard (*Brassica* sp.) yields by up to 65% (Yadav et al., 1993) under *A. nilotica* trees. In environments where seasonal rainfall (above approximately 800 mm during the cropping season) or irrigation provide adequate water for crop growth, reduction of crop yields under trees is mainly due to reduced light. A significant negative correlation between yield and light under néré and karité canopies indicated that shade was the major factor for the 50%

to 70% reduction of sorghum yields under those trees (Figure 5). Increased humidity under trees might favor the development of fungal diseases which may affect crop stand and contribute to yield reduction (Kater et al., 1992). Pruning reduces light competition to some extent, and may provide tree products such as fuelwood and fodder, and additional benefits through enhanced fruit production in certain tree species (Timmer et al., 1996).

The effect of boundary plantings of trees on crop yields depends on the site and age of the trees. At Hissar, India, 3.5-year-old *Eucalyptus tereticornis* on field boundaries reduced crop yields by 41% up to 10 m from the tree row (Malik and Sharma, 1990). Similarly, crop yield losses were observed at Machakos, Kenya, beginning two years after tree establishment, but different tree species showed distinctly different patterns of competition depending on growth rates and rooting pattern (Figure 6). The fast-growing grevillea and the slow-growing croton (*Croton megalocarpus*) depressed crop yields only over a short distance from the trees (< 3 m). At 3 m from the tree row, crop yields in the grevillea and croton plots approached that of the control plot, but gliricidia adversely affected yields for up to 4 m from the tree row. Although *Senna spectabilis* had faster height growth than the other tree species, it depressed yields least, probably because of its limited canopy and deeper rooting. *Melia volkensii*, which produced the most growth, was also the most competitive species; it depressed crop yields for up to 5 m from the tree row because of its large canopy and probably its lateral spread of roots near the soil surface. At 6 m from the tree row, there was little impact of any of the tree species on crop yields. It is interesting to note that contrary to the

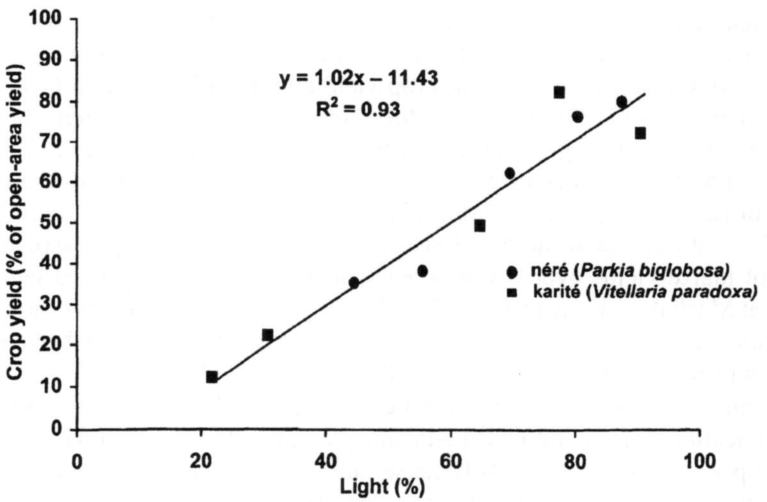

Fig. 5. Relationship between photosynthetically active radiation (PAR) transmitted and sorghum (*Sorghum bicolor*) yield under néré (*Parkia biglobosa*) and karité (*Vitellaria paradoxa*) canopies, measured at different distances from the base of trees, in Burkina Faso. (*Source:* adapted from Kessler, 1992).

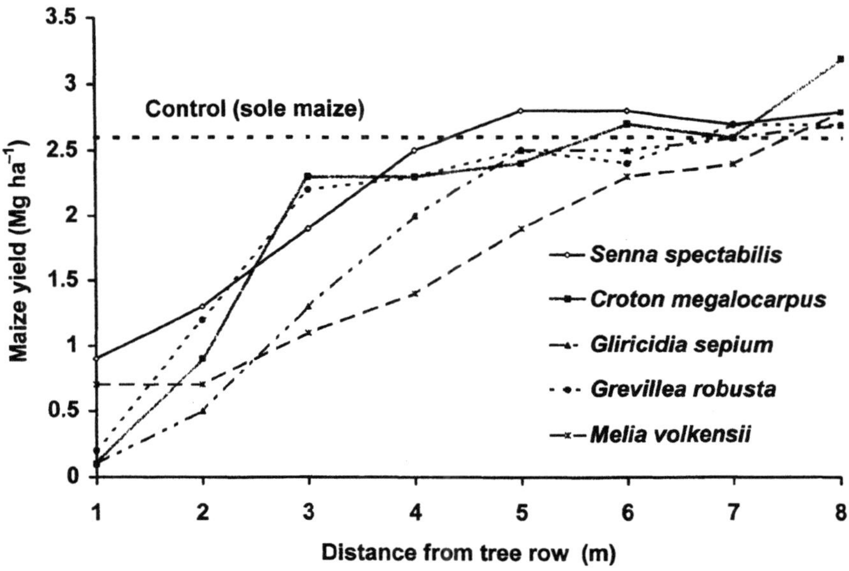

Fig. 6. Maize yield measured at different distances from tree rows of five different tree species in the 1995 first rainy season, two years after the establishment of trees, at Machakos, Kenya. (*Source:* ICRAF, 1996)

general view, the indigenous tree melia (*M. volkensii*) was the most competitive and the exotic species, grevillea and senna, the least competitive. Large differences among species for competitiveness were also observed three years after trees were established on field borders in Uganda (Okorio et al., 1994). Over 3 to 4 m from the tree row, crop yield effects ranged from a 60% yield loss adjacent to the fast-growing *Maesopsis eminii* to a 19% yield increase adjacent to the slow-growing and nitrogen-fixing *Alnus acuminata.* These studies point out the need for careful selection of tree species based on their growth rates, canopy and rooting characteristics. Trees become more competitive with age (and increased size), as illustrated by a progressive reduction of maize yields over 7.5 m on each side of a row of unpruned leucaena trees at Machakos, Kenya (Figure 7). After a certain number of years, however, the competitive effect of trees might be independent of tree age because all trees will have attained a large size (Khybri et al., 1992).

It must be recognized that in spite of large yield increases or decreases under scattered trees and near boundary plantings, the overall effect of trees on crop yields could be small because only a small proportion of the area is subjected to tree/crop interactions. Consider a scattered-tree system that contains 20 trees ha^{-1} with each tree affecting crops over a 100 m^2 area. If the trees increase or decrease crop yields by 50% over that area, the overall yield of the system will be only 10% higher or lower than that of sole crop. Similarly, if a boundary planting of a 100-m tree row reduces crop yield by

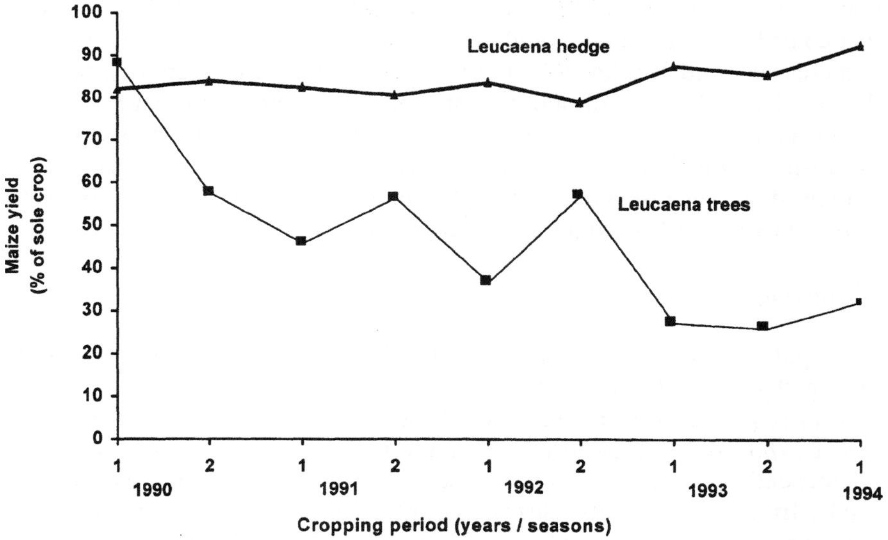

Fig. 7. Maize yield as affected by competition of unpruned *Leucaena leucocephala* trees and periodically pruned hedgerows during five years, since the establishment of *Leucaena* in November 1989 at Machakos, Kenya. Crop yields in the first and second rainy seasons in each year are indicated by 1 and 2 respectively. (*Source:* Rao MR and Mathuva MN, unpublished)

50% over a 10-m distance on either side, then the yield of the system as a whole over one hectare is reduced by only 10% (or 5% to each neighboring farmer).

Conclusions

Trees improve soil fertility and modify the microclimate both favorably and unfavorably under their canopies depending on tree species; the magnitude of change depends on canopy and root characteristics, age and size of trees, and density. Boundary plantings generally reduce crop yields but the effect extends over a relatively small area. The overall productivity of simultaneous systems depends on: (1) the complementarity of resource use by the components, especially by trees drawing resources from zones unexploited by crops, (2) the efficiency of nutrient cycling, and (3) the net value of harvested tree products relative to the net value of crop products that could be produced with the resources used by trees. Furthermore, the prospects for regular availability of tree products could reduce farmers' risk in the event of crop failures, for example, during drought which affects crops more seriously than trees.

Competition of trees can be managed to some extent by pruning, maintaining the appropriate density and sequential thinning; however, the scope for management of belowground competition is limited to the choice of trees which have deep and compact root systems. Whether tree density can be increased in parklands where crop yields are higher under trees than in the open (e.g., under faidherbia trees) depends on the nature of the processes that contribute to increased yields. The scope for higher stand density exists if increased yield is related to improved microclimate or increased fertility due to BNF and uptake of nutrients unexploited by crops, but not if it is due to concentration of nutrients by lateral exploitation. There is little scope for increasing faidherbia density in parklands beyond about 30 trees ha^{-1} at the mature stage because yield increases under faidherbia are due mostly to improved soil fertility, which is caused not solely by BNF.

Allelopathy

Allelopathy is discussed here as applicable to all agroforestry systems. If allelopathy exists, its effects could be more important in simultaneous systems than in sequential systems because of the permanent presence of trees and direct exposure of crops to the continuous release of chemicals. Although allelochemicals may be present in practically all tree parts (leaves, flowers, seeds, fruits, stems, roots, rhizomes), information on the nature of active chemicals and their mode of action is lacking (Rice, 1984). The effects of these chemicals on other plants are known to be dependent on the concentration as well as the combination in which one or more of these substances are released into the environment (Putnam and Tang, 1986). Considerable progress has been made in understanding the nature of allelochemicals and the extent of negative effects and potential benefits in annual cropping systems.

Several studies have reported the deleterious allelopathic effects of certain agroforestry trees on crops (Bhatt and Todaria, 1990; Willis, 1991; Nandal et al., 1994). Most allelopathy literature is based on laboratory studies or bioassays conducted under controlled conditions, using unrealistic concentrations of aqueous leachates or extracts of tree parts (Bhatt and Todaria, 1990; Nandal et al., 1994; Rice, 1984), and it is not clear to what extent these results have practical value. Even though certain trees may release harmful chemicals into soil through leaf litter or root secretion, they may have only temporary or no effect on crops under field conditions because of rapid elimination of chemicals by soil microorganisms and/or leaching, particularly in the humid environments. The practical significance of allelopathy cannot be ascertained unless the effect of allelochemicals on crop yields is demonstrated under field conditions, independent of the competition of trees for water and nutrients (Willis, 1991), and independent of the temporary immobilization of plant nutrients due to the addition of low quality tree litter (Sanginga and Swift, 1992).

The few field studies that reported allelopathic effects under field conditions were mostly concerned with poor germination and early growth of seedlings either under trees or after the application of tree prunings. Plant stand of many field crops (maize, pigeonpea (*Cajanus cajan*), beans, and groundnut (*Arachis hypogaea*)) was poor in the vicinity of *Gmelina arborea* in Malawi because of suspected allelopathy (J. Maghembe, pers. comm.). Boundary plantings of eucalyptus have been widely reported to affect adjacent crops due to allelopathy (Willis, 1991). Lal (1989a) reported poor germination and seedling establishment of cowpea in leucaena and gliricidia hedgerow systems in Ibadan, Nigeria, which often resulted in lower yields of cowpea under HI than in sole crop control. Interestingly, maize germination was not affected by the hedgerows. Similar observations have been made on cowpea at ICRAF's Machakos research station in Kenya (authors' unpublished data). Tian and Kang (1994) reported that gliricidia prunings at 5 Mg ha^{-1} caused 90% leaf chlorosis on maize and cowpea but the deleterious effect disappeared very soon or could be avoided by incorporating the material into the soil one week before crop sowing. However, the extent of allelopathic interaction cannot be determined from such studies because the observed results could have been caused by nutrient-related (especially N) soil factors. Early growth of maize seedlings was poor on plots mulched with grevillea at 4.6 Mg ha^{-1} in Machakos, Kenya, probably due to the same manganese toxicity that was observed in pot studies (Yobterik et al., 1994). Compared with the positive soil fertility effect of prunings or the negative competitive effect of trees for growth resources, the effect of allelopathy is probably of much less importance. Allelopathy could be beneficial through reduced weed emergence but the scope for selective reduction of weeds without affecting crops has not been investigated in agroforestry systems.

Sequential systems

Improved fallows

In improved fallows, selected tree (or herbaceous) species are either planted or retained from natural regeneration with some management. We restrict our discussion in this paper to planted tree fallows. Two categories of planted tree fallows can be distinguished: (1) short-duration fallows with fast-growing, leguminous trees or shrubs established primarily to replenish soil fertility to support food crop production, and (2) medium- to long-duration fallows with diverse species established for amelioration of degraded and abandoned lands as well as for utilization of tree products. Planted fallows are an improvement over natural fallows in that the objectives for which natural fallows have traditionally been used are attained in a shorter time through manipulation of management operations such as choice of tree species, spacing, density, establishment, and pruning.

In the rotation system of tree fallows followed by crops, three distinct phases can be distinguished based on the major soil changes that occur (Figure 8): (1) restoration (or fallow) phase, when the tree fallow accumulates nutrient stocks in the standing biomass through gains from natural processes (e.g., BNF) and recovery of lost nutrients (e.g., uptake from deeper soil layers), reduces weeds and pathogens, and improves other soil parameters, (2) nutrient transfer (or fallow clearing) phase when nutrient stocks in the vegetation are added to soil through fallow clearing, and (3) degradation (or cropping) phase, when nutrients are depleted through crop harvests, weeds and pests may increase, and soil conditions deteriorate. Tree fallows can be cleared for the subsequent cropping phase by either the slash-and-burn or slash-and-mulch technique, but these procedures have distinctly different effects on soil chemical, physical and biological processes (Figure 8). Nutrient cycling in the traditional slash-and-burn systems has been the subject of discussion of many papers (e.g., Sanchez and Hailu, 1996). We discuss in the following sections the tree–soil interactions in planted fallow systems and their effect on later crop yields.

Soil fertility: chemical

The major soil chemical changes that take place under tree fallows are increases of labile pools of SOM, N stocks, exchangeable cations and extractable P. Details of the mechanisms of soil improvement by tree fallows (as outlined in Table 1) are reviewed by Buresh and Tian (1997) and Szott et al. (1998). Planted fallows are expected to improve soil faster than HI and natural fallows because the land is completely occupied by fast-growing tree species that are specifically chosen for this purpose. However, the magnitude of improvement depends on tree species, length of fallow, soil and climatic conditions.

Significant increases in soil organic C, total N and exchangeable cations were observed in the topsoil even after short-duration fallows. For example, these variables were increased relative to initial soil conditions on nutrient-depleted soils in western Kenya, following one-year fallows of pigeonpea (*Cajanus cajan*) and leucaena (Onim et al., 1990), and in a one-year tephrosia (*Tephrosia vogelii*) fallow compared with grass fallow in Cameroon (Prinz, 1986). In Nigeria, two-year-old *Tephrosia candida* and pigeonpea fallows increased soil organic C and total N over natural bush in the surface 0–5 cm soil layer (Gichuru, 1991). In Zambia, a two-year sesbania fallow added 94 to 135 kg N and 10 to 15 kg P ha^{-1} through 3.4 to 4.5 Mg ha^{-1} of leaves, twigs and litterfall and 1.7 to 2.9 Mg ha^{-1} of belowground biomass; while N addition was through BNF and recycling from depth, all of the P was through recycling (Torquebiau and Kwesiga, 1996). An 18-month-old sesbania fallow in the highlands of western Kenya has recycled 135 kg N and 4.5 kg P ha^{-1} through 2 Mg ha^{-1} of leaves and pods and 3.3 Mg ha^{-1} of litterfall (ICRAF, 1997). In the subsequent cropping phase, the plots planted with sesbania fallow recorded significantly higher mineral N, higher rates of N mineralization and higher N associated with light fraction SOM than plots with grass fallow; these three indices of N availability were highly correlated with subsequent maize yields (ICRAF, 1996).

In the lowlands of Costa Rica, significant soil improvements were observed under short- and long-term woodlot plantations, irrespective of whether the

Phases	Major processes/changes
a) Tree fallow phase 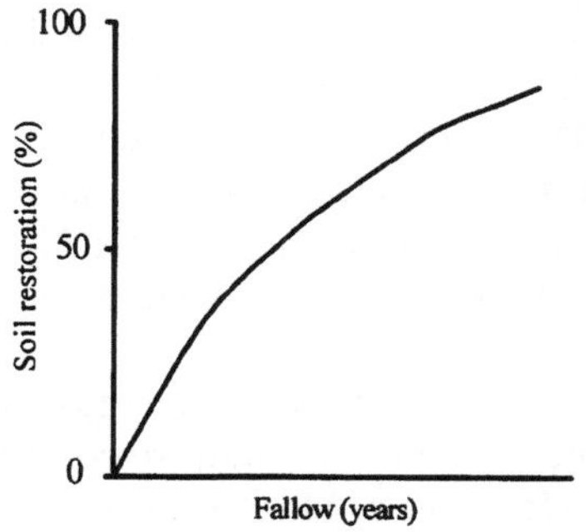	• Accumulation of nutrient stocks in fallow vegetation: -input through natural processes -capture from depth -acquisition of unavailable forms • Improved soil physical conditions • Improved soil biological activity • Reduced weed, insect and pathogen populations
b) Fallow clearing (i) Slash-and-burn	• Increased availability of nutrients (N, P & basic cations) and increased soil pH • Reduced Al, pests and beneficial soil biota • Loss of N and C through volatilization • Increased potential for nutrient loss through erosion and leaching
(ii) Slash-and-mulch	• Increase in SOM • Reduced nutrient availability in short-term • Prolonged nutrient release • Reduced nutrient losses • Enhanced soil biological activity • Longer term weed control
c) Cropping phase 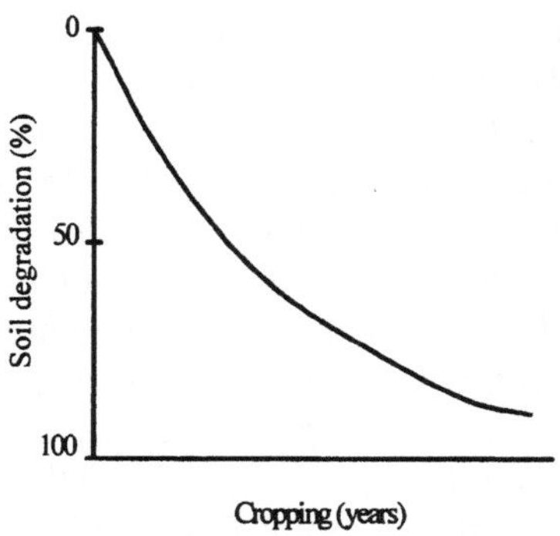	• Mineralization of organic inputs and SOM • Nutrient depletion through crop harvests • Loss of nutrients through erosion and leaching • Soil acidification • Weed invasion and build up of pests and diseases

Fig. 8. Summary of soil processes/changes as a result of tree–soil–crop interactions during different phases of tree fallow–crop rotation cycles.

trees were N_2 fixing (Montagnini and Sancho, 1990); total N, Ca and P were increased, relative to a grass fallow, under 30-month-old stands of a number of valuable indigenous trees. However, soil nitrate-N content was higher only under leguminous trees. On acrisols in Togo, Drechsel et al. (1991) recorded 40% to 60% improvement of SOM, CEC and exchangeable cations in the 0–15 cm soil layer under five-year-old trees of *Acacia auriculiformis*, *S. siamea* and *Albizia* sp., and 115% increase of SOM under neem, compared with grass fallows. On N-deficient soils in Tanzania, Jonsson et al. (1996) also noted significant species differences in effects on soil characteristics; N_2-fixing leucaena and *Prosopis chilensis* showed greater positive changes than non-N_2-fixing eucalyptus or *S. siamea*.

There are also several reports on tree stands causing changes in soil-reaction. Topsoil pH decreased under the fallows of *Acacia auriculiformis* (Drechsel et al., 1991) and eucalyptus (Jonsson et al., 1996), it remained unaffected under leucaena and *P. chilensis* (Jonsson et al., 1996) and it increased under neem, *S. siamea* and albizia (Drechsel et al., 1991) and leucaena, *Sesbania sesban* and pigeonpea (Onim et al., 1990). The pH changes are mostly the result of relocation of basic cations within the soil profile. The small changes that may take place after fallows will have little effect on crop yields in base-rich soils. However, the significant changes that occur on acidic, low-activity-clay soils in the humid tropics during tree fallow and the subsequent clearing process will have a major effect on crop growth. The decreases of pH during the fallow phase, which may be accentuated by fallow length (Adejuwan and Adesina, 1990), were probably related to leaching of basic cations as a result of leaching of excess nitrates and hydrogen released during BNF, and to uptake of cations from depth and their storage in the standing biomass. However, soil pH increases following the slash-and-burn of fallow vegetation. Any increase in pH under tree fallows was probably due to transfer of cations from depth to the soil surface via deep root uptake and litterfall.

Planted tree fallows have also been useful in ameliorating saline and alkali soils by leaching of salts beyond the crop root zone. On such soils in India, 30-month-old *S. sesban* decreased soil pH, and increased soil organic C (40%), total N (15%), and available N (53%) in the 0–30 cm soil layer (Rao et al., 1990). Production of a large quantity of biomass (30 to 40 Mg ha^{-1}) with high potential for BNF (312 to 466 kg ha^{-1} yr^{-1}) and its *in situ* decomposition were the major avenues for substantially reducing soil pH and ameliorating the saline and alkali soils. Long-term tree stands of a number of species (*Prosopis juliflora*, *Eucalyptus tereticornis*, *Terminalia arjuna*, *Albizia lebbeck* and *Acacia nilotica*) were also found to improve the soil chemical characteristics of salt-affected soils in India (Singh et al., 1994). Obviously, the extent of soil improvement depends on the length of fallow, with greater improvements following longer fallows (Adejuwan and Adesina, 1990; Mulongoy et al., 1993).

Compared with other nutrients, soil P did not change consistently under tree fallows. Onim et al. (1990) noted 85% lower extractable P compared to the initial level, one year after shrub fallows, on a P-deficient site in Kenya. Drechsel et al.(1991) observed higher soil P under cassia and neem than under albizia and *A. auriculiformis*. Adejuwan and Adesina (1990) did not find any relationship between length of gliricidia fallows and available P in the soil. Jonsson et al. (1996) reported a decrease of extractable soil P under eight-year-old tree stands. The decrease in extractable soil P resulted from uptake by trees and its storage in the standing tree biomass. Although trees have little scope to increase the supply of total soil P, they can acquire some unavailable forms of inorganic soil P by mechanisms such as mycorrhizal association (Bolan, 1991; Mason and Wilson, 1994) and secretion of weak acids (e.g., by pigeonpea, Ae et al., 1990), and thereby increase the plant-available forms of organic P following the return of tree residues to the soil (Tiessen et al., 1992; Buresh and Tian, 1997). The inability of fallow systems to meet the P requirements of crops highlights the need for P fertilizer use, especially on P-deficient soils, to ensure efficient fallow function and realize high productivity of crops (Palm, 1995). Fallow clearing by slash-and-burn does not lose phosphorus accumulated in the fallow vegetation.

Soil fertility: physical

Tree fallows improve soil physical properties due to the addition of large quantities of litterfall and root biomass (Schroth et al., 1995a), root activity of trees, biological activity and tree roots leaving macropores following their decomposition (van Noordwijk et al., 1991). Planted fallows of leucaena and sesbania improved soil physical conditions, compared with natural fallows or continuous cropping, as was evident from lower bulk density and reduced resistance to penetration in the surface soils at the end of the fallow phase. The consequence of these changes was greater infiltration and higher water storage under fallows (Juo et al., 1996; Torquebiau and Kwesiga, 1996). Tree roots can reach deep depths, penetrate compacted soil layers and improve subsoil structure and, thus, increase water infiltration and water storage. Roots of two-year-old sesbania were found to penetrate up to 5-m depth, and root mass in the 2 m soil profile accounted for 37% of the total biomass added (litterfall, foliage and roots) to the system (Torquebiau and Kwesiga, 1996). Species may differ greatly in their ability to root deeper and penetrate hard layers. Tree species with high root mass should be preferred for rehabilitation of physically degraded soils (Schroth et al., 1996).

Soil fertility: macrofauna

Accumulation of litter on the soil surface and microclimatic changes might lead to increased activity of soil macrofauna under tree fallows, particularly in the subhumid and humid environments. However, only a few studies have measured soil faunal activity under planted fallows, compared with natural fallows or continuous cropping. Observations made at Muguga, Kenya under natural forest, continuously cropped maize, one-year-old sesbania fallow, and grass fallow indicated that sesbania fallows restored the soil biological activity to the same level as in natural forest (Table 4). Although the diversity of faunal groups was not high, the total biomass of all groups under sesbania was several-fold higher than in the cropped fields or grass fallows. Of particular interest was the large increase under sesbania fallows of faunal groups such as earthworms and termites. Tian et al. (1993) working under subhumid conditions in Nigeria reported similar results and suggested that the increased faunal activity was due to the high quality of sesbania litter. Cases of increased soil faunal activity under planted fallows are also reviewed by Buresh and Tian (1997). The increased soil fauna will play a significant positive role in litter decomposition, mineralization of nutrients and improvement of soil structure by incorporating SOM into stable soil aggregates.

Weed control

Shade-based control of weeds is an essential aspect of the traditional bush fallow and planted-tree fallow systems. Two distinct phases can be recognized: initial suppression of aboveground weed growth, and subsequent depletion of the weed seed-bank. Natural fallows may control aboveground weeds within six years (Ramakrishnan and Misra, 1981), but a continuous canopy of 10 years or more is required to reduce the viable weed seed-bank (De Rouw, 1995; Staver, 1991). If selective weeding is done to retain certain species of natural vegetation, the weed suppression phase may be reduced to, for example, 0.5 to 1 year, as in the Täi farm system, Cote d'Ivoire (De Rouw, 1995). Such a practice will help increase land-use intensity.

Planted fallows can smother both annual and perennial weeds by the combined effect of reduced light and a thick mulch layer formed by canopy

Table 4. Biomass of macrofaunal groups under different land-use systems at Muguga, Kenya.

Group	Natural forest	Sesbania fallow	Grass fallow	Maize monocrop
	—— Macrofauna biomass (g m^{-2} of 0–30 cm soil depth) ——			
Earthworms	3.54	9.58	3.34	0.92
Woodlice	0.37	0	0.01	0
Millipedes	15.42	0	2.19	0.53
Centipedes	0.05	0.03	0.22	0
Termites	0.06	3.77	2.41	0.14
Cockroaches	0.11	0	0.05	0
Crickets	0.08	0	0	0
Beetle larvae	0.66	3.91	3.04	2.44
Beetle adults	0.78	0.17	0.53	0.12
Ants	0.11	0.05	0.25	0.05
Spiders	0.02	0.10	0.02	0
Others	0.05	2.20	0.02	0.01
Total	21.25	19.81	12.08	4.21

Source: J. M. Dangerfield, 1993 (unpublished).

leaf litter. The potential of fallow, however, depends on the weed species, fallow species, tree density, length of fallow and management in the initial establishment phase. Although tree fallows may take more time than shrubs and prostrate plants to suppress aboveground weeds, they could be equally effective on a longer term in controlling perennial weeds and reducing the weed seed-bank. In the humid tropics of Peru, the herbaceous fallow of *Centrosema macrocarpum* reduced weed populations to a very low level within a few months of its establishment. Although the slow-growing pigeonpea and inga tree fallows recorded higher weed growth than natural fallows or *C. macrocarpum* during the first 18 months after establishment, the weed population was similar to that under natural fallows by 29 months (Figure 9). After 32 months, weed growth was similar in all systems, and continued to be similar during the remaining fallow period (21 months) and the subsequent first crop. Weed growth in all systems started to increase during the cropping phase, but more rapidly in pigeonpea than in the others. In southern Africa, two-year fallows of sesbania and *Acacia angustissima* were found to reduce annual weeds effectively (K. Kwesiga and B. Dzowela, pers. comm., 1996). Peltophorum and chromolaena (*Chromolaena odorata*) fallows were found to smother annual weeds similarly in north Lampung, Indonesia (M. van Noordwijk, pers. comm., 1996).

It is also feasible that perennial grass weeds such as imperata (*Imperata cylindrica*) can be controlled through planted tree fallows, once the trees are well established (Garrity, 1997). On Alfisols in Nigeria, ten months of fallow growth of well-established leucaena and gliricidia hedgerows reduced the population of imperata by 51% to 67%, aboveground biomass by 78% to

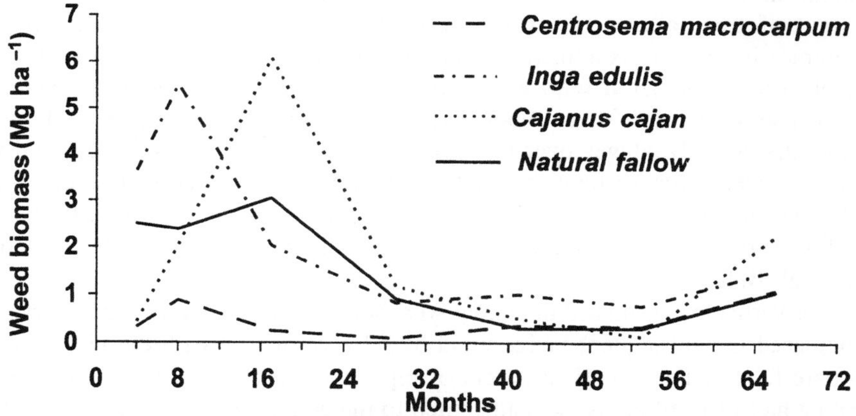

Fig. 9. Weed biomass in herbaceous legume (*Centrosema macrocarpum*), woody shrub (*Cajanus cajan*) and tree (*Inga edulis*) fallows compared with natural fallow during a 53-month fallow phase and a 12-month subsequent cropping phase at Yurimaguas, Peru (*Source:* Szott LT, Palm CA and Davey CB, unpublished).

81% and belowground rhizomes by 90% to 96% (Anoka et al., 1991). On Ultisols in north Lampung, Indonesia, the shade of unpruned hedgerows in a rotational fallow–HI system suppressed imperata population to a 'manageable' level; the hedgerows of gliricidia, peltophorum, or a combination of the two species were the most effective in controlling the grass (ICRAF, 1996). Fallow stands of *Acacia mangium* eliminated imperata in four to five years in Sabah, Malaysia (Miller and Hepburn, 1991). There seems to be a good prospect of controlling imperata in even a shorter period with closely planted peltophorum fallows (M. van Noordwijk, pers. comm., 1996) or by properly managing chromolaena (Akobundu and Ekeleme, 1996). For successful shade-based control of imperata, the tree species employed should be fire-tolerant and capable of regenerating quickly, as imperata lands are prone to fire hazards. The planted fallow technologies might be the most economical and effective strategy for small farmers to bring into cultivation the abandoned grasslands in the humid tropics.

On-going research at ICRAF on short-duration planted fallows has indicated the potential of certain fallow species to reduce infestations of the parasitic weed, striga. Sesbania fallows of 1.5 to 2 years duration have reduced infestations of *Striga asiatica* in Zambia (ICRAF, 1995) and *S. hermonthica* in western Kenya (H. Sjögren, pers. comm., 1996) on the subsequent maize crop. The decrease of striga could be partly due to increased soil fertility, which has a negative effect on striga incidence, and to sesbania acting as a false host (Oswald et al., 1996).

Effect of planted tree fallows on crop yields

Sequential systems using planted tree fallows have not received the same attention as HI, but the few studies undertaken so far illustrate the positive effect of tree fallows on crops. Short-duration fallows with herbaceous legumes have been examined widely and found to increase yields of subsequent crops compared with grass fallows or continuous cropping systems (Drechsel et al., 1996; van Noordwijk et al., 1995). However, tree fallows have distinct advantages over herbaceous fallows, particularly in seasonally dry climates, because they may take up nutrients from deep soil layers and recycle them better, and accumulate a large quantity of biomass (Szott et al., 1994); furthermore, leguminous tree fallows may add nitrogen to the system through BNF.

On Vertisols in semiarid India, one-year sole pigeonpea, harvested for grain, increased the subsequent maize yield by 57% compared with the yield after a bare fallow. The beneficial effect of pigeonpea, equivalent to the effect of 38 kg ha^{-1} of fertilizer N, was attributed to the enhanced mineralization of its post-harvest residues (Kumar Rao et al., 1983). In Malawi, a three-year-old pigeonpea fallow increased the yield of the following maize crop by 55%, compared with natural fallow (Prinz, 1986). However, on Ultisols in Nigeria, while a two-year fallow of *T. candida* increased the yield of subsequent maize by 1.5 Mg ha^{-1} (or 157%) compared with bush fallow, a two-year pigeonpea fallow did not show any benefit (Gichuru, 1991). *Tephrosia vogelii* was found to be promising in Rwanda, where its one-year fallow increased yield of the first sequential crop (maize) by 72% over control and the second crop (bean) by 96% (Balasubramanian and Sekayange, 1992). In researcher-managed plots, residual effects of fallows ranged from 40% to 170% in the first season and 42% to 84% in the second season (Drechsel et al., 1996). However, yield increases on farms following one-year fallows did not compensate for the loss of two crops during the fallow period (Drechsel et al., 1996).

In Zambia, one-year sesbania fallows increased the subsequent maize yields by 50% to 80%, and two-year fallows by 150% to 270% over the control maize yields after a grass fallow or continuous cropping (Kwesiga and Coe, 1994; Torquebiau and Kwesiga, 1996). The residual effect of both one- and two-year fallows was observed even four years after clearing the fallow, with three times higher yield than in monocropped maize (Kwesiga et al., 1998). Such a large residual effect more than compensates for the loss of production during the fallow period. In western Kenya, 18 months of sesbania fallow has produced in the subsequent three seasons 9.7 Mg ha^{-1} of maize compared with 6.9 Mg ha^{-1} after a grass fallow and 4.9 Mg ha^{-1} after continuous maize cropping. The fallow effect at this P-deficient site was substantially higher after applying P fertilizer to maize (B.A. Jama, pers.comm., 1997).

Pruning of trees during the fallow phase might increase the production of foliar biomass for greater *in situ* nutrient cycling at the expense of wood, which is normally taken out of the field. In western Kenya, Onim et al. (1990) pruned the fallows of sesbania, leucaena and pigeonpea at two-month intervals and incorporated the prunings into the soil. Following a one-year fallow, they recorded yield increases of 67% in the first crop of maize, compared with yields after a natural fallow. There was no residual effect in the second crop of maize, but a 26% yield increase was noted in the third crop of beans after leucaena and sesbania fallows; this could be an effect of the tree–crop rotation. Root biomass of fallow trees might contribute substantially to the residual effect of fallows over the years. Sanginga et al. (1988) estimated the N contribution of roots, nodules and, probably, a small percent of leaf litter of leucaena fallow to be equivalent to 32 kg N ha^{-1} on maize, similar to that of aboveground prunings in HI. Inoculated leucaena exhibited a greater effect than uninoculated leucaena.

The potential of short-duration tree fallows to restore soil productivity and increase yields of food crops has not been adequately explored in the humid tropics. Because soil degradation occurs rapidly in the humid tropics, longer duration fallows may be required to achieve the desired effect on subsequent crops. On Typic Paleudults in Peru, the total yield of three continuous crops (two rice and one cowpea) following 4.5-year inga and desmodium (*Desmodium ovalifolium*) fallows was 34% and 21% higher, respectively, than the yield after secondary forest of the same duration, due to higher rates of N mineralization following planted fallows (Szott and Meléndez, 1991).

A rotational woodlot-intercrop system has been developed in the Shinyanga region of Tanzania, in which the woodlot is intercropped with annual crops for two to three years in the early stage of tree growth and again after the harvest of trees two to three years later, keeping the stumps very low. Following a four-year-old mixed stand of *Acacia polyacantha* and leucaena, the first crop of maize gave 100% and the second crop 16% higher yields than the crops after a grass fallow (ICRAF, 1995). When yields decline to an uneconomical level, the stumps can be allowed to regrow into a woodlot.

Increased maize yields for one or two years have been reported following three to four years of *Chromolaena odorata* fallows in the Täi region of Ivory Coast (Slaats, 1996). This beneficial 'weed fallow,' which establishes rapidly by natural regeneration, could help intensify cropping in the humid tropics of Central and West Africa and Southeast Asia, where slash-and-burn is currently in vogue. However, it is not known if the positive effects of fallows with nonleguminous species such as chromolaena are short-lived or if they will sustain increased crop yields over successive fallow–cropping cycles. The issue of BNF may not be important for fallow species where the soils contain large quantities of accumulated nutrients at depth or are inherently fertile except for physical degradation. In these situations, tree species that efficiently recycle nutrients from depth, improve soil physical and biological conditions rapidly, and suppress weeds quickly are appropriate for use in tree fallows. In nitrogen-depleted soils, however, species that have high BNF ability should be preferred.

Conclusions

In the absence of direct competition between trees and crops for growth resources, sequential agroforestry systems offer an opportunity for exploiting the potential soil ameliorative attributes of trees for enhancing crop production. Soil amelioration is based primarily on N replenishment through BNF and recycling of N, P and basic cations to the crop root zone, and secondarily on improved soil physical conditions and biological activity. While chemical fertility is important in nutrient-depleted soils, the physical and biological aspects of soil fertility become more important in degraded soils. Tree fallows reduce weed populations through shading, which helps in the rehabilitation of abandoned grasslands.

On inherently fertile but nutrient-depleted soils, one- to two-year fallows with fast-growing leguminous shrubs, such as sesbania, can replenish soil N stocks to a level adequate for the production of 3 to 4 Mg ha^{-1} of maize during the first cropping season and for substantially increased yields of subsequent crops, compared with grass fallow or continuous cropping. The residual effect of such fallows can extend for up to four years after the removal of fallow, offsetting the loss of production during the fallow period and increasing the attractiveness of tree fallows to farmers. On infertile soils, tree fallows require a longer period to improve soil fertility. Tree fallows, however, do not increase the supply of P in the soil, although they may increase P availability within the system. Therefore, crops that follow tree fallows on P-deficient soils will need P fertilizers to fully exploit the potential benefits resulting from the fallow. Tree species that grow fast, fix nitrogen, root deeply, produce high root biomass and high quality foliar biomass, and provide economically valuable byproducts (e.g., fodder, fuelwood) are well suited for short-duration fallows.

Future research directions

This review has described the progress that has been made during the past 10 to 15 years in research efforts to increase and apply our knowledge of biophysical interactions in major tropical agroforestry systems and technologies. Obviously, all systems and technologies have not received equal attention during that period; for example, while HI has been explored widely and intensively since the early 1980s, planted fallows have been researched only since the late 1980s, and multistrata systems and others involving tree–livestock interactions have been little studied. Nevertheless, we have a much better understanding today than we had ten years ago with respect to the broad geographical and ecological limits and constraints of some of the evolving technologies, as well as the major interactions in each of them. Further effort is required to develop management practices in participation with farmers that will maximize complementary interactions and resolve the location-specific constraints to widespread adoption of the technologies. Intermediate agroforestry technologies might be more appropriate than straightforward HI or tree fallow systems for certain conditions. For example, HI could be rotated with tree fallows, or intercropping systems using coppiced trees or widely spaced rows of trees could be rotated with tree fallows or woodlots (using the already established trees).

As research in tropical agroforestry takes new directions, research on interactions will follow suit. Tropical agroforestry research has so far placed high emphasis on technologies that stress the service functions of trees for soil fertility improvement and/or soil conservation. Too little attention has been given to high-value or income-generating trees. Future research will concentrate on technologies involving high-value fruit, pole, and medicinal trees that can be integrated with crops, and biophysical interaction research will adopt the same focus. Secondly, there is need to examine the potential of exploiting inter- and intraspecies genetic variation in trees. The choice of tree species is an important management decision that determines the success or failure of a technology at a given place. Although the number of tree species explored in agroforestry research has increased substantially in recent years, much remains to be done. The usefulness of mixing species for hedgerow and fallow technologies needs to be evaluated. Mixing of species with different growth habits, rooting patterns, and quality of leaf residues may lead to higher biomass productivity by enhancing complementary use of growth resources, reducing pest problems, and promoting greater synchrony in the release of nutrients from residues in relation to crop requirements. Similarly, the range of crops used for testing the potential of agroforestry technologies meant for soil fertility improvement has been limited almost exclusively to cereals (mostly maize), despite the fact that even these crops, in practice, are grown either in rotation with or mixed with legumes. Continuous cultivation of cereal crops (in control treatments) depletes soil nutrients much faster than systems with cereal-legume rotations, and may lead to overestimation of the benefits of agroforestry treatments.

There is still inadequate understanding of belowground biophysical interactions in tropical agroforestry systems, but this area is now receiving considerable research attention. Progress has been made in recent years in the development of both the concepts and the techniques for belowground-interaction studies. Conceptually, a so-called 'central' hypothesis to justify agroforestry is that trees must utilize resources such as water and nutrients that are not utilized by crops (Cannell et al., 1996). This means that if the existing resources are under-utilized or unutilized in a crop-only system because they are inaccessible to crop roots, then spatial and/or temporal differences in rooting and root dynamics in an agroforestry system (complementarity) would be an advantage. For example, research on improved fallows in western Kenya showed that when growth of striga-infested maize was poor, large amounts of N remained unutilized and was eventually leached from the root zone (ICRAF, 1996). In terms of techniques for belowground-interaction studies, some of the recent developments such as simple allometric or fractal methods for quantifying root architecture (van Noordwijk and Purnomosidhi, 1995) and the low-cost heat-pulse technique for measuring root functioning (Khan and Ong, 1996) will be of value in the future. These techniques should be helpful in efforts to screen conveniently a large number of tree species or provenances suitable for simultaneous systems. A related development is in the area of modeling of biophysical interaction processes in agroforestry systems (Lawson et al., 1995). Further progress in model development and validation will, however, depend very much on the availability of physiological data (especially root structure and functioning) for the trees that are used in the systems. The various modelling efforts currently underway should help integrate the large body of data that is now available and increase our predictive understanding of agroforestry systems to guide future research and delineate biophysical limits for promising technologies.

Few studies have been published on the role of soil fauna, the functions of specific groups, and the scope for their manipulation for improving soil physical properties and nutrient availability to crops. With regard to sequential systems, although research has demonstrated the potential of fallows to improve soil fertility, knowledge on how fallows work in different soil and climatic conditions is still inadequate. Some issues that deserve further attention are (1) the relative importance of soil chemical, physical and biological changes that tree fallows cause in different soil and climatic conditions; (2) the length of residual effects of fallows of different durations; (3) the 'fate' of nitrogen in the system; (4) the external nutrients required for the sustainable function of fallows; and (5) the specific 'rotational effects' of fallows, for example, the control of pests and diseases. Short-term fallows may not have any specific advantage over annual crops in terms of carbon sequestration, but medium- to long-term tree fallows may have a favorable effect on the soil C budget and it needs to be quantified.

Another aspect that has not received adequate attention is the time and spatial dimension of biophysical interactions in agroforestry. Most studies have been conducted for only a few years, whereas agroforestry technologies must be evaluated on a long-term basis. Similarly, most information on biophysical interactions in agroforestry, as reviewed here, is based on plot-level studies. The interactions of systems at farm and landscape level should receive attention in the future because the lateral flow of resources (e.g., water and nutrients) becomes important at such spatial scales and because trees can exploit resources from larger areas than allocated (van Noordwijk and Ong, 1996). Knowledge of the larger-scale interactions is important in order to realistically assess the community and global benefits of agroforestry.

To date, most knowledge of biophysical interactions in agroforestry systems has come from studies of simultaneous or sequential systems which involve only two sets of components – the tree and the crop. The complexity and enormity of the task of studying biophysical interactions will be increased many-fold when research efforts are extended to multistrata systems such as homegardens and whole farms, including livestock. Furthermore, although it has not been discussed in this paper, the adoption of technologies that exploit interactions is an important issue. It is well known that the extent of adoption of any technology will depend not only on the magnitude of positive biophysical benefits, but also on a number of socioeconomic factors. Biophysical (interaction) research in agroforestry will inevitably and increasingly be linked to socioeconomic (interaction) research in the future.

Acknowledgments

The authors sincerely thank Pedro Sanchez, Meine van Noordwijk and John Beer for providing valuable comments on the drafts of this paper.

References

Ae N, Arihara J, Okada K, Yoshihara T and Johansen C (1990) Phosphorus uptake by pigeonpea and its role in cropping systems of the Indian subcontinent. Science 248: 477–480

Adejuwan JO and Adesina FA (1990) Organic matter and nutrient status under cultivated fallows: an example of *Gliricidia sepium* fallows from South Western Nigeria. Agroforestry Systems 10: 23–32

Akobundu IO and Ekeleme FE (1996) Potentials for *Chromolaena odorata* (L.) R. M. King and H. Robinson in fallow management in West and central Africa. In: Prasad UK,

Muniappan R, Ferrar P, Aechliman JP and de Foresta H (eds) Distribution, Ecology and Management of *Chromolaena odorata*, pp 88–98, Publication No. 202, University of Guam, Mangilao, Guam, USA

Aggarwal RK (1980) Physicochemical status of soils under khejri (*Prosopis cineraria* Linn.). In: Mann HS and Saxena SK (eds), Khejri (*Prosopis cineraria*) in the Indian Desert, pp 32–37, Central Arid Zone Research Institute, Jodhpur, India

Akinnifesi FK, Kang BT, Sanginga N and Tijani-Eniola H (1997) Nitrogen use and N-competition between *Leucaena* hedgerows and maize in an alley cropping system. Nutrient Cycling in Agroecosystems 47: 71–80.

Akyeampong E, Duguma B, Heineman AM, Kamara CS, Kiepe P, Kwesiga F, Ong CK, Otieno HJ and Rao MR (1995a) A synthesis of ICRAF's research on alley cropping. In: Kang BT, Osiname AO and Larbi A (eds) Alley Farming Research and Development, pp 40–51, IITA, Ibadan, Nigeria

Akyeampong A and Hitimana L (1996) Agronomic and economic appraisal of alley cropping with *Leucaena diversifolia* on an acid soil in the highlands of Burundi. Agroforestry Systems 33: 1–11

Akyeampong A, Hitimana L, Franzel S and Munyemana PC (1995b) The agronomic and economic performance of banana, bean and tree intercropping in the highlands of Burundi: an interim assessment. Agroforestry Systems 31: 199–210

Akondé TP, Leihner DE and Steinmüller N (1996) Alley cropping on an Ultisol in subhumid Benin: Part 1: Long-term effect on maize, cassava and tree productivity. Agroforestry Systems 34: 1–12

Alegre JC and Rao MR (1996) Soil and water conservation by contour hedging in the humid tropics of Peru. Agriculture, Ecosystems and Environment 57: 17–25

Anderson LS and Sinclair FL (1993) Ecological interactions in agroforestry systems. Agroforestry Abstracts 6: 57–91

Anoka UA, Akobundu IO and Okonkwo SNC (1991) Effects of *Gliricidia sepium* (Jacq.) Steud and *Leucaena leucocephala* (Lam.) de Wit on growth and development of *Imperata cylindrica* (L.) Raeuschel. Agroforestry Systems 16: 1–12

Balasubramanian V and Sekayange L (1991) Effects of tree legumes in hedgerows on soil fertility changes and crop performance in the semi-arid highlands of Rwanda. Biological Agriculture and Horticulture 8: 17–32

Balasubramanian V and Sekayange L (1992) Five years of research on improved fallow in the semi-arid highlands of Rwanda. In: Mulongoy K, Gueye M and Spencer DSC (eds) Biological Nitrogen Fixation and Sustainability of Tropical Agriculture, pp 405–422, John Wiley & Sons and Sayce Publishing, Chichester, UK

Banda AZ, Maghembe JA, Ngugi DN and Chome VA (1994) Effect of intercropping of maize and closely spaced leucaena hedgerows on soil conservation and maize yield on a steep slope at Ntcheu, Malawi. Agroforestry Systems 27: 17–22

Beer JW, Muschler RG, Kass DCL and Somarriba EJ (1997) Shade management in coffee and cacao plantations. Agroforestry Systems 38: 139–164

Belsky AJ, Mwonga SM and Duxbury JM (1993) Effects of widely spaced trees and livestock grazing on understory environments in tropical savannas. Agroforestry Systems 24: 1–20

Bhatt BP and Todaria NP (1990) Studies on the allelopathic effects of some agroforestry tree crops of Garhwal Himalaya. Agroforestry Systems 12: 251–255

Bolan NS (1991) A critical review on the role of mycorrhizal fungi in the uptake of phosphorus by plants. Plant and Soil 134: 189–207

Breman H and Kessler JJ (1995) Woody Plants in Agro-ecosystems of Semi-arid Regions. Springer-Verlag, Berlin Heidelberg

Brouwer J, Geiger SC and Vandenbeldt RJ (1992) Variability in the growth of *Faidherbia albida*: a termite connection. In: Vandenbeldt RJ (ed) *Faidherbia albida* in the West African Semi-arid Tropics, pp 131–138. ICRISAT, Patancheru-502 324, A.P., India

Bunderson WT (1992) Final report of the agroforestry technical assistance to the national agroforestry team, MARE project, Chitedze Agricultural Research Station, Malawi
Buresh RJ and Tian G (1997) Soil improvement by trees in sub-Saharan Africa. Agroforestry Systems 38: 51–76
Campbell BM, Frost P, King JA, Mawanza M and Mhlanga L (1994) The influence of trees on soil fertility on two contrasting semi-arid soil types at Matopos, Zimbabwe. Agroforestry Systems 28: 159–172
Cannell MGR, van Noordwijk M and Ong CK (1996) The central hypothesis: trees must acquire resources that the crop would not otherwise acquire. Agroforestry Systems 33: 27–31
Charreau C and Vidal P (1965) Influence de l'*Acacia albida* Del. sur le sol: nutrition minéral et rendements des mils *Pennisetum* au Sénégal. Agronomie Tropicale 6–7: 600–626
Chiyenda S and Materechera SA (1989) Some results from alley cropping *Leucaena leucocephala*, *Cassia siamea* and *Cajanus cajan* with maize at Bunda college of agriculture. In: Trees for Development in Sub-Saharan Africa, pp 135–142, IFS, Stockholm, Sweden
Cooper PJM, Leakey RRB, Rao MR and Reynolds L (1996) Agroforestry and the mitigation of land degradation in the humid and subhumid tropics of Africa. Experimental Agriculture 32: 235–290
Craswell E (1997) Agroforestry in the management of sloping lands in Asia and the Pacific. Agroforestry Systems 38: 121–137
CTFT (1988) *Faidherbia albida* (Del.) A. Chev. (synonyme *Acacia albida* Del.). Centre Technique Forestier Tropical, Nogent-sur-Marne, France
Dalland A, Våje PI, Mathews RB and Singh BR (1993) The potential of alley cropping in improvement of cultivation systems in the high rainfall areas of Zambia. III. Effects on soil chemical and physical properties. Agroforestry Systems 21: 117–132
Depommier D, Janodet E and Oliver R (1992) *Faidherbia albida* parks and their influence on soils and crops at Watinoma, Burkina Faso. In: Vandenbeldt RJ (ed) *Faidherbia albida* in the West African Semi-arid Tropics, pp 111–116, ICRISAT, Patancheru-502 324, A.P., India
De Rouw A (1995) The fallow period as a weed-break in shifting cultivation (tropical wet forests). Agriculture, Ecosystems and Environment 54: 31–43
Drechsel P, Glaser B and Zech W (1991) Effect of four multipurpose tree species on soil amelioration during tree fallow in central Togo. Agroforestry Systems 16: 193–202
Drechsel P, Steiner KG and Hagedorn F (1996) A review on the potential of improved fallows and green manure in Rwanda. Agroforestry Systems 33: 109–136
Dommergues YR (1995) Nitrogen fixation by trees in relation to soil nitrogen economy. Fertilizer Research 42: 215–230
Duguma B, Mollet M and Tiki Manga T (1994) Annual Progress Report, Institute of Agronomic Research (IRA) and International Centre for Research in Agroforestry (ICRAF), AFRENA Report, ICRAF, Nairobi, Kenya
Evensen CI, Dierolf TS and Yost RS (1995) Decreasing rice and cowpea yields in alley cropping on highly weathered Oxisol in West Sumatra, Indonesia. Agroforestry Systems 31: 1–19
Fernandes ECM, Davey CB and Nelson LA (1993) Alley cropping on an acid soil in the upper Amazon: mulch, fertilizer, and hedgerow root pruning effects. In: Technologies for Sustainable Agriculture in the Tropics, pp 77–96, ASA Special Publication 56, American Society of Agronomy, Madison, WI 53711, USA
Garrity, DP (ed) (1997) Agroforestry innovations for *Imperata* grassland rehabilitation. Agroforestry Systems (Special Issue) 36: 1–274
Geiger SC, Vandenbeldt RJ and Manu A (1994) Variability in the growth of *Faidherbia albida*: the soils connection. Soil Science Society of America Journal 58: 227–231
Groot JJR and Soumaré A (1995) The roots of the matter: soil quality and tree roots in the Sahel. Agroforestry Today 7(1): 9–11
Gichuru MP (1991) Residual effects of natural bush, *Cajanus cajan* and *Tephrosia candida* on the productivity of an acid soil in southeastern Nigeria. Plant and Soil 134: 31–36

Govindarajan M, Rao MR, Mathuva MN and Nair PKR (1996) Soil-water and root dynamics under hedgerow intercropping in semiarid Kenya. Agronomy Journal 88: 513–520

Haggar JP, Warren GP, Beer JW and Kass D (1991) Phosphorus availability under alley cropping and mulched and unmulched sole cropped systems in Costa Rica. Plant and Soil 137: 275–283

Hands MR, Harrison AF and Bayliss-Smith T (1995) Phosphorus dynamics in slash-and-burn and alley cropping systems of the humid tropics. In: Tiessen H (ed) Phosphorus in the Global Environment: Transfers, Cycles and Management, pp 155–170, John Wiley, New York

Hauser S (1993) Distribution and activity of earthworms and contribution to nutrient cycling in alley cropping. Biology and Fertility of Soils 15: 16–20

Hauser S and Kang BT (1993) Nutrient dynamics, maize yield and soil organic matter dynamics in alley cropping with *Leucaena leucocephala*. In: Mulongoy K and Merckx R (eds) Soil Organic Matter Dynamics and Sustainability of Tropical Agriculture, pp 215–222. John Wiley & Sons and Sayce Publishing, Chichester, UK

Hulugalle NR and Kang BT (1990) Effect of hedgerow species in alley cropping systems on surface soil physical properties of an Oxic Paleustalf in southwestern Nigeria. Journal of Agricultural Science (Cambridge, UK) 114: 301–307

Hulugalle NR and Ndi JN (1993) Effects of no-tillage and alley cropping on soil properties and crop yields in a Typic Kandiudult of southern Cameroon. Agroforestry Systems 22: 207–220

Huxley PA, Pinney A, Akunda E and Muraya P (1994) A tree/crop interface orientation experiment with a *Grevillea robusta* hedgerow and maize. Agroforestry Systems 26: 23–45

ICRAF (1995) Annual Report for 1994, pp 91–93, 153–154. International Centre for Research in Agroforestry, Nairobi, Kenya

ICRAF (1996) Annual Report for 1995, pp 83–88, 64–69, 172–180, 207–210. International Centre for Research in Agroforestry, Nairobi, Kenya

ICRAF (1997) Annual Report for 1996. International Centre for Research in Agroforestry, Nairobi, Kenya

Jama B, Getahun A and Ngugi DN (1991) Shading effects of alley cropped *Leucaena leucocephala* on weed biomass and maize yield at Mtwapa, Coast Province, Kenya. Agroforestry Systems 13: 1–11

Jama B, Nair PKR and Rao MR (1995) Productivity of hedgerow shrubs and maize under alleycropping and block planting systems in semiarid Kenya. Agroforestry Systems 31: 257–274

Jonsson K, Ong CK and Odongo JCW (1997) Tree–crop interactions in a parkland system with millet, néré and karité in Burkina Faso. Experimental Agriculture (in press)

Jonsson K, Ståhl L and Hägberg P (1996) Tree fallows: a comparison between five tropical tree species. Biology and Fertility of Soils 23: 50–56

Juo ASR, Franzluebbers K, Dabiri A and Ikhile B (1996) Soil properties and crop performance on a kaolinitic Alfisol after 15 years of fallow and continuous cultivation. Plant and Soil 180: 209–217

Kamara CS and Haque I (1992) *Faidherbia albida* and its effects on Ethiopian highland Vertisols. Agroforestry Systems 18: 17–29

Kang BT (1993) Alley cropping: past achievements and future directions. Agroforestry Systems 23: 141–155

Kang BT, Akinnifesi FK and Pleysier JL (1994) Effect of agroforestry woody species on earthworm activity and physicochemical properties of worm casts. Biology and Fertility of Soils 18: 193–199

Kang BT, Reynolds L and Atta-Krah AN (1990) Alley farming. Advances in Agronomy 43: 315–359

Kass DL, Araya JS, Sanchez JO, Pinto LS and Ferreira P (1995) Ten years experience with alley farming in Central America. In: Kang BT, Osiname AO and Larbi A (eds) Alley Farming Research and Development, pp 393–402, IITA, Ibadan, Nigeria

Kater LJM, Kante S and Budelman A (1992) Karité (*Vitellaria paradoxa*) and néré (*Parkia biglobosa*) associated with crops in South Mali. Agroforestry Systems 18: 89–105

Kessler JJ (1992) The influence of karité (*Vitellaria paradoxa*) and néré (*Parkia biglobosa*) trees on sorghum production in Burkina Faso. Agroforestry Systems 17: 97–118

Kiepe P and Rao MR (1994) Management of agroforestry for the conservation and utilization of land and water resources. Outlook in Agriculture 23(1): 17–25

Khan AAH and Ong CK (1996) A low cost heat pulse method for measuring tree root water uptake. Agroforestry Forum 7(2): 19–22

Khanna PK (1997) Nutrient cycling under mixed-species tree systems in southeast Asia. Agroforestry Systems 38: 99–120

Khybri ML, Gupta RK, Sewa Ram and Tomar HPS (1992) Crop yields of rice and wheat grown in rotation as intercrops with three tree species in the outer hills of western Himalaya. Agroforestry Systems 17: 193–204

Kumar Rao JVDK, Dart PJ and Sastry PVSS (1983) Residual effect of pigeonpea (*Cajanus cajan*) on yield and nitrogen response of maize. Experimental Agriculture 19: 131–141

Kwesiga F and Coe R (1994) The effect of short rotation *Sesbania sesban* planted fallows on maize yield. Forest Ecology and Management 64: 199–208

Kwesiga F, Franzel S, Place F, Phiri D and Simwanza CP (1998) *Sesbania sesban* improved fallows in eastern Zambia: their inception, development and farmer enthusiasm. Agroforestry Systems (this issue)

Laike A (1992) *Faidherbia albida* in the traditional farming systems of central Ethiopia. In: Vandenbeldt RJ (ed) *Faidherbia albida* in the West African Semi-arid Tropics, pp 39–42, ICRISAT, Patancheru-502 324, A.P., India

Lal R (1989a) Agroforestry systems and soil surface management of a tropical alfisol. I. Soil moisture and crop yields. Agroforestry Systems 8: 7–29

Lal R (1989b) Agroforestry systems and soil surface management of a tropical alfisol. II. Water runoff, soil erosion and nutrient loss. Agroforestry Systems 8: 97–111

Lal R (1989c) Agroforestry systems and soil surface management of a tropical alfisol: III. Changes in soil chemical properties. Agroforestry Systems 8: 113–132

Lal R (1989d) Agroforestry systems and soil surface management of a tropical alfisol: IV. Effects on soil physical and mechanical properties. Agroforestry Systems 8: 197–215

Lal R (1989e) Agroforestry systems and soil surface management of a tropical alfisol: V. Water infiltrability, transmissivity and soil water sorptivity. Agroforestry Systems 8: 217–238

Lawson TL and Kang BT (1990) Yield of maize and cowpea in an alley cropping system in relation to available light. Agricultural and Forest Meteorology 52: 347–357

Lawson GJ, Crout NMJ, Levy PE, Mobbs DC, Wallace JS, Cannell MGR and Bradley RG (1995) The tree–crop interface: representation by coupling of forest and crop process models. Agroforestry Systems 30: 199–221

Malik RS and Sharma SK (1990) Moisture extraction and crop yield as a function of distance from a row of *Eucalyptus tereticornis*. Agroforestry Systems 12: 187–195

Mapa RB and Gunasena HPM (1995) Effects of alley cropping on soil aggregate stability of a tropical Alfisol. Agroforestry Systems 32: 237–245

Mason PA and Wilson J (1994) Harnessing symbiotic associations: vesicular-arbuscular mycorrhizas. In: Leakey RRB and Newton AC (eds) Tropical Trees: the Potential for Domestication and the Rebuilding of Forest Resources, pp 165–175, HMSO, London, UK

Mathuva MN, Rao MR, Smithson PC and Coe R (1997) Improving maize yields in semi-arid highlands of Kenya: agroforestry or fertilizers? Field Crops Research (in press)

Matthews RB Holden ST, Volk J and Lungu S (1992a) The potential of alley cropping in improvement of cultivation systems in the high rainfall areas of Zambia. I. *Chitemene* and *Fundikila*. Agroforestry Systems 17: 219–240

Matthews RB, Lungu S, Volk J, Holden ST and Solberg K (1992b) The potential of alley cropping in improvement of cultivation systems in the high rainfall areas of Zambia. II. Maize production. Agroforestry Systems 17: 241–262

McIntyre BD, Riha SJ and Ong CK (1997) Competition for water in a hedge–intercrop system. Field Crops Research 52: 151–160

Miller RR and Hepburn AJ (1991) A review of the growth of *Acacia mangium* at the Bengkoka afforestation and settlement project North Sabah. In: Abod SA, Tahir PM, Tsai LM, Shukor NAA, Sajap AS and Manikam D (eds) Recent Developments in Tree Plantations of Humid/Subhumid Tropics of Asia, pp 93–99, Universiti Pertanian Malaysia

Mittal SP and Singh P (1989) Intercropping field crops between rows of *Leucaena leucocephala* under rainfed conditions in northern India. Agroforestry Systems 8: 165–172

Montagnini F and Sancho F (1990) Impacts of native trees on tropical soils: a study in the Atlantic lowlands of Costa Rica. Ambio 19 (8): 386–390

Monteith JL, Ong CK and Corlett JE (1991) Microclimate interactions in agroforestry. Forest Ecology and Management 45: 31–44

Mugendi DN, Mochoge BO, Coulson CL and Sang FK (1997) Effect of incorporating *Cassia siamea* prunings on maize yield in semiarid Kenya. African Crop Science Journal 5: 271–277

Mulongoy K, Kunda KN and Chiang CNK (1993) Effect of alley cropping and fallowing on some soil fertility parameters in southern Nigeria. In: Mulongoy K and Merckx R (eds) Soil Organic Matter Dynamics and Sustainability of Tropical Agriculture, pp 47–56, John Wiley & Sons and Sayce Publishing, Chichester, UK

Mulongoy K and van der Meersch MK (1988) Nitrogen contribution by leucaena (*Leucaena leucocephala*) prunings to maize in an alley cropping system. Biology and Fertility of Soils 6: 282–285

Mureithi JG, Taylor RS and Thorpe W (1994) The effects of alley cropping with *Leucaena leucocephala* and of different management practices on the productivity of maize and soil chemical properties in lowland coastal Kenya. Agroforestry Systems 27: 31–52

Nair PKR (ed) (1989) Agroforestry Systems in the Tropics. Kluwer Academic Publishers, Dordrecht, The Netherlands

Nair PKR (1990) The Prospects for Agroforestry in the Tropics. Technical Paper No. 131: World Bank, Washington, DC, USA

Nair PKR (1991) State-of-the-art of agroforestry systems. Forest Ecology and Management 45: 5–29

Nair PKR (1993) An Introduction to Agroforestry. Kluwer Academic Publishers, Dordrecht, The Netherlands

Nair PKR, Rao MR and Fernandes ECM (1994) Tree–crop interactions in sustainable agroforestry systems. In: Soil Technology for Sustainable Agriculture, Symposia VIa, ISSS Commission VI: Soil Technology, pp 110–137, 15th International Congress of Soil Science, 10–16 July 1994, Acapulco, Mexico

Nandal DPS, Bisla SS, Narwal SS and Kaushik JC (1994) Allelopathic interactions in agroforestry systems. In: Narwal SS and Tauro P (eds) Allelopathy in Agriculture and Forestry, pp 92–130, Scientific Publishers, Jodhpur, India

Nygren P and Jiménez JM (1993) Radiation regime and nitrogen supply in modelled alley cropping systems of *Erythrina poeppigiana* with sequential maize–bean cultivation. Agroforestry Systems 21: 271–285

Obondo L (1987) Potencial alelopatico de *Gliricidia sepium* (Jacq.) Walp. sobre los cultivo de maiz y frijol y las malezas predominantes. In: Withington D, Glover N and Brewbaker JL (eds), pp 59–60, *Gliricidia sepium* (Jacq.) Walp. Management and Improvement, Nitrogen Fixing Tree Association, Waimanalo, Hawaii, USA

Okorio J, Byenkya S, Wajja N and Peden D (1994) Comparative performance of seventeen upperstorey tree species associated with crops in the highlands of Uganda. Agroforestry Systems 26: 185–203

Okorio J and Maghembe JA (1994) The growth and yield of *Acacia albida* intercropped with maize (*Zea mays*) and beans (*Phaseolus vulgaris*) at Morogoro, Tanzania. Forest Ecology and Management 64: 183–190

Ong CK and Huxley P (eds) (1996) Tree–Crop Interactions: A Physiological Approach. CAB International, Wallingford, UK

Ong CK and Black CR (1994) Complementarity in resource use in intercropping and agroforestry systems. In: Monteith JL, Scott RK and Unsworth MH (eds) Resource Capture by Crops, pp 255–278, Nottingham University Press, Loughborough, UK

Ong CK, Corlett JE, Singh RP and Black CR (1991) Above and below ground interactions in agroforestry systems. Forest Ecology and Management 45: 45–57

Onim JFM, Mathuva M, Otieno K and Fitzhugh HA (1990) Soil fertility changes and response of maize and beans to green manures of leucaena, sesbania and pigeonpea. Agroforestry Systems 12: 197–215

Oswald A, Frost H, Ransom JK, Kroschel J, Shepherd KD and Sauerborn J (1996) Studies on the potential for improved fallow using trees and shrubs to reduce striga infestations in Kenya. In: Moreno MT, Cubero JI, Berner D, Joel D, Musselman LJ and Parker C (eds) Advances in Parasitic Plant Research, pp 795–780, Junta de Andalucia, Codoba, Spain

Palm CA (1995) Contribution of agroforestry trees to nutrient requirements of intercropped plants. Agroforestry Systems 30: 105–124

Palm CA, Salazar AA, Szott LT and Fernandes ECM (1995) Long-term performance of alley cropping on acid soils of the Amazon basin. In: Kang BT, Osiname AO and Larbi A (eds) Alley Farming Research and Development, pp 379–392, IITA, Ibadan, Nigeria

Poschen, P (1986) An evaluation of the *Acacia albida*-based agroforestry practices in the Hararghe highlands of eastern Ethiopia. Agroforestry Systems 4: 129–143

Prinz D (1986) Increasing the productivity of smallholder farming systems by introduction of planted fallows. Plant Research and Development 24: 31–56

Puri S and Bangarwa KS (1992) Effects of trees on the yield of irrigated wheat crop in semi-arid regions. Agroforestry Systems 20: 229–241

Putnam AR and Tang CS (1986) The Science of Allelopathy, Wiley, New York, USA

Ralhan PK, Singh A and Dhanda RS (1992) Performance of wheat as intercrop under poplar (*Populus deltoides* Bartr.) plantations in Punjab (India). Agroforestry Systems 19: 217–222

Ramakrishnan PS and Misra BK (1981) Population dynamics of *Eupatorium adenophorum* Spreng. during secondary succession after slash and burn agriculture (jhum) in northeastern India. Weed Research 22: 77–84

Rao DLN, Gill HS and Ibrol IP (1990) Regional experience with perennial *Sesbania* in India. In: Macklin B and Evans DO (eds), Perennial *Sesbania* Species in Agroforestry, pp 189–198, NFTA, Waimanalo, Hawaii, USA

Rao MR, Muraya P and Huxley PA (1993) Observations of some tree root systems in agroforestry intercrop situations, and their graphical representation. Experimental Agriculture 29: 183–194

Rao MR, Sharma MM and Ong CK (1990) A study of the potential of hedgerow intercropping in semi-arid India using a two-way systematic design. Agroforestry Systems 11: 243–258

Rao MR, Ong CK, Pathak P and Sharma MM (1991a) Productivity of annual cropping and agroforestry systems on a shallow Alfisol in semi-arid India. Agroforestry Systems 15: 51–63

Rao MR, Sharma MM and Ong CK (1991b) A tree/crop interface design and its use for evaluating the potential of hedgerow intercropping. Agroforestry Systems 13: 143–158

Rhoades C (1995) Seasonal pattern of nitrogen mineralization and soil moisture beneath *Faidherbia albida* (syn. *Acacia albida*) in central Malawi. Agroforestry Systems 29: 133–145

Rice EL (1984) Allelopathy. Academic Press, New York, USA

Rippen M, Hagger JPH, Kass D and Köpke U (1994) Alley cropping and mulching with *Erythrina poeppigiana* (Walp.) O. F. Cook and *Gliricidia sepium* (Jacq.) Walp.: effects on maize/weed competition. Agroforestry Systems 23: 119–134

Rosecrance RC, Rogers S and Tofinga M (1992) Effects of alley cropped *Calliandra calothyrsus* and *Gliricidia sepium* hedgerows on weed growth, soil properties and taro yields in Western Samoa. Agroforestry Systems 19: 57–66

Ruhigwa BA, Gichuru MP, Mambani B and Tariah NM (1992) Root distribution of *Acioa barteri*, *Alchornea cordifolia*, *Cassia siamea* and *Gmelina arborea* in an acid Ultisol. Agroforestry Systems 19: 67–78

Saka AR, Bunderson WT, Itimu OA, Phombeya HSK and Mbekeani Y (1994) The effects of *Acacia albida* on soils and maize grain yields under smallholder farm conditions in Malawi. Forest Ecology and Management 64: 217–230

Salazar A, Szott LT and Palm CA (1993) Crop–tree interactions in alley cropping systems on alluvial soils of the Upper Amazon Basin. Agroforestry Systems 22: 67–82

Sanchez PA (1995) Science in agroforestry. Agroforestry Systems 30: 5–55

Sanchez PA and Hailu M (1996) (eds) Alternatives to slash-and-burn agriculture. Agriculture, Ecosystems and Environment 58: 1–87

Sanginga NK, Mulongoy K and Ayanaba A (1988) Nitrogen contribution of leucaena/rhizobium symbiosis to soil and subsequent maize crop. Plant and Soil 112: 137–141

Sanginga NK and Swift MJ (1992) Nutritional effects of *Eucalyptus* litter on the growth of maize (*Zea mays*). Agriculture, Ecosystems and Environment 41: 55–65

Schroth G and Zech W (1995) Above- and below-ground biomass dynamics in sole cropping and alleycropping system with *Gliricidia sepium* in semi-deciduous rainforest zone of West Africa. Agroforestry Systems 31: 181–198

Schroth G, Kolbe D, Pity B and Zech W (1995a) Searching for criteria for the selection of efficient tree species for fallow improvement, with special reference to carbon and nitrogen. Fertilizer Research 42: 297–314

Schroth G, Oliver R, Balle P, Gnahoua GM, Kanchenkanti N, Leduc B, Mallet B, Peltier R and Zech W (1995b) Alley cropping with *Gliricidia sepium* on a high base status soil following forest clearing: effects on soil conditions, plant nutrition and crop yields. Agroforestry Systems 32: 261–276

Schroth G, Kolbe D, Pity B, and Zech W (1996) Root system characteristics with agroforestry relevance of nine leguminous tree species and a spontaneous fallow in a semi-deciduous rainforest area of West Africa. Forest Ecology and Management 84: 199–208

Shannon DA, Vogel WO and Kabaluaba KN (1994) The effects of alley cropping and fertilizer application on continuously cropped maize. Tropical Agriculture 71: 163–169

Siaw DEKA, Kang BT and Okali OUU (1991) Alley cropping with *Leucaena leucocephala* (Lam.) De Wit and *Acioa barteri* (Hook. f.) Engl. Agroforestry Systems 14: 219–231

Singh A, Dhanda RS and Ralhan PK (1993) Performance of wheat varieties under poplar (*Populus deltoides* Bartr.) plantations in Punjab (India). Agroforestry Systems 22: 83–86

Singh G, Singh NT and Abrol IP (1994) Agroforestry techniques for the rehabilitation of degraded salt-affected lands in India. Land Degradation & Rehabilitation 5: 223–242

Singh RP, Ong CK and Saharan N (1989) Above and below ground competitions in alley-cropping in semi-arid India. Agroforestry Systems 9: 259–274

Singh Rathore MP (1995) Insect Pests in Agroforestry. Working Paper No. 70, ICRAF, Nairobi, Kenya

Slaats JJP (1996) The use of *Chromolaena odorata* as fallow in a semi-permanent system in South-West Côte d'Ivoire. In: Prasad UK, Muniappan R, Ferrar P, Aeschliman JP and de Foresta H (eds) Distribution, Ecology and Management of *Chromolaena odorata*, pp 68–75, Publication No. 202, University of Guam, Mangilao, Guam, USA

Soman P, Stomph TJ, Bidinger F and Fussel LK (1987) Improvement in stand establishment in pearl millet. In: Food Grain Production in the Semi-Arid Africa: Proceedings of the OAU/STRC-SAFGRAD International Drought Symposium, 17–23 May 1986, Nairobi, Kenya, pp 159–171, OAU/STRC-SAFGRAD, Ouagadougou, Burkina Faso

Staver C (1991) The role of weeds in the productivity of Amazonian bush fallow agriculture. Experimental Agriculture 27: 287–304

Szott LT and Meléndez G (1991) Crop yields, soil nitrogen mineralization and soil chemical properties following 4.5 years of managed leguminous fallows. In: TropSoils Technical Report, 1988–1989, pp 234–236, North Carolina State University, Raleigh, USA

Szott LT, Palm CA and Buresh RJ (1998) Ecosystem fertility and fallow function. Agroforestry Systems (in press)

Szott LT, Palm CA and Davey CB (1994) Biomass and litter accumulation under managed and natural tropical fallows. Forest Ecology and Management 67: 177–190

Szott LT, Palm CA and Sanchez PA (1991) Agroforestry in acid soils of the humid tropics. Advances in Agronomy 45: 275–301

Tejwani KG (1994) Agroforestry in India. Oxford & IBH Publishing Co Pvt Ltd, New Delhi, India

Tian G and Kang BT (1994) Evaluation of phytotoxic effects of *Gliricidia sepium* (Jacq.) Walp. prunings on maize and cowpea seedlings. Agroforestry Systems 26: 249–254

Tian G, Brussaard L and Kang BT (1993) Biological effects of plant residues with contrasting chemical composition under humid tropical conditions: effects on soil fauna. Soil Biology and Biochemistry 25: 731–738

Tiessen H, Salcedo IH and Sampaio EVSB (1992) Nutrient and soil organic matter decomposition dynamics under shifting cultivation in semi-arid northeastern Brazil. Agriculture, Ecosystems and Environment 38: 139–151

Tilander Y, Ouedraogo G and Yougma F (1995) Impact of tree coppicing on tree–crop competition in parkland and alley farming systems in semiarid Burkina Faso. Agroforestry Systems 30: 363–378

Timmer LA, Kessler JJ and Slingerland M (1996) Pruning of néré trees (*Parkia biglobosa* (Jacq.) Benth.) on the farmlands of Burkina Faso, West Africa. Agroforestry Systems 33: 87–98

Tomlinson H, Teklehaimanot Z, Traoré A and Olapade E (1995) Soil amelioration and root symbioses of *Parkia biglobosa* (Jacq.) Benth. in West Africa. Agroforestry Systems 30: 145–159

Torquebiau EF and Kwesiga F (1996) Root development in *Sesbania sesban* fallow-maize system in eastern Zambia. Agroforestry Systems 34: 193–211

Vandenbeldt RJ (ed.) (1992) *Faidherbia albida* in the West African Semi-Arid Tropics. ICRISAT, Patancheru-502 234, Andhra Pradesh, India and ICRAF, Nairobi, Kenya

Vandenbeldt RJ and Williams JH (1992) The effect of soil surface temperature on the growth of millet in relation to the effect of *Faidherbia albida* trees. Agriculture and Forest Meteorology 60: 93–100

van Noordwijk M and Ong CK (1996) Lateral resource flow and capture – the key to scaling up agroforestry results. Agroforestry Forum 7(3): 29–31

van Noordwijk M and Purnomosidhi P (1995) Root architecture in relation to tree–crop–soil interactions and shoot pruning in agroforestry. Agroforestry Systems 30: 161–173

van Noordwijk M, Widianto M, Heinen M and Hairiah K (1991) Old tree roots channels in acid soils in the humid tropics: important for crop root penetration, water infiltration and nitrogen management. Plant and Soil 134: 37–44

van Noordwijk M, Sitompul SM, Hairiah K, Listyarini E and Syekhfani MS (1995) Nitrogen supply from rotational or spatially zoned inclusion of Leguminosae for sustainable maize production on an acid soil in Indonesia. In: Date RA, Grundon NJ, Rayment GE and Probert ME (eds) Plant–Soil Interactions at Low pH, pp 779–784, Kluwer, Dordrecht, The Netherlands

Wallace JS (1996) The water balance of mixed tree–crop systems. In: Ong CK and Huxley PA (eds) Tree–Crop Interactions: A Physiological Approach, pp 189–233, CAB International, Wallingford, UK

Wallace JS, Jackson NA and Ong CK (1997) Soil evaporation in a *Grevillea robusta* agroforestry system. 2. Modeling the effects of a tree canopy. Agricultural and Forest Meteorology (in press).

Weston LA (1996) Utilization of allelopathy for weed management in agroecosystems. Agronomy Journal 88: 860–866

Willis RJ (1991) Research on allelopathy on *Eucalyptus* in India and Pakistan. Commonwealth Forest Review 70: 279–289

Wong MTF, Akyeampong A, Nortcliff S, Rao MR and Swift RS (1995) Initial responses of maize and beans to decreased concentrations of monomeric inorganic aluminium with application of manure or tree prunings to an Oxisol in Burundi. Plant and Soil 171: 275–282

Woomer P, Bajah AN, Atta-Krah and Sanginga N (1995) Analysis and interpretation of alley farming network data from tropical Africa. In: Kang BT, Osiname AO and Larbi A (eds), Alley Farming Research and Development, pp 189–20, IITA, Ibadan, Nigeria

Xu ZH, Myers RJK, Saffinga PG and Chapman AL (1993a) Nitrogen fertilizer in leucaena alley cropping. II. Residual value of nitrogen fertilizer and leucaena residues. Fertilizer Research 34: 1–8

Xu ZH, Myers RJK, Saffinga PG and Chapman AL (1993b) Nitrogen cycling in leucaena (*Leucaena leucocephala*) alley cropping in semi-arid tropics. II. Response of maize growth to addition of nitrogen fertilizer and plant residues. Plant and Soil 148: 73–82

Yadav JP, Sharma KK and Khanna P (1993) Effect of *Acacia nilotica* on mustard crop. Agroforestry Systems 21: 91–98

Yamoah CF, Agboola AA and Mulongoy K (1986a) Decomposition, nitrogen release and weed control by prunings of selected alley cropping shrubs. Agroforestry Systems 4: 239–246

Yamoah CF, Agboola AA, Wilson,GF and Mulongoy K (1986b) Soil properties as affected by the use of leguminous shrubs for alley cropping with maize. Agriculture, Ecosystems and Environment 18: 167–177

Yobterik AC, Timmer VR and Gordon AM (1994) Screening agroforestry tree mulches for corn growth: a combined soil test, pot trial and plant analysis approach. Agroforestry Systems 25: 153–166

Young A (1997) Agroforestry for Soil Management. CAB International, Wallingford, UK and ICRAF, Nairobi, Kenya

Agroforestry Systems **38**: 51–76, 1998.

Soil improvement by trees in sub-Saharan Africa

R. J. BURESH[1,] * and G. TIAN[2]
[1] *International Centre for Research in Agroforestry (ICRAF), P.O. Box 30677, Nairobi, Kenya; E-mail: r.buresh@cgnet.com;* [2] *International Institute of Tropical Agriculture (IITA), P.M.B. 5320, Ibadan, Nigeria (* Corresponding author)*

Key words: nitrogen, nutrient cycling, phosphorus, soil fertility, soil organic matter

Abstract. Trees can influence both the supply and availability of nutrients in the soil. Trees increase the supply of nutrients within the rooting zone of crops through (1) input of N by biological N_2 fixation, (2) retrieval of nutrients from below the rooting zone of crops and (3) reduction in nutrient losses from processes such as leaching and erosion. Trees can increase the availability of nutrients through increased release of nutrients from soil organic matter (SOM) and recycled organic residues.

Roots of trees frequently extend beyond the rooting depth of crops. Research on a Kandiudalfic Eutrudox in western Kenya showed that fast-growing trees with high N demand (*Calliandra calothyrsus*, *Sesbania sesban* and *Eucalyptus grandis*) took up subsoil nitrate that had accumulated below the rooting depth of annual crops. *Sesbania sesban* was also more effective than a natural grass fallow in extracting subsoil water, suggesting less leaching loss of nutrients under *S. sesban* than under natural uncultivated fallows.

Nutrient release from SOM is normally more dependent on the portion of the SOM in biologically active fractions than on total quantity of SOM. Trees can increase inorganic soil N, N mineralization and amount of N in light fraction SOM. Among six tree fallows of 2- and 3-year duration on an Ustic Rhodustalf in Zambia, inorganic N and N mineralization were higher for the two tree species with lowest (lignin + polyphenol)-to-N ratio (mean = 11) in leaf litter than for the two tree species with highest ratio (mean = 20) in leaf litter. Trees can also restore soil fauna, which are important for SOM and plant residue decomposition.

Some agroforestry trees have potential to provide N in quantities sufficient to support moderate crop yields through (i) N inputs from biological N_2 fixation and retrieval of nitrate from deep soil layers and (ii) cycling of N from plant residues and manures. The cycling of P from organic materials is normally insufficient to meet the P requirements of crops. Sustained crop production with agroforestry on P-deficient soils will typically require external P inputs.

Introduction

Increasing population in sub-Saharan Africa (SSA) has led to a replacement of traditional systems of shifting cultivation with generally unsustainable systems such as shorter-duration fallows, sedentary agriculture on smallscale land holdings and expansion of agriculture into marginal areas. Many smallscale farmers lack the financial means and appropriate incentives to purchase sufficient fertilizer to balance the nutrients that are removed with harvested plant products (Cooper et al., 1996). The result has been widespread 'mining' of soil nutrients and depletion of soil fertility in SSA (Stoorvogel et al., 1993).

Sub-Saharan Africa is dominated by Alfisols, Ultisols and Oxisols with low activity clay (Deckers, 1993). Alfisols are widely distributed in the subhumid and semiarid tropics. They have low water retention and are susceptible to soil erosion and soil compaction. Deficiencies of N and P are common, while localized deficiencies of other nutrients occur under intensive cultivation. Alfisols acidify rapidly under continuous cultivation without proper nutrient management. Ultisols and Oxisols cover considerable areas of the subhumid and humid tropics. They have major chemical constraints including low nutrient reserves, multiple nutrient deficiencies, high P fixation and Al toxicity. Ultisols, like Alfisols, also have major physical constraints of low water retention and susceptibility to erosion and compaction.

The main sources of plant-available N are mineralization of soil organic matter (SOM), biological N_2 fixation, fertilizers and organic inputs (e.g., plant residues, composts and manures) (Giller et al., 1997). The main sources of plant-available P are weathering of soil minerals, mineralization of SOM, fertilizers and organic materials (Buresh et al., 1997). Phosphorus, unlike N, is not biologically fixed from air, and the P content of plant residues and manures is normally insufficient to meet the requirements for sustained crop production (Palm, 1995). Sustainable crop production in many soils of SSA requires P inputs because the soils are either derived from parent material with low levels of P or have been depleted of plant-available P through continuous cropping with insufficient P inputs (Sanchez et al., 1997a).

Trees can potentially improve soils through numerous processes including maintenance or increase of SOM, biological N_2 fixation, uptake of nutrients from below the reach of crop roots, increased water infiltration and storage, reduced loss of nutrients by erosion and leaching, improved soil physical properties, reduced soil acidity and improved soil biological activity (Young, 1997). These processes are generally more researched and reviewed for simultaneous agroforestry systems, such as hedgerow intercropping (alley cropping) (Juo et al., 1994; Kang, 1997) and parklands (Breman and Kessler, 1995), than for sequential systems, such as planted tree fallows (Sanchez, 1995). Fernandes et al. (1994) reviewed the desirable characteristics of soil-improving trees, and Szott et al. (1991, 1998) concluded that the ability of trees to enhance nutrient availability was greater on high base status soils than low base status soils. Recent reviews (Rhoades, 1997; Young, 1997) on the soil improvement effects of trees have largely concentrated on studies of soils under forest stands or along transects under individual trees. Moreover, previous reviews have generally not given much attention to the relationships between tree-induced soil changes and the production of crops.

Soil improvement by trees will be simplified for this review by grouping potential benefits into three categories: (i) increased supply of nutrients through increased inputs and reduced outputs; (ii) increased availability of nutrients, through enhanced nutrient cycling and conversion of nutrients to more labile forms; and (iii) a more favorable environment for plant growth, through improved soil chemical and physical properties. Nutrient inputs are additions to a soil-plant system, which includes soil to the depth of crop rooting. Inputs include biological N_2 fixation, retrieval of nutrients from below the rooting depth of crops, organic materials produced outside the system and fertilizers. Nutrient outputs include removal through plant harvest, soil erosion, leaching and gaseous losses (Smaling et al., 1996). Nutrient outputs are increased by soil improvements that increase production of harvested plant products which are removed from the soil-plant system. Nutrient cycling refers to the transfers of nutrients already within the soil-plant system (Sanchez and Palm, 1996). In our review, 'soil-plant system' will refer to the scale of a farmer's field.

In order to complement other recent reviews on soil improvement by trees, our review will (i) focus on recent research findings and (ii) highlight relationships between tree-induced soil changes and the production of crops. The review will include both (i) sequential systems in which trees and crops are either grown in rotation or maximum growth rates of the tree and crop components occur at different times and (ii) simultaneous systems in which trees and crops are grown together. The tree and crop components compete for nutrients and water in simultaneous systems, but competition for growth resources is absent or minimal in sequential systems (Rao et al., 1997). A large reduction in nutrient losses can be achieved through control of erosion by agroforestry (Kiepe and Rao, 1994; Nair et al., 1995; Craswell et al., 1997). This process, however, lies outside the scope of this review.

Increased nutrient supply

Deep capture and uptake of nutrients

Trees with deep roots can potentially intercept nutrients leaching down soil profiles and 'capture' nutrients accumulated in subsoil below the rooting depth of annual crops (van Noordwijk et al., 1996). Nutrients taken up by trees from below the rooting zone of annual crops become an input when transferred to surface soil in the form of leaf litter, roots and prunings of tree leaves and branches (Schroth, 1995). The lateral extension of tree roots can be considerable, particularly in semiarid areas (Breman and Kessler, 1995). The lateral capture and uptake of nutrients from within the rooting zone of crops, however, represents a redistribution of nutrients within the soil-plant system rather than an input.

There is no doubt that tree roots can extend beyond the rooting depth of annual crops (Stone and Kalisz, 1991; Torquebiau and Kwesiga, 1996; Mekonnen et al., 1997) and that the spatial distribution and temporal patterns of root growth vary among tree species (Jonsson et al., 1988; Ruhigwa et al., 1992; Hauser, 1993a). The extent to which tree roots in deep soil layers contribute to the overall uptake of nutrients, however, is less clear. Buresh (1995) concluded from a review of the literature that the potential of trees to retrieve subsoil nutrients is generally greatest when (i) trees have deep rooting systems and high demand for nutrients, (ii) water and/or nutrient stress occurs in the surface soil and (iii) considerable reserves of plant-available nutrients or weatherable minerals occur in the subsoil. Greater capture of subsoil resources by roots would be expected for water and mobile nutrients, such as nitrate, than for less mobile nutrients such as P. There is typically little potential of trees to capture P from below the rooting depth of crops (IAEA, 1975) because plant extractable P is normally low in subsoil (Breman and Kessler, 1995).

Acid subsoils rich in iron oxides and low in organic matter can have appreciable anion sorption capacity, which enables the soil to retain leached nitrate (Cahn et al., 1992). Hartemink et al. (1996) reported that about 60% of the nitrate at 1- to 2-m depth on a Kandiudalfic Eutrudox in western Kenya was sorbed on soil surfaces. Sorption of nitrate delays its downward movement (Wong et al., 1987) and results in nitrate accumulation in the subsoil (Wild, 1972). Michori (1993) observed 2200 kg nitrate-N ha^{-1} at 1- to 5-m depth under fertilized coffee in Kenya. The peak in subsoil nitrate corresponded to a soil layer with low pH, high positive surface charge and 1:1 clay minerals. Kihanda (1996) found > 400 kg N ha^{-1} as nitrate + ammonium at 0.6- to 2.4-m depth on an Ando-Humic Nitisol in Kenya after 7 years of maize (*Zea mays*) cropping with annual application of 50 kg N ha^{-1}. Measurement of N leaching with ^{15}N-labeled fertilizer in cores indicated 10% recovery of added N at 0.8- to 2.4-m depth after 1.5 years.

An accumulation of nitrate at 0.5- to 2.0-m depth has been observed under unfertilized maize on acid soil (Kandiudalfic Eutrudoxs or Kandiudalfs, depending on topsoil clay content) in western Kenya (Table 1). Mean nitrate levels in six farmers' fields before the initiation of researcher-managed experiments ranged from 70 to 315 kg N ha^{-1} at 0.5- to 2.0-m depth, with relatively high standard deviations. Aluminum saturation to 2-m depth at the sites was < 10%, and growth of maize was severely limited by low available soil P and parasitism by *Striga hermonthica*. The accumulation of subsoil nitrate is attributed to greater formation of nitrate by mineralization of SOM than uptake of N by unfertilized maize (Mekonnen et al., 1997). The excess

Table 1. Nitrate at 0.5- to 2.0-m depth on acid soils in farmers' fields in Vihiga and Siaya Districts in western Kenya.

Location	Cropping history	Number of plots	Nitrate–N (kg ha^{-1})	
			Mean	Standard deviation
Ochinga	Weed fallow	12	70	27
Ochinga	Maize	30	155	96
Ochinga	Maize	20	210	67
Abuneri	Maize	32	198	102
Julius	Maize and fallow	60	133	61
Nyabeda	Maize and fallow	4	315	111

nitrate can then leach to the subsoil where it is sorbed on positively charged soil surfaces (Hartemink et al., 1996), which act as a 'chemical safety net' retarding the downward movement and leaching loss of nitrate. Unfertilized maize because of shallow rooting and poor growth is not able to take up this subsoil nitrate.

Hartemink et al. (1996) and Mekonnen et al. (1997) showed that *Sesbania sesban* grown in rotation with maize in western Kenya can rapidly root into this 'chemical safety net' and take up the sorbed nitrate that is inaccessible to unfertilized maize. Nitrate to 4-m depth was 51 kg N ha^{-1} for a 15-month *S. sesban* fallow, as compared to 199 kg N ha^{-1} for unfertilized maize (Mekonnen et al., 1997). The maximum rooting depth was 1.2 m for maize, whereas roots in a 15-month-old *S. sesban* fallow extended below 4 m.

Research in western Kenya on an acid soil, without chemical and physical barriers to rooting in the top 4 m, suggests a direct relationship between the demand of trees for N and the magnitude of nitrate uptake from the subsoil (Jama et al., 1998). Fast-growing trees, such as *S. sesban* and *Calliandra calothyrsus*, reduced soil nitrate in the top 2 m by 150 to 200 kg N ha^{-1} by 11 months after establishment from seedlings (Figure 1). The reduction in soil nitrate at 0.5- to 2-m depth was about 75 kg N ha^{-1} under *S. sesban* and about 100 kg N ha^{-1} under *C. calothyrsus*. Van Noordwijk (1989) reported that root length densities of about 0.1 cm cm^{-3} in the subsoil are sufficient for utilizing virtually all available nitrate. Root length densities for *S. sesban* and *C. calothyrsus* were $\geq$ 0.1 cm cm^{-3} to below 1.5-m depth at 11 months. For slower growing *Grevillea robusta*, with low demand for N, root length density at 11 months was $\geq$ 0.1 cm cm^{-3} to only 0.45-m depth, and soil nitrate increased rather than decreased during the 11 months after establishment (Figure 1).

The procedures for accurately determining spatial variability of subsoil nitrate concentrations as influenced by trees have not yet been fully developed. For example, at two locations in Kenya, subsoil nitrate below *S. sesban* was highly variable among experimental plots and with distance from a tree (Mekonnen, 1996; Kipleting, 1996). This high spatial variability will lead to experimental challenges in (i) the selection of sampling locations that are representative of the unit soil surface area in agroforestry systems and (ii) the quantification of subsoil N capture and recycling by trees.

There is increasing evidence that capture of subsoil nitrate by trees might be an important process in subhumid and humid areas on deep soils with relatively high base status and anion sorption capacity. Many of these soils, however, are P deficient. The effective utilization by crops of the N transferred from subsoil to topsoil by trees will require P fertilization of crops on P-deficient soils. Rotation of *S. sesban* with maize at the study site in western Kenya, for example, did not eliminate P deficiency in the subsequent maize crop (ICRAF, 1996). Furthermore, soil chemical and physical barriers to deep rooting will reduce the potential of trees to capture and recycle subsoil nutrients. High aluminum saturation in subsoil can markedly restrict deep

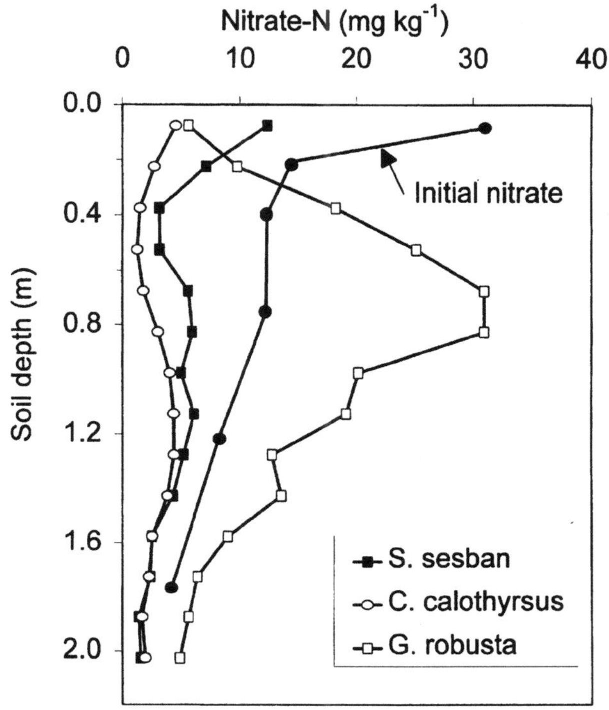

Fig. 1. Effect of trees on soil nitrate eleven months after planting as seedlings in western Kenya. *Source:* Adapted from Jama et al. (1998).

rooting and retrieval of deep soil nutrients on Ultisols and Oxisols (van Noordwijk, 1989).

In semiarid areas, the lateral capture of soil nutrients by trees is considered to be much more important than the capture from deep soil layers (Breman and Kessler, 1995). Based on a review of research in West Africa, Kessler and Breman (1991) concluded that deep rooting in the Sahelian zone (mean annual rainfall ≤ 600 mm) is limited by low levels of available nutrients and water in subsoil. In the Sudanian zone (savanna, mean annual rainfall = 600 to 1200 mm) deep rooting is limited by shallow soils and compacted subsoil layers. Deans et al. (1994), nonetheless, found fine roots of 9-year-old *Acacia senegal* and neem (*Azadirachta indica*) down to the water table at 32-m depth in the Sahel in Senegal. Subsoil nitrate was very high below *A. senegal* and *Faidherbia albida* and exceeded the European limit in drinking water (11 mg N L^{-1}) in many soil layers down to the water table. High subsoil nitrate was also observed below an open field frequently cropped with groundnut (*Arachis hypogaea*). Deans et al. (1994) speculated that N_2 fixation was the likely source of the subsoil nitrate. The potential role of trees in reducing the downward movement of nitrate and in capturing the accumulated subsoil nitrate in semiarid areas is uncertain.

Reduced leaching

Indirect evidence of reduced leaching loss of nutrients under trees was provided by an 18-month study on a Kandiudalfic Eutrudox in western Kenya (ICRAF, 1996). Subsoil water (1- to 2-m depth) was lower under a *S. sesban* fallow than under a natural uncultivated fallow and unfertilized maize monoculture (Figure 2). Subsoil water in the S. sesban fallow seldom exceeded field capacity despite a mean annual rainfall of about 1800 mm. Subsoil water in the natural uncultivated fallow and maize monoculture, on the other hand, occasionally exceeded field capacity, indicating that mobile water was present to transport nitrate downward. The duration and potential for leaching were underestimated in this study because soil water was measured intermittently rather than continuously. The relatively low soil water (Figure 2) and low soil nitrate (Braun, 1995; van Bodegom, 1995; Hartemink et al., 1996) under *S. sesban*, nonetheless, provide indirect evidence for less leaching loss. The low subsoil water and nitrate content under *S. sesban* can be attributed to high water and N demand by the fast-growing *S. sesban*.

Hedgerow intercropping of *Leucaena leucocephala* with a maize-cassava (*Manihot esculenta*) rotation reduced nitrate leaching on a Ferralic-haplic Acrisol with 87% sand content in the Benin Republic (Horst et al., 1995). Nitrate leaching was estimated from nitrate concentration in soil solution and water flux below 1.5-m soil depth. The *L. leucocephala* hedgerows increased

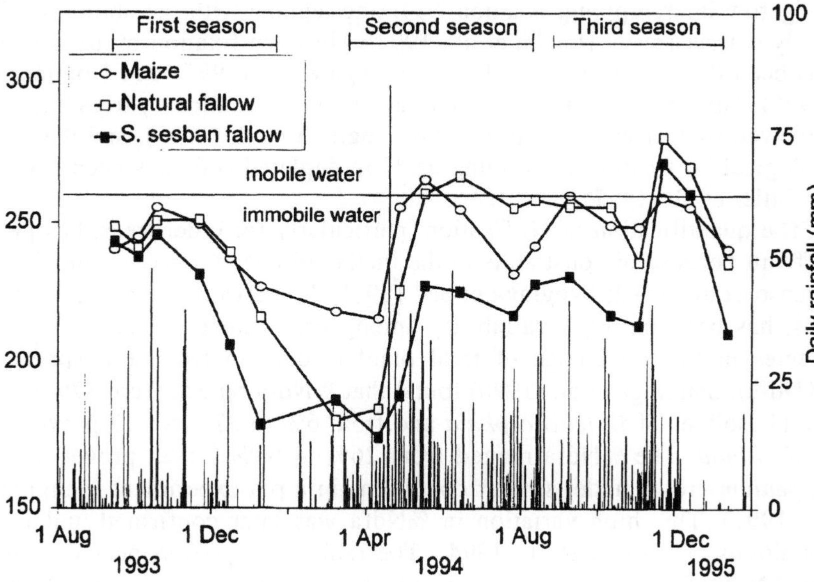

Fig. 2. Effect of land-use system on soil water at 1- to 2-m depth on a Kandiudalfic Eutrudox in western Kenya. *Source:* ICRAF (1996).

total root length density under both maize and cassava. Crop yields, however, were not increased by hedgerow intercropping.

The role of trees in reduction of leaching might be particularly important during periods of reduced ground cover by herbaceous plants, such as between cropping seasons. At the beginning of rainy seasons, root length density increased at 0.3- to 0.5-m depth in *Gliricidia sepium* hedgerows in central Côte d'Ivoire (Schroth and Zech, 1995). The authors speculated that this subsoil rooting might reduce nutrient leaching at the onset of rains when, as reported by Wong and Nortcliff (1995), N release from the flush in mineralization of SOM can exceed N demand by crops and weeds.

Leguminous trees can effectively exploit soil nitrate, thereby resulting in N inputs to the soil-plant system through retrieval of deep soil N and reduction in N leaching. These inputs of N, however, might occur to the detriment of N input by biological N_2 fixation (Giller, 1997). Effective net increases in N inputs by N_2-fixing leguminous trees require that they meet a substantial proportion of their N requirements through biological N_2 fixation, particularly when uptake of subsoil nitrate is low.

Biological nitrogen fixation

The N contribution of leguminous trees through N_2 fixation is well recognized, although it is important to note that not all legumes fix N_2 (MacDicken, 1994). *Senna spectabilis*, for example, which has been widely tested in hedgerow intercropping, is a non-N_2-fixing legume with the ability to effectively extract soil N (Ladha et al., 1993). Nitrogen fixation by trees in SSA has been thoroughly reviewed by Sanginga et al. (1995) and Dommergues (1995). The effects of environmental factors on N_2 fixation by trees have been reviewed by Danso et al. (1992) and Sanginga et al. (1995), and the role of biological N_2 fixation in building up N in depleted soils has been reviewed by Giller et al. (1997).

The quantification of N_2 fixation, particularly for older trees, has proven difficult because of constraints in the methodologies for measuring N_2 fixed (Danso et al., 1992; Sanginga et al., 1996). A noteworthy finding, nonetheless, has been the high variability among provenances or isolines of tree species in the percentage of total plant N derived from atmospheric N_2 (%Ndfa). Sanginga et al. (1990) found that %Ndfa ranged from 37% to 74% for 11 isolines of *L. leucocephala* and from 6% to 37% for 13 provenances of *F. albida*. The %Ndfa ranged from 26% to 68% for 25 provenances of *G. sepium* grown under similar conditions in a pot experiment (Sanginga et al., 1991). This high variation in %Ndfa was later confirmed under field conditions (Sanginga et al., 1994). The ranking of provenances or isolines for %Ndfa, however, is very dependent on growth stage, and differences observed in short-duration pot studies may not reflect the long-term N_2 fixation in the field (Sanginga et al., 1995).

The deficiency of available P, which is characteristic of many SSA soils, can limit N_2 fixation and growth of N_2-fixing trees. Large differences in growth and P use efficiency, nonetheless, occur among and within N_2-fixing tree species (Sanginga et al., 1994; 1995). Three provenances of *G. sepium*, for example, differed significantly in %Ndfa on P-deficient soil (Sanginga et al., 1994). Nodulation and symbiotic N_2 fixation were stimulated relatively more by 20 kg P ha^{-1} than by higher P rates. Differences in tolerance to low soil P were critical for early tree growth, but they tended to disappear with increasing tree age. These results highlight the merit of selecting provenances of N_2-fixing trees that are tolerant to low available P at an early growth stage.

Enhancement in N inputs from N_2 fixation requires the effective integration of appropriate tree and herbaceous legumes within farming systems (Giller and Cadisch, 1995). Increases in N_2 fixation by trees can likely be achieved through careful selection of tree germplasm and proper fertilization, particularly with P. Inoculation with ***Rhizobium*** or ***Frankia*** has greatest potential on specific (non-promiscuous) N_2-fixing trees grown on sites with low or negligible density of compatible ***Rhizobium*** or ***Frankia*** (Dommergues, 1995). The potential for adoption of inoculants, however, might be limited considering that adoption of rhizobial inoculants for food legumes in developing countries remains low despite years of research.

Input of organic materials

Nutrients released from the biomass of trees grown within a soil-plant system constitute internal cycling, except for the nutrients added by N_2 fixation, retrieval from below the rooting depth of crops and lateral retrieval from outside the soil-plant system. The addition of organic materials originating from outside the soil-plant system is a nutrient input. Examples include biomass from trees and shrubs, farmyard manure and composts. Their effectiveness as nutrient sources depends upon their composition (Mafongoya et al., 1997).

Leaf biomass from trees and shrubs in the miombo woodland of southern Africa (Nyathi and Campbell, 1993) or from along field and farm borders is a potential source of nutrients for crops. The transfer of biomass to fields, however, involves the 'mining' of nutrients from outside the field in order to add nutrients to the field. The transfer of nutrients either directly through tree biomass or indirectly through foraging livestock can contribute to microvariability in soil fertility and plant growth, which is particularly prevalent in the Sahel (Brouwer et al., 1993; Brouwer and Bouma, 1997). Nutrients from farmyard manure are not a net input to a soil-plant system if they were originally exported from the system in plant biomass to feed the animals (Haque et al., 1995). Considerable nutrients, including P, can be lost by leaching from farmyard manure applied to sandy soil (Brouwer and Powell, 1995).

One potentially promising shrub for biomass transfer in humid and subhumid areas is *Tithonia diversifolia*, a non-N_2-fixer found along field borders and roads (Gachengo, 1996; ICRAF, 1997). Analysis of 100 samples of leaves + tender stems of *T. diversifolia* in Sri Lanka revealed ranges of 32 to 55 g N kg^{-1}, 2.0 to 5.2 g P kg^{-1} and 23 to 55 g K kg^{-1} on a dry weight basis (Nagarajah and Nizar, 1982). Fresh leaf biomass of *T. diversifolia* decomposes rapidly and is an effective source of N, P and K for maize (Gachengo, 1996) and vegetables (ICRAF, 1997). The cutting of *T. diversifolia* biomass from hedges and transfer to fields is financially more attractive for higher valued crops, such as vegetables, than for maize (ICRAF, 1997).

Jama et al. (1997) compared leaf biomass from *C. calothyrsus*, manure from dairy cattle fed *C. calothyrsus* as a protein supplement and commercial triple superphosphate (TSP) fertilizer as sources of P for maize on P-deficient acid soil at two sites in western Kenya. Application of all sources at 10 kg P ha^{-1}, either individually or as mixtures of organic (manure or *C. calothyrsus*) and inorganic (TSP + urea) sources, significantly increased maize yield and had a residual benefit on the following maize crop. Financial analysis indicated that *C. calothyrsus* was much more attractive as a protein supplement for dairy cattle than as a P source for maize production. These results indicate that tree biomass with value as a protein supplement for dairy cattle is better fed to cattle than used as a P source for maize; the resulting manure can then be used as a nutrient source for soil fertility improvement.

Leaf biomass of trees frequently has a higher N-to-P ratio than the ratio of N to P required by crops (Palm, 1995). Therefore, the quantities of biomass that are sufficient to supply N to a crop can be insufficient to supply P. Jama et al. (1997) concluded that rather than using organic material with a high N-to-P ratio to supply all of the required P, it would be financially more attractive to integrate an inorganic P source with the organic material, whereby the organic material would provide the required N for the crop and the inorganic P source would meet the additional requirement for P.

Increased nutrient availability

Scattered trees, particularly in semiarid areas, have long been recognized to represent islands of improved soil conditions within landscapes (Breman and Kessler, 1995; Rhoades, 1997). *Faidherbia albida* is well known for the so-called 'albida effect' which refers to greater growth of crops or herbaceous plants under a tree canopy than in an open field. Increased growth of herbaceous plants has been observed under the canopy of a number of tree species, including non-N_2 fixers. This increased growth under tree canopies can result from a combination of factors such as (i) increased nutrient inputs including manure and urine from livestock 'camping' under the tree (Brouwer and Bouma, 1997), (ii) increased nutrient availability through enhanced soil biological activity and rates of nutrient turnover and (iii) improved microclimate and soil physical properties. There is, however, increasing evidence that trees selectively establish at sites with favorable soil properties (Adderley et al., 1997). At least part of the difference in soil properties that has been observed in comparisons of soil from beneath the canopy of *F. albida* and in open fields can result from preexisting differences (Geiger et al., 1994). Reports of improved soil properties under tree canopies must, therefore, be interpreted with caution.

The decomposition of organic materials from trees not only supplies nutrients but can also increase plant availability of nutrients in soil by processes such as (i) supplying an energy source to soil organisms thereby enhancing nutrient cycling through soil organisms, (ii) reducing P-sorption capacity of soil and (iii) stimulating plant root growth (Palm et al., 1997). Relatively high rates of fresh *T. diversifolia* biomass (5 Mg ha^{-1} on a dry weight basis), for example, increased soil biological activity and reduced P sorption on an acid soil in western Kenya (ICRAF, 1997; Nziguheba et al., 1998). These effects may lead to enhanced plant availability of soil P.

Soil organic matter

Numerous studies have reported higher SOM in topsoil under trees than in open areas (Mulongoy et al., 1993; Breman and Kessler, 1995; Young, 1997). Mordelet et al. (1993), for example, reported higher soil organic C under tree clumps than in open grassland in a humid savanna in Côte d'Ivoire. Trouve et al. (1994) found a progressive increase in SOM under *Eucalyptus* sp. and *Pinus caribaea* in plantations on sandy soil in the Congo, and Belsky et al. (1993) showed an accumulation of SOM at 0- to 15-cm depth under *Acacia tortilis* and *Adansonia digitata* in a savanna in Kenya. Kang (1997) reported that soil organic C at 0- to 15-cm depth after 5-year hedgerow intercropping with *L. leucocephala* was 12.3 g kg^{-1} under the hedgerow and 9.4 g kg^{-1} between hedgerows, as compared to 5.9 g kg^{-1} in the control without a hedgerow.

The effect of trees on SOM varies among tree species and soils (Kang et al., 1994). Organic C at 0- to 15-cm depth was generally higher under 10-year-old trees than under adjacent grass vegetation in Nigeria. The relative difference between tree fallows and grass vegetation, however, was greater on an Oxic Paleustalf in the forest-savanna transition zone than on a Typic Paleudult in the forest zone. Nyberg and Högberg (1995) used ^{13}C natural abundance to demonstrate varying contributions of 4- to 5-year-old trees of six species to soil organic C on farms in Kenya. A lower ^{13}C abundance in soils under a tree than outside the tree canopy reflected a shift in inputs of C from C_4 grasses to C_3 trees. A significant decrease ($P < 0.05$) in ^{13}C abundance was observed under four of the six species, whereas the soil organic C was significantly higher under only two of the species.

Total SOM frequently does not relate to crop yields (Sanchez and Miller, 1986), and nutrient release from SOM is normally more dependent on the portion of the SOM in biologically active fractions than on total quantity of SOM. Recent research has focused on the identification of biologically active

SOM fractions that are important sources of nutrients (Bernhard-Reversat, 1993; Vanlauwe et al., 1996). The amount of N in the sand-associated fraction of SOM that floats in a dense liquid (referred to as light fraction N) was recently identified as a sensitive measure of SOM differences among cropping systems (Hassink, 1995; Barrios et al., 1996a). The N in this light fraction (particle size > 150 μm, density < 1.37 g cm^{-3}) also correlated with N mineralization of the whole soil (Hassink, 1994; Barrios et al., 1996b).

Light fraction SOM under maize can be increased by addition of tree biomass (Barrios et al., 1996a) and by the rotation of maize with planted tree fallows (Barrios et al., 1997). In these studies, the tree biomass and tree fallows had no effect on total soil organic C and N as compared to a control with continuous unfertilized maize. The quantity of light fraction SOM can differ among tree species (Bernhard-Reversat, 1987). *Sesbania sesban* was more effective than five other tree species (*C. calothyrsus*, *Flemingia macrophylla*, *G. sepium*, *L. leucocephala* and *Senna siamea*) in increasing light fraction SOM at 0- to 15-cm depth after 2- and 3-year fallows on an Ustic Rhodustalf in Zambia (Barrios et al., 1997). The quantity of light fraction SOM after the 3-year tree fallows was not correlated with grain yield of maize during the following season. The amount of N in the light fraction, however, correlated ($P < 0.01$) with maize yield (E. Barrios, personal communication, 1997).

Soil biological activity

The decomposition of SOM and plant residues is largely controlled by soil biota, particularly macrofauna (Tian et al., 1992). These organisms are important for the availability of nutrients in low input systems, where crops largely depend upon nutrients released from organic materials rather than from inorganic fertilizers. Adejuyigbe et al. (1998) and Tian et al. (unpub.) in monthly monitoring of the population dynamics of soil fauna, found higher earthworm and microarthropod populations under planted woody fallows than under continuous cropping with a maize/cassava intercrop (Table 2). The relatively low earthworm population under *L. leucocephala* can be attributed to relatively lower soil water content under *L. leucocephala* (Tian et al., unpub.). Microarthropod populations were positively correlated with lignin content of the leaf litter in the natural fallow and planted fallow plots (Adejuyigbe et al., 1998).

Tian et al. (1993a) found that tree residues applied as a soil mulch increased earthworm, termite and ant populations. Earthworm populations were inversely correlated with the lignin-to-N ratio of plant residues, and ant populations were positively related to N content of residues. Tian et al. (1995) integrated the C-to-N ratio, lignin content and polyphenol content of plant residues into a quality index. Ant density and decomposition of the residues correlated positively to the plant quality index, whereas termite population correlated negatively to the plant quality index.

In hedgerow intercropping, Kang et al. (1990) reported a higher number

Table 2. Effects of tree fallows on mean earthworm population at 0- to 30-cm depth, and soil microarthropod population at 0- to 6-cm depth during the wet season in Nigeria.

Treatment	Earthworm population (no. m^{-2})[1]	Microarthropod population (no. m^{-2})[2]
Continuous cropping	58 b	21900 c
Natural bush fallow	194 a	32600 ab
Leucaena leucocephala	118 a	27900 bc
Acacia leptocarpa	100 a	45000 a
Senna siamea	142 a	45300 a

Notes: Analysis of variance based on log $(x + 1)$ transformed data. Values in a column followed by the same letter are not significantly different at $P = 0.05$.
[1] May to October 1994.
[2] May to September 1995.
Sources: Earthworm population: G. Tian, personal observation, 1997; Microarthropod population: Adejuyigbe et al., 1998.

of earthworm casts with *Dactyladenia (Acioa) barteri, Alchornea cordifolia, G. sepium* and *L. leucocephala* than in the control with no trees. Hauser (1993b) found that mean dry matter of earthworm casts was 117 Mg ha^{-1} yr^{-1} under a *L. leucocephala* hedgerow as compared to 28 Mg ha^{-1} yr^{-1} in control plots without trees. Increased microbial biomass was also found with hedgerow intercropping in southwest Nigeria (van der Meersch et al., 1993).

Nitrogen

Barrios et al. (1997) measured availability of soil N following 2- and 3-year fallows on a N-deficient Ustic Rhodustalf in eastern Zambia. Nitrogen availability, as determined by N in light fraction SOM, inorganic soil N (nitrate + ammonium) and aerobic N mineralization at 0- to 15-cm depth differed among the fallows (Table 3). Among the trees, all measures of N availability were higher for S. sesban than for the mean for five other tree species and for the natural grass fallow. The measures of N availability were higher for the two tree species with lowest (lignin + polyphenol)-to-N ratios in leaf litter (*S. sesban* and *G. sepium*) than for the two species with highest ratios in leaf litter (*C. calothyrsus* and *F. macrophylla*). *Sesbania sesban* fallow and fertilized maize monoculture resulted in similar soil inorganic N levels, but N mineralization and light fraction N were greater after *S. sesban* (data not shown). These results confirm that certain trees can increase N availability as compared to continuous cropping without fertilization. Tree species, however, vary greatly in effect on N availability and hence their suitability for soil fertility replenishment (Schroth et al., 1995; Barrios et al., 1997).

Few studies have examined the relationship between tree-induced soil changes in nutrient availability and yield of crops. In the tree fallow study in eastern Zambia mentioned above, Barrios (personal communication, 1997)

Table 3. Effects of fallows on N availability on an Ustic Rhodustalf in eastern Zambia.

Treatment	(L + PP)/N in leaf litter	Light fraction N (mg kg^{-1} soil)	Inorganic N (mg kg^{-1})	Aerobic N mineralization (mg kg^{-1} day^{-1})
Natural grass fallow	–	15	11	0.45
Sesbania sesban	8.5	22	23	0.60
Gliricidia sepium	13	15	16	0.38
Leucaena leucocephala	16	13	10	0.36
Flemingia macrophylla	21	16	13	0.30
Senna siamea	16	13	7	0.27
Calliandra calothyrsus	20	16	12	0.27
Single df contrasts				
S. sesban vs. other trees		***	***	***
S. sesban vs. natural fallow		***	***	**
Low vs. high (L+PP)/N[a]		*	***	***

Notes: All values are means for 2- and 3-year fallows.
*, **, and *** = significant at the 0.05, 0.01, 0.001 levels, respectively.
L + PP = lignin plus polyphenols.
[a] *S. sesban* and *G. sepium* vs. *C. calothyrsus* and *F. macrophylla.*
Source: Adapted from Barrios et al. (1997).

found that yield of unfertilized maize following nine 3-year systems (treatments in Table 3 plus a groundnut/maize/soybean [*Glycine max*] rotation and unfertilized maize monoculture) was not related to total SOM or total soil N. Soil inorganic N, aerobic N mineralization and N in light fraction SOM at 0- to 15-cm depth, however, correlated positively ($P < 0.05$) with maize grain yield at the N-deficient site. Soil inorganic N before maize planting correlated better than N mineralization and N in light fraction SOM with maize yield (Figure 3). Initial soil inorganic N, however, might not be as well related to crop yield in areas with higher rainfall and appreciable leaching of nitrate during early crop growth.

Maize yield after 3-year fallows in eastern Zambia was well predicted ($r^2 = 0.65$) with a multiple regression that included soil inorganic N before maize planting and population of *Striga asiatica* during maize growth (Barrios et al., unpub.). *Striga asiatica* is a parasitic weed that can be controlled to varying degrees by different tree fallows. This observation suggests that when trees differ greatly in their effect on a major crop pest, the incidence of the pest should be considered along with soil nutrient pools in the prediction of crop yield.

Schroth et al. (1995) reported that N mineralization following 5-year-old tree fallows in Côte d'Ivoire differed significantly among nine tree species. Aerobic N mineralization, after an initial flush following soil wetting, correlated positively with dry matter production of rice in a bioassay in pots with soil collected from the tree fallows. Their results suggest that N availability could be increased by selecting trees with high litter production and high root mass. Jonsson et al. (1996), however, showed that the quality of tree

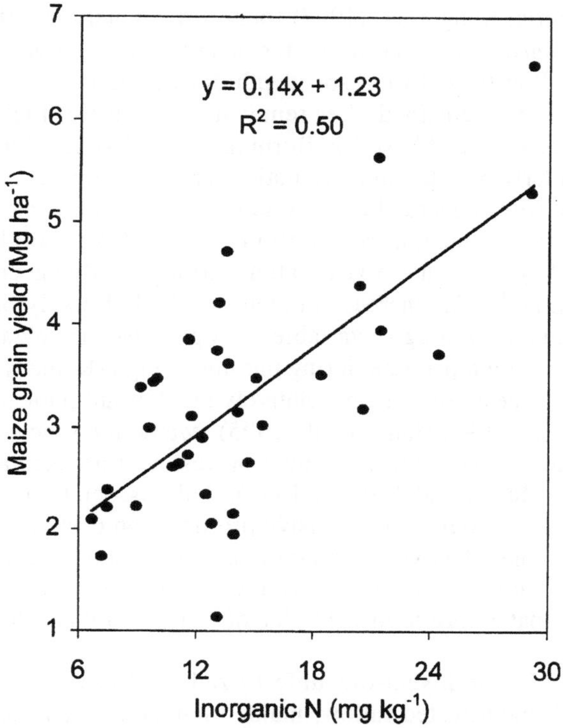

Fig. 3. Relationship between soil inorganic N and maize yield on an Ustic Rhodustalf in eastern Zambia. *Source:* Adapted from ICRAF (1996).

leaves was also related to crop growth. They found that maize growth in a bioassay in pots with soil from five fallows (each with a different tree species) in Tanzania was inversely related to the (lignin + phenolics)-to-N ratio of the tree leaves.

Phosphorus

Most studies have found little or no benefit of trees on extractable inorganic soil P. Siaw et al. (1991) found no significant difference in available soil P under hedgerow intercropping with *L. leucocephala* and *D. barteri* as compared to a control without trees. Weil and Mughogho (1993) found no consistent and significant effect of *F. albida* on Bray 1 extractable P in Malawi. Kang et al. (1994) observed greater Bray 1 extractable P under 10-year-old trees than under adjacent grass vegetation on a Typic Paleudult but not on an Oxic Paleustalf in Nigeria. The Paleudult, however, was not deficient in plant-available P. Some studies have reported a decrease in extractable inorganic P under trees, presumably because of sequestration of P in tree biomass (Haggar et al., 1991).

In research on the use of woody species to regenerate degraded soil in southwest Nigeria, Kang et al. (1997) found little beneficial effect of 4-year-old trees (*S. siamea*, *L. leucocephala*, *Acacia leptocarpa* and *A. auriculiformis*) on extractable P at 0- to 15-cm soil depth during the fallow phase. Burning of tree residues that remained after removal of the useful fuelwood, considerably increased extractable soil P. Burning of piled vegetation in a miombo woodland similarly resulted in a dramatic increase in extractable inorganic P to 0.4-m depth (Strømgaard, 1992). Acid extractable inorganic P at 0- to 10-cm soil depth, for example, increased from 1–12 mg P kg^{-1} soil before burning to 19–24 mg P kg^{-1} soil at 3.5 years after burning piled vegetation (50 or 63 Mg dry weight ha^{-1}). The increase in plant-available P likely originates from the burnt biomass, although available P might also be released from soil organic P when the temperature during burning is high (Romanya et al., 1994).

Trees might access soil P from relatively recalcitrant pools not utilized by crops (Palm et al., 1991; Hands et al., 1995). Many tree species in the tropics form associations with ectomycorrhizal or vesicular-arbuscular mycorrhizal (VAM) fungi (Mason and Wilson, 1994). This association can enhance P uptake by the tree through the extensive proliferation of mycorrhizal hyphae, which results in the effective exploitation of a larger soil volume. Mycorrhizal fungi may also contribute to the production of phosphatase enzyme and organic acids that improve availability of soil P (Bolan, 1991; Handreck, 1997).

The effect of four 2-year-old unfertilized land-use systems (*S. sesban* fallow, natural uncultivated fallow, maize monoculture and bare fallow) on soil inorganic and organic P fractions was recently examined at two P-deficient sites (Kandiudalfic Eutrudox and Kandic Paleustalf) in Kenya. Conventional measurements of extractable inorganic soil P (acid extractable P, bicarbonate extractable P, P extracted with iron-oxide impregnated paper, resin extractable P and P desorbable by repeated resin extractions) at 0- to 15-cm depth did not detect differences in soil P availability. The amount of P in light fraction SOM, however, was greater following *S. sesban* and natural fallows than unfertilized maize monoculture at both sites (J. Maroko, personal communication, 1996). The amount of P in light fraction SOM correlated with grain yield of the subsequent maize crop (ICRAF, 1996). Chloroform extractable P, which is directly related to microbial biomass P, also correlated with maize grain yield on the Eutrudox (Figure 4), which contained only 2 mg kg^{-1} bicarbonate extractable P. These results highlight the importance of SOM as a source of plant-available P in unfertilized tropical soil with low extractable inorganic P (Friend and Birch, 1960).

The above study on an Eutrudox in western Kenya showed that rotation of unfertilized maize with *S. sesban* rather than continuous maize cropping can slightly increase availability of soil P, as determined by chloroform extractable P, P in light fraction SOM and sodium hydroxide-extractable organic P (Buresh et al., 1997). The *S. sesban* fallow also increased maize yield on this P-deficient soil, but it did not eliminate P deficiency in subsequent maize crops. High yields of maize following the *S. sesban* fallow required P inputs (ICRAF,

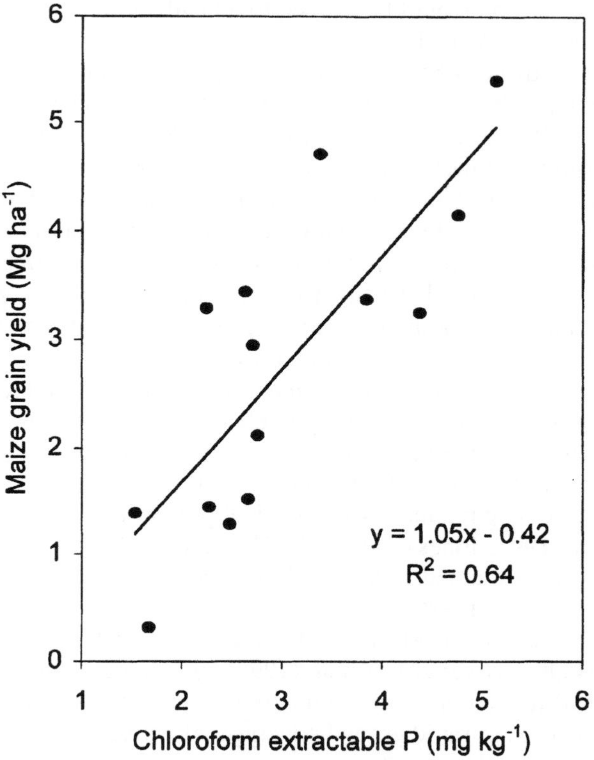

Fig. 4. Relationship between soil chloroform extractable P, which is directly related to microbial biomass P, and maize yield on a Kandiudalfic Eutrudox in western Kenya. *Source:* J. Maroko, personal communication, 1997, unpublished.

1996). The rotation of *S. sesban* with maize became much more financially attractive when P fertilizer was applied to maize (ICRAF, 1997).

Improved soil physical and chemical environment

The ability of trees and biomass from trees to maintain or improve soil physical properties is well documented (Young, 1997). Higher availability of soil water can be maintained under trees because of interception and redistribution of rainwater within the system, reduced evapotranspiration and increased infiltration (Breman and Kessler, 1995; Wallace, 1996). Torquebiau and Kwesiga (1996) showed that 2-year fallows with *S. sesban* can decrease soil bulk density and resistance to penetration and increase water infiltration on an Alfisol in eastern Zambia. Kang et al. (1997), however, reported that four years of fallow with various woody species (*S. siamea*, *L. leucocephala*, *A. leptocarpa* and *A. auriculiformis*) did not improve bulk density and aggregate stability in a degraded Alfisol in southwest Nigeria. They suggested that a longer fallow period would be needed to amend soil physical conditions of the highly degraded Alfisol.

The effect of trees in hedgerow intercropping on soil physical properties has been studied in detail in southwest Nigeria by Lal (1989a; b) and Hulugalle and Kang (1990), and reviewed by Rao et al. (1997). Hulugalle and Kang (1990), in a study that included *L. leucocephala, G. sepium, A. cordifolia, and D. barteri* on an Oxic Paleustalf, found that particle size distribution, bulk density, apparent pore size distribution and water infiltration were superior in the hedgerow plots as compared to the control without trees. Hulugalle and Ndi (1993) reported that hedgerow intercropping with *S. spectabilis* on a Typic Kandiudult in southern Cameroon reduced dry season soil temperature and surface seal formation and increased infiltration rate. Whereas high quality (low C-to-N ratio and lignin content) tree residues can directly enhance crop production through relatively rapid release of nutrients, low quality (high C-to-N ratio and lignin content) residues can benefit crops through mulching effects on the microclimate (Tian et al., 1993b).

Trees can also help maintain soil chemical properties (Dalland et al., 1993; Ruhigwa et al., 1993; Jones et al., 1996; Wendt et al., 1996). Juo et al. (1995, 1996) found that a planted fallow of *L. leucocephala*, but not of pigeonpea (*Cajanus cajan*), was as effective as a natural bush fallow in maintaining soil pH, exchangeable Ca and Mg and effective cation exchange capacity on an Oxic Kandiustalf in Nigeria. Juo et al. (1995) hypothesized that planted fallow systems should include several contrasting species in order to combine the beneficial attributes of rapid ground coverage, effective biological N_2 fixation and cycling of subsoil nutrients.

Implications for sustainable soil management

Smallholder farmers in SSA are frequently constrained by low income from on-farm and off-farm sources. Small farm size, low crop yields, climate- and pest-related risks, and lack of high-value enterprise options can all contribute to low farm income. As a consequence of low income, farmers have frequently not purchased sufficient fertilizer inputs to offset the outputs of nutrients with harvested plant products (Cooper et al., 1996). Sanchez and Leakey (1997) and Sanchez et al. (1997b) advocated (i) investment in soil fertility replenishment (ii) intensive use of agroforestry trees to produce diverse high-value products and (iii) enabling policies, as three crucial ingredients for the transformation of smallholder farms in SSA to productive and sustainable enterprises. High-value trees (e.g., for indigenous fruits, medicinal products, fodder and high-grade timber) can fit specific niches on farms while leaving land for production of staple food crops and income-generating crops, such as vegetables. Trees can also be vital for soil fertility replenishment, particularly when their contribution to soil improvement involves increased farm income originating from either a reduction in costs or an increase in benefits.

Planted tree fallows have demonstrated potential for soil improvement on

N-deficient soil (Kwesiga and Coe, 1994; Barrios et al., 1997) where farmers perceive declining soil fertility, and fallowing is common but fallow periods are decreasing (Franzel, 1998). An example of such an area is eastern Zambia (population density = 25 to 40 persons km^{-2}) where *S. sesban* fallows are financially attractive and are being evaluated by thousands of farmers (Kwesiga et al., 1998). The decomposing litter, leaves and roots from 2-year *S. sesban* dramatically increase N availability (Table 3) and yield of subsequent crops. The *S. sesban* also provides fuelwood.

Short duration fallows (< 14 months), in which crop production is foregone for only one season, may have some potential in areas with high population where cropping is often continuous, such as western Kenya (Franzel, 1998). The potential is greatest when returns to cropping are low, the opportunity cost of labor is high and the tree provides a valuable product such as fuelwood (ICRAF, 1997; Swinkels et al., 1997). On P-deficient soil, P fertilization during the crop phase might be essential for financially attractive tree fallow systems, as was shown with *S. sesban* in western Kenya (ICRAF, 1997).

The transfer to crops of either leaf biomass from trees grown outside the field or farmyard manure generated with feed sources from outside the field can be an effective source of nutrients. Biomass transfer has greatest potential when (i) the biomass has a high nutrient concentration and rapidly releases plant-available nutrients after addition to the field (Palm et al., 1997), (ii) the opportunity cost for labor is low, (iii) the value of the crop is high and (iv) the biomass does not have higher-valued uses than as a nutrient source (ICRAF, 1997; Jama et al., 1997). An example of such a situation is the transfer of fresh leaf biomass from *T. diversifolia* grown near fields in western Kenya (Gachengo, 1996; Nziguheba et al., 1998).

The quantity of available biomass is frequently insufficient to meet the nutrient requirements of crops, particularly for P on P-deficient soil (Palm, 1995). Biomass, may therefore need to be supplemented with inorganic fertilizers. The integration of organic materials with inorganic P fertilizers might have potential to improve the availability of P by enhanced cycling of P through labile organic P pools (Palm et al., 1997) and reduced P-sorption capacity of soil (Iyamuremye and Dick, 1996). When biomass has value as fodder for milk production, it is financially more attractive to use the biomass as fodder and apply the farmyard manure to fields rather than to directly apply the biomass to fields (ICRAF, 1993; Jama et al., 1997).

In semiarid areas, trees can contribute to the microvariability in soil properties and soil fertility. As a result of microvariability, different parts of a field will produce the best yield in different years of contrasting rainfall. Although this reduces the probability of a large crop harvest, it reduces the risk of crop failure by ensuring crop production in poor rainfall years (Brouwer et al., 1993). Given the unlikelihood of large-scale fertilizer inputs in semiarid areas in the near future, a critical challenge may be to increasingly utilize knowledge of soil microvariability to improve the use of plant growth resources and stabilize crop yields (Brouwer and Bouma, 1997).

Whereas trees have potential to supply N to crops through biological N_2 fixation and uptake of subsoil nitrate, the potential of trees to supply P to crops is limited (Sanchez and Palm, 1996). There is typically little potential of trees to retrieve P from below the rooting depth of crops. Trees essentially utilize plant-available P that exists in the topsoil and then recycle this P to crops through decomposition of litter, prunings and roots or through burning of tree biomass. Agroforestry will, therefore, not eliminate the need for P inputs on P-deficient soils (Sanchez and Palm, 1996; Buresh et al., 1997). A possible alternative to conventional water-soluble P fertilizers (i.e., TSP, single superphosphate and diammonium phosphate) is direct application of phosphate rock (PR) (Chien and Menon, 1995). As suggested by Sanchez et al. (1997a), the use of P fertilizers, including PR, merits careful assessment in an integrated strategy with improved fallows of N_2-fixing trees and biomass transfer for replenishing soil fertility in SSA.

Acknowledgments

We thank the following for helpful suggestions in reviews of drafts of this paper: Joost Brouwer, Peter Cooper, B. T. Kang, Roger Leakey, Chuck Rhoades, Pedro Sanchez, Meine van Noordwijk and Anthony Young.

References

Adejuyigbe CO, Tian G and Adeoye GO (1998) Potentials of woody fallows in restoration of soil microarthropods in a degraded tropical soil. Agroforestry Systems (in press)

Adderley WP, Jenkins DA, Sinclair FL, Stevens PA and Verinumbe I (1997) The influence of soil variability on tree establishment at an experimental agroforestry site in North East Nigeria. Soil Use and Management 13: 1–8

Barrios E, Buresh RJ and Sprent JI (1996a) Organic matter in soil particle size and density fractions from maize and legume cropping systems. Soil Biology and Biochemistry 28: 185–193

Barrios E, Buresh RJ and Sprent JI (1996b) Nitrogen mineralization in density fractions of soil organic matter from maize and legume cropping systems. Soil Biology and Biochemistry 28: 1459–1465

Barrios E, Kwesiga F, Buresh RJ and Sprent JI (1997) Light fraction soil organic matter and available nitrogen following trees and maize. Soil Science Society of America Journal 61: 826–831

Belsky AJ, Mwonga SM, Amundson RG, Duxbury JM and Ali AR (1993) Comparative effects of isolated trees on their undercanopy environments in high- and low-rainfall savannas. Journal of Applied Ecology 30: 143–155

Bernhard-Reversat F (1987) Litter incorporation to soil organic matter in natural and planted tree stands in Senegal. Pedobiologia 30: 401–417

Bernhard-Reversat F (1993) Dynamics of litter and organic matter at the soil-litter interface in fast-growing tree plantations on sandy ferrallitic soils (Congo). Acta Oecologica 14: 179–195

Bolan NS (1991) A critical review on the role of mycorrhizal fungi in the uptake of phosphorus by plants. Plant and Soil 134: 189–207

Braun A (1995) Nutrient flows in fallow systems and maize in western Kenya. MSc thesis. Wageningen Agricultural University, Wageningen, The Netherlands
Breman H and Kessler JJ (1995) Woody plants in agro-ecosystems of semi-arid regions. Springer-Verlag, Berlin, Germany
Brouwer J and Bouma J (1997) Soil and crop growth variability in the Sahel: highlights of research (1990–95) at ICRISAT Sahelian Center. Information Bulletin No. 49. ICRISAT, Patancheru, Andhra Pradesh, India and Agricultural University of Wageningen, Wageningen, The Netherlands
Brouwer J, Fussell LK and Herrmann L (1993) Soil and crop growth micro-variability in the West African semi-arid tropics: a possible risk-reducing factor for subsistence farmers. Agriculture, Ecosystems and Environment 45: 229–238
Brouwer J and Powell JM (1995) Soil aspects of nutrient cycling in a manure application experiment in Niger. In: Powell JM, Fernández-Rivera S, Williams TO and Renard C (eds) Livestock and Sustainable Nutrient Cycling in Mixed Farming Systems of Sub-Saharan Africa. Volume II. Technical Papers. ILCA, Addis Ababa, Ethiopia
Buresh RJ (1995) Nutrient cycling and nutrient supply in agroforestry systems. In: Dudal R and Roy RN (eds) Integrated Plant Nutrition Systems, pp 155–164. FAO Fertilizer and Plant Nutrition Bulletin No. 12. Rome, Italy
Buresh RJ, Smithson PC and Hellums DT (1997) Building soil P capital in sub-Saharan Africa. In: Buresh RJ and Sanchez PA (eds) Replenishing Soil Fertility in Africa. SSSA Special Publication 51. SSSA and ASA, Madison, WI, USA (in press)
Cahn MD, Bouldin DR and Cravo MS (1992) Nitrate sorption in the profile of an acid soil. Plant and Soil 143: 179–183
Chien SH and Menon RG (1995) Factors affecting the agronomic effectiveness of phosphate rock for direct application. Fertilizer Research 41: 227–234
Cooper PJM, Leakey RRB, Rao MR and Reynolds L (1996) Agroforestry and the mitigation of land degradation in the humid and sub-humid tropics of Africa. Experimental Agriculture 32: 235–290
Craswell ET, Sajjipongse A, Howlett DJB and Dowling AJ (1997) Agroforestry in the management of sloping lands in Asia and the Pacific. Agroforestry Systems 38: 121–137
Dalland A, Våje PI, Matthews RB and Singh BR (1993) The potential of alley cropping in improvement of cultivation systems in the high rainfall areas of Zambia. III. Effects on soil chemical and physical properties. Agroforestry Systems 21: 117–132
Danso SKA, Bowen GD and Sanginga N (1992) Biological nitrogen fixation in trees in agroecosystems. Plant and Soil 141: 177–196
Deans JD, Lindley DK and Munro RC (1994) Deep beneath the trees in Senegal. In: Annual Report of the Institute of Terrestrial Ecology 1993–1994, pp 12–14. NERC, Swindon, UK
Deckers J (1993) Soil fertility and environmental problems in different ecological zones of the developing countries of Sub-Saharan Africa. In: van Reuler H and Prins WH (eds) The Role of Plant Nutrients and Sustainable Food Production in Sub-Saharan Africa, pp 37–52. Vereniging van Kunstmest Producenten, Leidschendam, The Netherlands
Dommergues YR (1995) Nitrogen fixation by trees in relation to soil nitrogen economy. Fertilizer Research 42: 215–230
Fernandes ECM, Garrity DP, Szott LT and Palm CA (1994) Use and potential of domesticated trees for soil improvement. In: Leakey RRB and Newton AC (eds) Tropical Trees: The Potential for Domestication and the Rebuilding of Forest Resources, pp 137–147. HMSO, London, UK
Franzel S (1998) Socioeconomic factors affecting the adoption potential of improved fallows in Africa. Agroforestry Systems (in press)
Friend MT and Birch HF (1960) Phosphate responses in relation to soil tests and organic phosphorous. Journal of Agricultural Science 54: 341–347
Gachengo CN (1996) Phosphorus release and availability on addition of organic materials to phosphorus fixing soils. M Phil thesis, Moi University, Eldoret, Kenya

Geiger SC, Vandenbeldt RJ and Manu A (1994) Variability in the growth of *Faidherbia albida*: the soils connection. Soil Science Society of America Journal 58: 227–231

Giller KE (1997) Tropical legumes: providers and plunderers of nitrogen. In: Bergström L and Kirchmann H (eds) Carbon and Nitrogen Cycling in Natural and Agricultural Tropical Ecosystems. CAB International, Wallingford, UK (in press)

Giller KE and Cadisch G (1995) Future benefits from biological nitrogen fixation: An ecological approach to agriculture. Plant and Soil 174: 255–277

Giller KE, Cadisch G, Ehaliotis C, Adams E, Sakala WD and Mafongoya PL (1997) Building soil nitrogen capital in sub-Saharan Africa. In: Buresh RJ and Sanchez PA (eds) Replenishing Soil Fertility in Africa. SSSA Special Publication 51. SSSA and ASA, Madison, WI, USA (in press)

Haggar JP, Warren GP, Beer JW and Kass D (1991) Phosphorus availability under alley cropping and mulched and unmulched sole cropping systems in Costa Rica. Plant and Soil 137: 275–283

Handreck KA (1997) Phosphorus requirements of Australian native plants. Australian Journal of Soils Research 35: 241–289

Hands MR, Harrison AF and Bayliss-Smith T (1995) Phosphorus dynamics in slash-and-burn and alley cropping systems of the humid tropics. In: Tiessen H (ed) Phosphorus in the Global Environment; Transfers, Cycles and Management. Scope 54, pp 155–170. John Wiley and Sons Ltd, New York

Haque I, Powell JM and Ehui SK (1995) Improved crop-livestock production strategies for sustainable soil management in tropical Africa. In: Lal R and Stewart BA (eds) Soil Management: Experimental Basis for Sustainability and Environmental Quality, pp 293–345. CRC Press, Boca Raton, FL

Hartemink AE, Buresh RJ, Jama B and Janssen BH (1996) Soil nitrate and water dynamics in sesbania fallow, weed fallows, and maize. Soil Science Society of America Journal 60: 568–574

Hassink (1994) Active organic matter fractions and microbial biomass as predictors of N mineralization. In: Neeteson JJ and Hassink J (eds) Nitrogen Mineralization in Agricultural Soils, pp 1–15. AB-DLO Thema's 1. DLO Research Institute for Agrobiology and Soil Fertility, Wageningen/Haren, The Netherlands

Hassink J (1995) Decomposition rate constants of size and density fractions of soil organic matter. Soil Science Society of America 59: 1631–1635

Hauser S (1993a) Root distribution of *Dactyladenia (Acioa) barteri* and *Senna (Cassia) siamea* in alley cropping on Ultisol. I. Implication for field experimentation. Agroforestry Systems 24: 111–121

Hauser S (1993b) Distribution and activity of earthworms and contribution to nutrient recycling in alley cropping. Biology and Fertility of Soils 15: 16–20

Horst WJ, Kühne R and Kang BT (1995) Nutrient use in *Leucaena leucocephala* and *Cajanus cajan* in maize/cassava alley cropping on Terre de Barre, Benin Republic. In: Kang BT, Osiname AO and Larbi A (eds) Alley Farming Research and Development, pp 122–136. Alley Farming Network for Tropical Africa, Ibadan, Nigeria

Hulugalle NR and Kang BT (1990) Effect of hedgerow species in alley cropping systems on surface soil physical properties of an Oxic Paleustalf in southwestern Nigeria. Journal of Agricultural Science (Cambridge, UK) 114: 301–307

Hulugalle NR and Ndi JN (1993) Effects of no-tillage and alley cropping on soil properties land crop yields in a Typic Kandiudult of southern Cameroon. Agroforestry Systems 22: 207–220

IAEA (1975) Root activity patterns of some tree crops. Technical Report Series No. 170. International Atomic Energy Agency, Vienna

ICRAF (1993) Annual report 1992. International Centre for Research in Agroforestry, Nairobi, Kenya

ICRAF (1996) Annual report 1995. International Centre for Research in Agroforestry, Nairobi, Kenya

ICRAF (1997) Annual report 1996. International Centre for Research in Agroforestry, Nairobi, Kenya

Iyamuremye F and Dick RP (1996) Organic amendments and phosphorus sorption by soils. Advances in Agronomy 56: 139–185

Jama B, Buresh RJ, Ndufa JK and Sheperd KD (1998) Vertical distribution of roots and soil nitrate: Tree species and phosphorus effects. Soil Science Society of America Journal (in press)

Jama B, Swinkels RA and Buresh RJ (1997) Agronomic and economic evaluation of organic and inorganic sources of phosphorus in western Kenya. Agronomy Journal 89: 597–604

Jones RB, Wendt JW, Bunderson WT and Itimu OA (1996) Leucaena + maize alley cropping in Malawi. Part 1: Effects of N, P, and leaf application on maize yields and soil properties. Agroforestry Systems 33: 281–294

Jonsson K, Fidjeland L, Maghembe JA and Högberg P (1988) The vertical distribution of fine roots of five tree species and maize in Morogoro, Tanzania. Agroforestry Systems 6: 63–69

Jonsson K, Ståhl L and Högberg P (1996) Tree fallows: a comparison between five tropical tree species. Biology and Fertility of Soils 23: 50–56

Juo ASR, Caldwell JO and Kang BT (1994) Place for alley cropping in sustainable agriculture in the humid tropics. 15th World Congress of Soil Science, Acapulco, Mexico, 7a: 98–109

Juo ASR, Franzluebbers K, Dabiri A and Ikhile B (1995) Changes in soil properties during long-term fallows and continuous cultivation after forest clearing in Nigeria. Agriculture, Ecosystems and Environment 56: 9–18

Juo ASR, Franzluebbers K, Dabiri A and Ikhile B (1996) Soil properties and crop performance on a kaolinitic Alfisol after 15 years of fallow and continuous cultivation. Plant and Soil 180: 209–217

Kang BT (1997) Alley cropping–soil productivity and nutrient recycling. Forest Ecology and Management 91: 75–82

Kang BT, Akinnifesi FK and Ladipo DO (1994) Performance of selected woody agroforestry species grown on an Alfisol and an Ultisol in the humid lowland of West Africa, and their effects on soil properties. Journal of Tropical Forest Science 7: 303–312

Kang BT, Reynolds L and Atta-Krah AN (1990) Alley Farming. Advances in Agronomy 43: 315–359

Kang BT, Salako FK, Akobundu IO, Pleysier JL and Chianu JN (1997) Amelioration of a degraded Oxic Paleustalf by leguminous and natural fallows. Soil Use and Management 13: 130–135

Kessler JJ and Breman H (1991) The potential of agroforestry to increase primary production in the Sahelian and Sudanian zones of West Africa. Agroforestry Systems 13: 41–62

Kiepe P and Rao MR (1994) Management of agroforestry for the conservation and utilization of land and water resources. Outlook in Agriculture 23(1): 17–25

Kihanda FM (1996) The role of farm yard manure in improving maize production in the sub-humid highlands of Central Kenya. PhD diss., University of Reading, Reading, UK

Kipleting KM (1996) Root and inorganic nitrogen distribution in an Alfisol under different landuse systems. MSc thesis. Moi University, Eldoret, Kenya

Kwesiga F and Coe R (1994) The effect of short rotation Sesbania sesban fallows on maize yield. Forest Ecology and Management 64: 199–208

Kwesiga FR, Franzel S, Place F, Phiri D and Simwanza CP (1998) Sesbania sesban improved fallows in eastern Zambia: their inception, development and farmer enthusiasm. Agroforestry Systems (in press)

Ladha JK, Peoples MB, Garrity DP, Capuno VT and Dart PJ (1993) Estimating dinitrogen fixation of hedgerow vegetation using the nitrogen-15 natural abundance method. Soil Science Society of America Journal 57: 732–737

Lal R (1989a) Agroforestry systems and soil surface management of a tropical Alfisol. IV: Effect on soil physical and mechanical properties. Agroforestry Systems 8: 197–215

Lal R (1989b) Agroforestry systems and soil surface management of a tropical Alfisol: V: Water infiltrability, transmissivity and soil water sorptivity. Agroforestry Systems 8: 217–238

MacDicken KG (1994) Selection and management of nitrogen-fixing trees. Winrock International, Morrilton, Arkansas and FAO, Bangkok

Mafongoya PL, Giller KE and Palm CA (1997) Decomposition and nitrogen release patterns of tree prunings and litter. Agroforestry Systems 38: 77–97

Mason PA and Wilson J (1994) Harnessing symbiotic associations: vesicular–arbuscular mycorrhizas. In: Leakey RRB and Newton AC (eds) Tropical Trees: The Potential for Domestication and the Rebuilding of Forest Resources, pp 165–175. HMSO, London, UK

Mekonnen K (1996) Rooting characteristics and inorganic nitrogen distribution in three landuse systems on an Oxisol in western Kenya. MSc thesis. Moi University, Eldoret, Kenya

Mekonnen K, Buresh RJ and Jama B (1997) Root and inorganic nitrogen distributions in sesbania fallow, natural fallow and maize fields. Plant and Soil 188: 319–327

Michori P (1993) Nitrogen budget under coffee. PhD diss., University of Reading, Reading, UK [British Thesis Service DX 175716]

Mordelet P, Abbadie L and Menaut JC (1993) Effects of tree clumps on soil characteristics in a humid savanna of West Africa (Lamto, Côte d'Ivoire). Plant and Soil 153: 103–111

Mulongoy K, Kunda KN and Chiang CNK (1993) Effect of alley cropping and fallowing on some soil fertility parameters in southern Nigeria. In: Mulongoy K and Merckx R (eds) Soil Organic Matter Dynamics and Sustainability of Tropical Agriculture, pp 47–55. John Wiley & Sons and Sayce Publishing, Chichester, UK

Nagarajah S and Nizar BM (1982) Wild sunflower as a green manure for rice in the mid country wet zone. Tropical Agriculturalist 138: 69–79

Nair PKR, Kang BT and Kass DCL (1995) Nutrient cycling and soil-erosion control in agroforestry systems. In: Juo ASR and Freed RD (eds) Agriculture and the Environment: Bridging Food Production and Environmental Protection in Developing Countries, pp 117–138. ASA Special Publication 60. ASA, CSSA and SSSA, Madison, WI, USA

Nyathi P and Campbell BM (1993) The acquisition and use of miombo litter by small-scale farmers in Masvingo, Zimbabwe. Agroforestry Systems 22: 43–48

Nziguheba G, Palm CA, Buresh RJ and Smithson PC (1998) Soil phosphorus fractions and adsorption as affected by organic and inorganic sources. Plant and Soil (in press)

Nyberg G and Högberg P (1995) Effects of young agroforestry trees on soils in on-farm situations in western Kenya. Agroforestry Systems 32: 45–52

Palm CA (1995) Contribution of agroforestry trees to nutrient requirements of intercropped plants. Agroforestry Systems 30: 105–124

Palm CA, McKerrow AJ, Glasener KM and Szott LT (1991) Agroforestry systems in lowland tropics: Is phosphorus important?. In: Tiessen H, Lopez-Hernandez D and Salcedo IH (eds) Phosphorus Cycles in Terrestrial and Aquatic Ecosystems, Regional Workshop 3: South and Central America, pp 134–141. Saskatchewan Institute of Pedology, University of Saskatchewan, Saskatoon, Canada

Palm CA, Myers RJK and Nandwa S (1997) Combined use of organic and inorganic nutrient sources for soil fertility maintenance and replenishment. In: Buresh RJ and Sanchez PA (eds) Replenishing Soil Fertility in Africa. SSSA Special Publication 51. SSSA and ASA, Madison, WI, USA (in press)

Rao MR, Nair PKR and Ong CK (1997) Biophysical interactions in tropical agroforestry systems. Agroforestry Systems 38: 3–50

Rhoades CC (1997) Single-tree influences on soil properties in agroforestry: lessons from natural and savanna ecosystems. Agroforestry Systems 35: 71–94

Romanya J, Khanna PK and Raison RJ (1994) Effects of slash burning on soil phosphorus fractions and sorption and desorption of phosphorus. Forest Ecology and Management 65: 89–103

Ruhigwa BA, Gichuru MP, Mambani B and Tariah NM (1992) Root distribution of *Acioa barteri*, *Alchornea cordifolia*, *Cassia siamea* and *Gmelina arborea* in an acid Ultisol. Agroforestry Systems 19: 67–78

Ruhigwa BA, Gichuru MP, Tariah NM, Isirimah NO and Douglas DC (1993) Spatial variability in soil chemical properties under *Dactyladenia barteri*, *Alchornea cordifolia*, *Senna siamea* and *Gmelina arborea* hedgerows on an acid Ultisol. Experimental Agriculture 29: 365–372

Sanchez PA (1995) Science in agroforestry. Agroforestry Systems 30: 5–55

Sanchez PA, Shepherd KD, Soule MJ, Place FM, Buresh RJ, Izac A-MN, Mokwunye AU, Kwesiga FR, Ndiritu CG and Woomer PL (1997a) Soil fertility replenishment in Africa: an investment in natural resource capital. In: Buresh RJ and Sanchez PA (eds) Replenishing Soil Fertility in Africa. SSSA Special Publication 51. SSSA and ASA, Madison, WI, USA (in press)

Sanchez PA, Buresh RJ and Leakey RRB (1997b) Trees, soils and security. Philosophical Transactions of the Royal Society, Series B 352: 949–961

Sanchez PA and Leakey RRB (1997) Land-use transformation in Africa: three determinants for balancing food security with natural resource utilization. European Journal of Agronomy 7: 1–9

Sanchez PA and RH Miller (1986) Organic matter and soil fertility management in acid soils of the tropics. Transactions 13th International Congress Soil Science, Hamburg, 6: 609–625

Sanchez PA and Palm CA (1996) Nutrient cycling and agroforestry in Africa. Unasylva 185 47: 24–28

Sanginga N, Bowen GD and Danso SKA (1990) Assessment of genetic variability for N_2 fixation between and within provenances of *Leucaena leucocephala* and *Acacia albida* estimated by ^{15}N labelling techniques. Plant and Soil 127: 169–178

Sanginga N, Danso SKA and Zapata F (1996) Field requirements of nitrogen fixation in leguminous trees used in agroforestry systems: Influence of ^{15}N-labeling approaches and reference trees. Biology and Fertility of Soils 23: 26–32

Sanginga N, Danso SKA, Zapata F and Bowen GD (1994) Field validation of intraspecific variation in phosphorus use efficiency and nitrogen fixation by provenances of *Gliricidia sepium* grown in low P soils. Applied Soil Ecology 1: 127–138

Sanginga N, Manrique K and Hardarson G (1991) Variation in nodulation and N_2 fixation by the *Gliricidia sepium/Rhizobium* spp. symbiosis in a calcareous soil. Biology and Fertility of Soils 11: 273–278

Sanginga N, Vanlauwe B and Danso SKA (1995) Management of biological N_2 fixation in alley cropping systems: Estimation and contribution to N balance. Plant and Soil 174: 119–141

Schroth G (1995) Tree root characteristics as criteria for species selection and systems design in agroforestry. Agroforestry Systems 30: 125–143

Schroth G, Kolbe D, Pity B and Zech W (1995) Searching for criteria for the selection of efficient tree species for fallow improvement, with special reference to carbon and nitrogen. Fertilizer Research 42: 297–314

Schroth G and Zech W (1995) Root length dynamics in agroforestry with *Gliricidia sepium* as compared to sole cropping in the semi-deciduous rainforest zone of West Africa. Plant and Soil 170: 297–306

Siaw DEKA, Kang BT and Okali DUU (1991) Alley cropping with *Leucaena leucocephala* (Lam.) De Wit and *Acioa barteri* (Hook. f.) Engl. Agroforestry Systems 14: 219–231

Smaling, EMA, Fresco LO and de Jager A (1996) Classifying, monitoring and improving soil nutrient stocks and flows in African agriculture. Ambio 25: 492–496

Stone EL and Kalisz PJ (1991) On the maximum extent of tree roots. Forest Ecology and Management 46: 59–102

Stoorvogel JJ, Smaling EMA and Janssen BH (1993) Calculating soil nutrient balances in Africa at different scales. I. Supra-national scale. Fertilizer Research 35: 227–235

Strømgaard P (1992) Immediate and long-term effects of fire and ash-fertilization on a Zambian miombo woodland soil. Agriculture, Ecosystems and Environment 41: 19–37

Swinkels RA, Franzel S, Shepherd KD, Ohlsson E and Ndufa JK (1997) The economics of short rotation improved fallows: evidence from areas of high population density in western Kenya. Agricultural Systems 55: 99–121

Szott LT, Fernandes ECM and Sanchez PA (1991) Soil-plant interactions in agroforestry systems. Forest Ecology and Management 45: 127–152

Szott LT, Palm CA and Buresh RJ (1998) Ecosystem fertility and fallow function. Agroforestry Systems (in press)

Tian G, Brussaard L and Kang BT (1993a) Biological effects of plant residues with contrasting chemical compositions under humid tropical conditions: effects on soil fauna. Soil Biology and Biochemistry 25: 731–737

Tian G, Brussaard L and Kang BT (1995) An index for assessing the quality of plant residues and evaluating their effects on soil and crop in the (sub-) humid tropics. Applied Soil Ecology 2: 25–32

Tian G, Kang BT and Brussaard L (1992) Biological effects of plant residues with contrasting chemical compositions under humid tropical conditions – decomposition and nutrient release. Soil Biology and Biochemistry 24: 1051–1060

Tian G, Kang BT and Brussaard L (1993b) Mulching effect of plant residues with chemically contrasting compositions on maize growth and nutrients accumulation. Plant and Soil 153: 179–187

Torquebiau EF and Kwesiga F (1996) Root development in a *Sesbania sesban* fallow-maize system in Eastern Zambia. Agroforestry Systems 34: 193–211

Trouve C, Mariott A, Schwartz D and Guillet B (1994) Soil organic carbon dynamics under *Eucalyptus* and *Pinus* planted on savannas in the Congo. Soil Biology and Biochemistry 26: 287–295

van Bodegom P (1995) Water, nitrogen and phosphorus dynamics in three fallow systems and maize in western Kenya. MSc thesis. Wageningen Agricultural University, Wageningen, The Netherlands

van der Meersch MK, Merckx R and Mulongoy K (1993) Evolution of plant biomass and nutrient content in relation to soil fertility changes in two alley cropping systems. In: Mulongoy K and Merckx R (eds) Soil Organic Matter Dynamics and Sustainability of Tropical Agriculture, pp 143–154. John Wiley & Sons and Sayce Publishing, Chichester, UK

Vanlauwe, B, Swift MJ and Merckx R (1996) Soil litter dynamics and N use in a leucaena (*Leucaena leucocephala* Lam. (De Witt)) alley cropping system in southwestern Nigeria. Soil Biology and Biochemistry 28: 739–749

van Noordwijk M (1989) Rooting depth in cropping systems in the humid tropics in relation to nutrient use efficiency. In: van der Heide J (ed) Nutrient Management for Food Crop Production in Tropical Farming Systems, pp 129–144. Institute for Soil Fertility, Haren, The Netherlands

van Noordwijk M, Lawson G, Soumare A, Groot JJR and Hairiah K (1996) Root distribution of trees and crops: competition and/or complementarity. In: Ong CK and Huxley P (eds) Tree-Crop Interactions, a Physiological Approach, pp 319–364. CAB International, Wallingford, UK

Wallace JS (1996) The water balance of mixed tree-crop systems. In: Ong CK and Huxley P (eds) Tree-Crop Interactions, a Physiological Approach, pp 73–158. CAB International, Wallingford, UK

Weil RR and Mughogho SK (1993) Nutrient cycling by Acacia albida (syn. Faidherbia albida) in agroforestry systems. In: Ragland J and Lal R (eds) Technologies for Sustainable Agriculture in the Tropics, pp 97–108. ASA Special Publication 56, ASA, CSSA and SSSA, Madison, WI

Wendt JW, Jones RB, Bunderson WT and Itimu OA (1996) Leucaena + maize alley cropping in Malawi. Part 1: Residual P and leaf management effects on maize nutrition and soil properties. Agroforestry Systems 33: 295–305

Wild A (1972) Nitrate leaching under bare fallow at a site in northern Nigeria. Journal of Soil Science 23: 315–324

Wong MTF and Nortcliff S (1995) Seasonal fluctuations of native available N and soil management implications. Fertilizer Research 42: 13–26

Wong MTF, Wild A and Juo ASR (1987) Retarded leaching of nitrate measured in monolith lysimeters in south-east Nigeria. Journal of Soil Science 38: 511–518

Young A (1997) Agroforestry for soil management. CAB International, Wallingford, UK and ICRAF, Nairobi, Kenya

Agroforestry Systems **38**: 77–97, 1998.

Decomposition and nitrogen release patterns of tree prunings and litter

P. L. MAFONGOYA[1,*], K. E. GILLER[2] and C. A. PALM[3]
[1] *Zambia-ICRAF, P.O. Box 510046, Chipata, Zambia;* [2] *Wye College, University of London, Wye, Ashford, Kent TN25 5AH, UK;* [3] *Tropical Soil Biology and Fertility Programme, P.O. Box 30592, Nairobi, Kenya (* Author for correspondence: E-mail: zamicraf@zamnet.zm)*

Key words: polyphenols, protein binding capacity, resource quality

Abstract. Many studies have shown that agroforestry tree prunings can supply sufficient nutrients to meet crop demand, with the exception of phosphorus. The potential of these organic inputs to supply nutrients depends on their resource quality. Various indices have been developed to predict decomposition and nitrogen release patterns of tree prunings. To date the (lignin + polyphenol):N ratio seems to be the most robust ratio for predicting mass loss and nitrogen release. However, no critical value can be given because of the different methods used to analyze polyphenols. Suggested areas of future research include development of robust indices for predicting plant litter quality, decomposition patterns of belowground litter (roots), residual effects of tree biomass additions, and effects of adding mixtures of organic materials of contrasting quality. The overall challenge is to develop ways of managing organic matter decomposition to optimize short- and long-term release of nutrients and the maintenance of soil organic matter.

Introduction

Most soils in the tropics are deficient in soil nitrogen (N), phosphorus (P), or both (Sanchez and Logan, 1992). Many of these soils are acid, infertile, and cannot support sustainable crop production without external inputs of inorganic fertilizers. Even some soils which were once fertile have become depleted of nutrients and can no longer sustain crop production. Smallscale farmers in the tropics have limited access to inorganic fertilizers due to high costs and, in some cases, unavailability of such fertilizers. In these situations, agroforestry systems are considered to be viable and sustainable land-use alternatives because of the benefits of trees in maintenance of soil fertility. Trees can maintain or enhance soil fertility by adding N through N_2-fixation, recycling nutrients through litterfall or prunings, or importing nutrients through biomass transfer systems. These nutrients become available to crops through decomposition of the tree prunings and litter. Trees can develop deep roots that absorb nutrients from below the rooting depths of most annual crops, thereby establishing a nutrient cycle that is more closed than those of most agricultural systems, which are characterized by leaky nutrient cycles (Nair, 1993). Belowground biomass accumulation by tree roots can be very high, 3 to 6 Mg ha^{-1} yr^{-1} (Sanchez, 1995). This biomass can make a substantial contribution to soil organic matter carbon and nutrient cycling.

Due to differences in the types and quality of inputs, the materials available for decomposition in agroforestry systems are not the same as in natural or agricultural systems. Senescent material (litter) dominates as an input in natural and agricultural systems, while inputs in agroforestry systems include both fresh and senescent material. Litter differs from green foliage in that it has lower nutrient content due to translocation of nutrients before litter fall. Hence the quality and decomposition rates of litter and green foliage from the same plant can be markedly different (e.g., Constantinides and Fownes, 1994a). Increased knowledge of the interactions that occur when residues of different quality are applied to crops can lead to various strategies for the management of tree residues and other nutrient inputs. These include the method and time of application, and the application of mixtures of different quality.

Only about 20% of the N released from tree prunings or litter is taken up by the current crop (Palm, 1995; Giller and Cadisch, 1995). Much of the remaining 40% to 80% of the applied organic N is incorporated into soil organic matter (SOM) (Haggar et al., 1993). Because of the slow release of N and P from SOM, organic inputs have a greater residual effect on soil fertility than do inorganic fertilizers. The quality of the prunings or litter has a crucial influence in determining their impact on soil fertility, especially with regard to nutrient supply and SOM formation. High-quality organic inputs (high N but low lignin and polyphenol contents) will release nutrients rapidly. In contrast, low-quality organic inputs (poor in N and high in polyphenols and lignin) will release nutrients slowly or even immobilize them (Palm, 1995). This has practical implications for attaining synchrony between nutrient release and crop nutrient demand and, hence, for efficiency of nutrient use. This paper attempts to review the current state of knowledge on the quality and decomposition patterns of organic inputs in agroforestry systems, and their management implications.

Resource quality and decomposition

Parameters for characterizing organic input quality

Decomposition and nutrient release rates are determined by the resource quality of the organic material, the environment, and the decomposer organisms present (Swift et al., 1979). The resource quality is defined by the organic constituents and nutrient contents of the material. Several factors must be considered when choosing parameters to describe plant quality: (1) the processes of decomposition and nutrient release may be controlled by different parameters; (2) the critical parameters will depend on the time frame of the study (Melillo et al., 1989); and (3) the importance of certain parameters changes with the type of plant material (Palm, 1995). The ultimate aim is to identify robust parameters that predict decomposition and nutrient release.

These parameters could be used to assess the potential decomposition and nutrient release patterns of organic materials, and replace the need for detailed decomposition studies for each plant material in each location.

The carbon quality of an organic material depends on the proportions of soluble carbon, cellulose (+hemicellulose), and lignin; quality, in this case, refers to the energy available for decomposer organisms. Soluble carbon, which includes metabolic and storage carbon, is of high quality and is primarily responsible for promoting microbial growth and activity (Smith, 1994). The amount of soluble carbon compounds is also the determinant of the immediate nutrient mineralization or immobilization pattern. If there is a large amount of soluble C but little soluble N and P in the decomposing plant materials, then net immobilization will occur (Aber and Melillo, 1991; Figure 1). The celluloses, or structural polysaccharides, are of intermediate quality to the decomposers and are attacked by microbes after the soluble carbohydrates have been depleted (Swift et al., 1979; Figure 1). Lignin is considered by many to be the most important component for determining the rate of decomposition (e.g., Meentemeyer, 1978); it is of low quality and provides little or no energy to decomposers until the late stages of decomposition. Lignin intertwines with the cell wall, physically protecting cellulose and other cell wall constituents from degradation (Chesson, 1997). A detailed description of the chemical structure of these organic constituents and their role in decomposi-

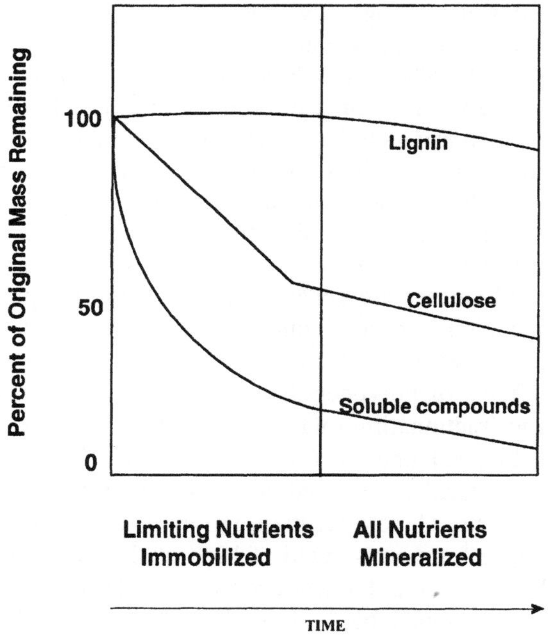

Fig. 1. Patterns of mass loss from the various carbon constituents during decomposition, and the associated mineralization/immobilization patterns (adapted from Aber and Melillo, 1991).

tion and nutrient release can be found in Swift et al. (1979), Aber and Melillo (1991) and Cadisch and Giller (1997).

Another organic component, polyphenols, comprises a relatively small percentage of the organic material but can have a disproportionate effect on decomposition (Vallis and Jones, 1973). Many recent studies with agroforestry species have definitively shown an inverse relationship between polyphenolic content and nitrogen release (Palm and Sanchez, 1990; Constantinides and Fownes, 1994a; Handayanto et al., 1994). Polyphenolics include a range of compounds differing in size, solubility, and reactivity. Polyphenols can serve as a carbon substrate for decomposers (Martin and Haider, 1980) but in general they inhibit the growth or function of decomposers and other organisms (Swift et al., 1979). It is not necessarily the quantity of polyphenolics that is important in decomposition, but the types of polyphenolic compounds (Handayanto et al., 1994; Mafongoya et al., 1997).

Condensed tannins, also known as proanthocyanidins, are the polyphenols most noted for their effects on decomposition and nutrient dynamics, which result from their reactions with proteins and nitrogen (Martin and Haider, 1980, Waterman and Mole, 1994). Some condensed tannins are soluble in polar extractants, while others are insoluble and bound to proteins or the cell wall (Jackson et al., 1996). The soluble condensed tannins and hydrolyzable tannins can react with nitrogen and amino groups in the plant or soil solution, reducing N release. In fact, Handayanto et al. (1994) found the relationship between polyphenols and nitrogen dynamics increased when the extract was analyzed for its protein-binding capacity, a means of assessing the reactivity of soluble polyphenols. The strength of the bonds formed between polyphenolics and nitrogen varies but can be quite strong (Martin and Haider, 1980), and there is increasing evidence that this nitrogen is unavailable indefinitely (Handayanto et al., 1997a; Cadisch and Giller, unpublished results).

The bound, or insoluble, condensed tannins may simply slow decomposition by binding to the cell wall and proteins, making them physically or chemically inaccessible to decomposers, an effect similar to that of lignin. Insoluble condensed tannins are generally included in the lignin fraction by most separation techniques so the 'lignin' content may be overestimated if the material has a high level of insoluble condensed tannins. To aid our understanding of the effects of polyphenols on decomposition, greater attention needs to be paid to fractionating total polyphenols into extractable (soluble) and bound (insoluble) polyphenols (cf. Jackson et al., 1996).

Nutrient quality is usually measured in terms of net nitrogen or phosphorus mineralization patterns. Generally, materials with nitrogen concentrations greater than 20 mg g^{-1} are considered of high quality although this can be modified by high lignin and polyphenol contents. In defining nutrient quality, it may be useful to distinguish between soluble organic N and N that is bound to the cell wall, particularly with green foliage (Mafongoya et al., 1997). Phosphorus contents greater than 2.5 mg g^{-1} can generally be considered of high quality.

Resource quality of agroforestry inputs

The development of a useful plant quality index requires standardization of methods or a means of comparing results obtained by different methods. Table 1 identifies the important parameters and recommended methods for characterizing organic materials from agroforestry species. Researchers are encouraged to provide this basic list of plant quality characteristics, in addition to information on the climate, ecosystem, and soil type where the study was conducted.

Decomposition in agroforestry systems differs from that in natural or agricultural systems because of differences in the types and quality of organic inputs. Litter is a general term for senescent plant parts. Litter differs from green foliage in that much of the soluble carbon compounds and nutrients are translocated from leaves during senescence, prior to abscision and leaf fall. As a result, the quality and decomposition rates of litter and foliage from the same plant can be quite different (Constantinides and Fownes, 1994a). A simplified comparison of the resource quality of green foliage and litter is presented in Figure 2.

Soluble carbon has seldom been measured in the foliage and litter of agroforestry systems, except for recent work by Vanlauwe et al. (1997b), yet it is the most important plant constituent in determining immediate patterns of nutrient release, microbial growth and activity, and perhaps rates of nutrient cycling (Smith, 1994). Soluble carbon varies depending on the species, plant part, and plant's stage of maturity (Amato et al., 1984). Green foliage usually contains 20% to 30% soluble carbon (Swift et al., 1979; Collins et al., 1990),

Table 1. Recommended parameters and methods for analysis of organic materials for quality characterization.

Quality parameter	Recommended method
Lignin	ADF-H_2SO_4 (Van Soest, 1963; Rowland and Roberts, 1994)
Soluble carbon	[a] Hot water extraction (TAPPI, 1988) followed by total carbon or simple sugar analysis (Dubois et al., 1956)
Soluble phenolics	[b] Aqueous methanol (50%) extraction 2mg/mL, followed by analysis with Folin-Ciolcalteu reagent (Constantinides and Fownes (1994b)
Protein-binding capacity[c]	Bovine serum albumin method (Dawra et al., 1988)
Total N	Kjeldahl (Anderson and Ingram, 1993)
Total P	Same digest as for N (Anderson and Ingram, 1993)
NDF-N[d]	micro-Kjeldahl on NDF (Van Soest, 1963)

Source: Adapted from Palm and Rowland (1997).
[a] This method also gives cellulose.
[b] Only if total N > 2%.
[c] Only if total soluble phenolics > 3%.
[d] Non-digestible fiber-N.

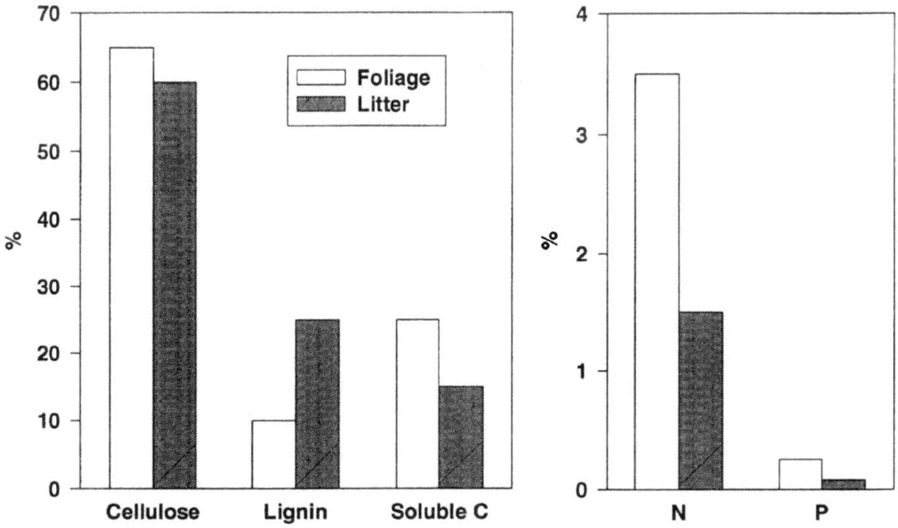

Fig. 2. Generalized comparison of the carbon and nutrient qualilty of green foliage and litter of the same material.

much of which can be polyphenols. Litter can also contain significant levels of soluble carbon compound materials (Trofymow et al., 1995), but there are few direct comparisons of green foliage and litter. Schlesinger and Hasey (1981) found 10% less soluble carbohydrate in the litter, compared to green foliage from the same plants.

Cellulose and hemicelluloses are the main components of the cell wall and constitute from 30% to 70% of the plant carbon (Theander and Westerlund, 1993; Rowland and Roberts, 1994). Cellulose is perhaps the least variable of the carbon constituents because its concentration does not change during leaf senescence. Its concentration increases due to loss of soluble compounds.

The percentage of lignin in the leaves of agroforestry species varies widely, ranging from 5% to 20% of dry weight in green foliage and from 10% to 40% in leaf litters (Constantinides and Fownes, 1994a). It has been suggested that 15% lignin is a critical level, above which decomposition is impaired because lignin protects the cellulose in the cell wall from attack (Chesson, 1997; Palm and Rowland, 1997). Many agroforestry trees also contain a large amount of insoluble condensed tannins in the cell wall (Reed, 1986; Mafongoya, 1995) which may result in a large fraction of the plant N being bound to the insoluble fraction. As mentioned previously, high contents of insoluble condensed tannins may result in overestimation of the lignin content.

Total soluble polyphenols content of green foliage of agroforestry species can be as high as 10% of dry weight, but is usually less than 5% (Constantinides and Fownes, 1994a). Care must be taken in comparisons between studies because the soluble polyphenols content is highly dependent on the extractant and tissue-to-solvent ratio (Constantinides and Fownes, 1994b). In comparisons of biomass from the same plant, Schlesinger and Hasey (1981) found soluble polyphenol contents were more than 20% higher in litter compared to green foliage. Constantinides and Fownes (1994a) found both increases and decreases.

Except for the work of Handayanto et al. (1994) and Mafongoya et al. (1997), little information has been published on the protein-binding capacity of polyphenols and condensed tannins in agroforestry species. There is no apparent correlation between total polyphenols and condensed tannins. Materials from some species, such as *Senna siamea*, have a fairly high total soluble polyphenol content (Constantinides and Fownes, 1994a) but a very low level of condensed tannins (Jackson et al., 1996), whereas materials from *Calliandra calothrysus* have a level of total soluble polyphenols that is similar to *S. siamea*, but also a high level of condensed tannins (Handayanto et al., 1995). This perhaps explains the different nitrogen dynamics following application of prunings from the two species: *Senna* exhibits a short-term immobilization followed by rapid N mineralization, and *Calliandra* shows slower N mineralization. Considerable information about polyphenolics in agroforestry trees and their role in decomposition might be gained by examining the wealth of data on condensed tannins in legume forages and their effect on feed digestibility (Jackson et al., 1996; Reed, 1995). To enable exchange of information for predicting plant quality, a parallel should be drawn between the roles of condensed tannins as an antinutritive factor and as an inhibitor to decomposition and nitrogen release.

The nutrient content of agroforestry trees depends on many factors, including the species, plant part, relative proportion of leaves and twigs, age of the tissue, frequency of pruning, and the soil and climate (Palm, 1995). Budelman (1989) suggests that soil fertility is the most important environmental or management factor affecting nutrient content in the plant tissues. As mentioned earlier, translocation affects the N and P content of litter and green foliage. The extent of translocation in green tissues before abscission depends on soil fertility, with the most limiting nutrient exhibiting the greatest translocation (Vitousek, 1984). The nitrogen and phosphorus contents of litter of several agroforestry trees were less than those found in the green foliage (Constantinides and Fownes, 1994a).

Because of the influence of studies from natural systems, lignin measurements have dominated decomposition studies in agroforestry systems, whereas soluble carbon compounds have been virtually ignored. However, in agroforestry systems where green foliage is applied to the soil, the soluble forms of carbon (including polyphenols) compared to the available nutrients (N and P) in organic materials will, in fact, have a dominating influence on microbial growth and activity, and net nutrient mineralization or immobilization.

Indices for predicting release patterns

Many researchers have developed predictors or indices of mass loss and N release (Fox et al., 1990; Palm and Sanchez, 1991; Oglesby and Fownes, 1992; Tian et al., 1992a; Tian et al, 1995; Handayanto et al., 1994, 1995; Constantinides and Fownes, 1994b; Mtambanengwe and Kirchman, 1995; Lehman et al., 1995; Mafongoya et al., 1997). These indices or predictors include ratios of carbon to nitrogen (C:N), polyphenol to nitrogen (PP:N), lignin to nitrogen (L:N), and polyphenol plus lignin to nitrogen (PP + L):N. All of the indices are apparently valid but depend on the time course under consideration and the type of materials. For example, nitrogen mineralization in the short term has been negatively correlated with PP:N when considering green foliage with high N content and also substantial concentrations of total soluble polyphenolics (Palm and Sanchez, 1991; Oglesby and Fownes, 1992). Over a longer time frame, however, (PP + L):N is a better predictor for these same materials (Constantinides and Fownes, 1994a; Tian et al., 1992a). For litters which generally have lower N and higher lignin contents than green biomass, L:N is a better predictor (Melillo et al., 1989). The (PP + L):N ratio has been found to be the most robust index to predict mass loss and N release pattern in most agroforestry systems (e.g., Figure 3), but

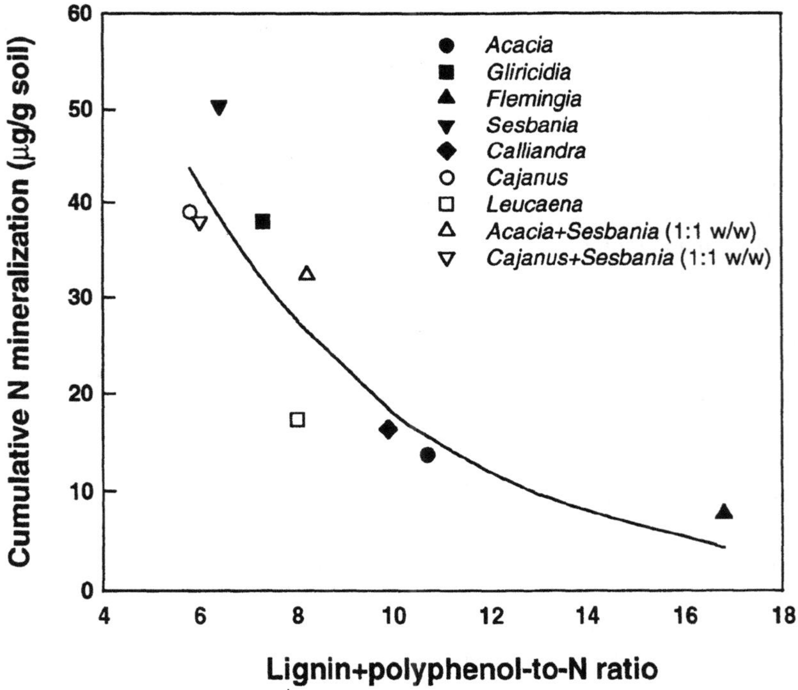

Fig. 3. The relationship between cumulative mineralized N and the (lignin + polyphenol):N ratio of leaves of multipurpose trees (Mafongoya et al., 1997).

a critical ratio cannot be recommended due to different methods of polyphenol analysis. As a general guideline, however, materials with a (PP + L):N ratio of less than 10 will release N. The predictive power of ratios that consider polyphenols is likely to be improved by including an estimate of their protein-binding capacity.

The main objectives of the development of these indices are to identify robust parameters that predict decomposition and nutrient release, and to avoid the need for detailed studies of decomposition of various plant materials in every location. The indices could also facilitate the screening of agroforestry trees for their nutrient-supplying capacity, and improve understanding of controls on nutrient cycling. In addition, the indices could be used to model soil-fertility changes, including SOM dynamics in systems where prunings are applied as a source of nutrients for a long time.

Traditionally, the C:N ratio has been used as the best index to predict decomposition and N mineralization in leguminous and nonleguminous green manures (Frankenberger and Abdelmagid, 1985). However, the C:N ratio has been found to be a very poor predictor of decomposition and nutrient release in most agroforestry studies, with the exception of studies by Sandhu et al. (1990), Tian et al. (1992a), and Mtambanengwe and Kirchman (1995). Litter used in these studies had a very wide range of C:N ratios, and included both herbaceous legumes and trees. Recent work has shown that, for predicting decomposition and N mineralization, the partitioning of total N and C into labile pools appears to be more important than C:N ratios in determining decomposition rates of agroforestry prunings and litter from trees (Mtambanengwe and Kirchman, 1995; Mafongoya et al., 1997). Fractionation of total N into soluble N and fiber-bound N (NDF-N) was more important than total N content in determining the rate of N mineralization from various tree prunings (Mafongoya et al., 1997). Similar partitioning of total C into labile fractions by Mtambanengwe and Kirchman (1995) improved the correlations of different C compounds with decomposition and N mineralization of diverse miombo litters.

A plant residue quality index (PRQI) was recently introduced by Tian et al. (1995). It basically combines the indices mentioned above and includes C:N ratio, lignin content, and total soluble polyphenol content. The PRQI was found to be correlated to decomposition rates, soil microclimate effects, and soil faunal populations. Unfortunately, critical levels or the utility of PRQI for selection of plant materials were not specified.

Indices for predicting decomposition and nutrient release rates are useful to an extent, but the types of plant materials for which they are useful must be kept in mind. For example, the C:N or L:N ratios are good predictive tools until residues with a N content greater than 3 g/100 g are considered. It then becomes necessary to ascertain whether the residues are rich in total soluble polyphenols. If the total polyphenol contents are high, it is then necessary to look at the reactivity or protein-binding capacity of the phenolics. The development of such a hierarchical (or decision tree) approach may simplify the characterization of materials, and the use of residue-quality parameters as tools for predicting decomposition.

Nutrient release patterns of biomass from common multipurpose trees

As described above, chemical qualities and decomposition characteristics differ widely among multipurpose tree species and this may largely determine their utility in supplying nutrients for crops. It has long been known that in multipurpose tree leaves, nutrient concentrations decline and the degree of lignification increases with age (e.g, Joachim, 1926; Joachim and Kandiah, 1936). Polyphenol contents are less predictable and may increase or decrease with age, depending on the species (Vanlauwe et al., 1997a). This may be due partly to a 'dilution' effect as leaves expand. Thus, the age of leaves or branches may significantly affect the quality and decomposition of prunings (Mafongoya et al., 1997; Vanlauwe et al., 1997b). Furthermore, decomposition characteristics will be strongly influenced by the prevailing weather conditions at the time biomass is added to the soil (Vanlauwe et al., 1997b). Although there is comparatively little information about nutrient concentrations in freshly fallen litter from agroforestry trees, litter is usually composed of old leaves, so will generally be highly lignified.

Categories of nutrient release patterns

Several factors, other than quality differences among leaves of the same species, complicate comparisons of data on the decomposition of leaf material from multipurpose tree species. Two examples of complicating factors are the use of different analytical methods to determine chemical qualities, and the choice of either laboratory or field incubations. However, as abundant data are now available for a number of species (Table 2), it is possible to make some fairly confident predictions as to the relative decomposition rates of their leaves.

Leaves that have high nitrogen contents but are low in lignin and polyphenols (e.g., leaves of *Gliricidia sepium* and *Sesbania* spp.) will decompose very quickly, releasing a large proportion of their N. Slow decomposition patterns can result from several different characteristics, generally related to large amounts of reactive polyphenols or structural lignin and associated insoluble proanthocyanidins. Some agroforestry trees have leaves that are generally lignified and will therefore decompose slowly, possibly immobilizing N for long periods (e.g., *Dactyladenia barteri* and *Flemingia macrophylla*). Species with high N contents and large polyphenol contents can be divided into two categories: (1) species such as *Leucaena leucocephala*, with a small capacity to bind protein and rapidly decomposing leaves that release a large proportion of their nitrogen, despite the presence of polyphenols, and (2) species such as *Calliandra calothyrsus* with a large capacity to bind protein

Table 2. Leaf and litter characteristics and resulting decomposition patterns of multipurpose agroforestry trees. Direct values are not given due to difficulties in comparing results obtained with different analytical methods. Instead, the likely relative leaf qualities and the probable decomposition characteristics are given, based on extrapolation from data for fully opened leaves (including pinnae). A higher number of crosses (+) indicates a higher relative value.

Species	N_2-fixer	C-to-N ratio	Lignin content	Polyphenol content	Protein-binding capacity	Decomposition rate	% N released in first month	Ref.[a]
Acacia angustissima	yes	18	+++	+++	++++	slow		7
A. auriculiformis	yes	20	++	+++	ND	ND	immobilizes	1
Apuleia leiocarpa	no	18	+	+	+	moderate	immobilizes	13
Dactyladenia barteri[b]	no	25–30	++++	+++	ND	slow		11
Cajanus cajan	ye	10–15	+ to ++	+ to ++	+	moderate		7, 9
Calliandra calothyrsus	yes	10–18	+	+++	++++	slow	15–30	3, 7, 8
Centrolobium ochroxylum	no	18	ND	+++	+++	moderate	immobilizes	13
Crotalaria anagyroides	yes	12	+	ND	ND	fast		5
Croton macrostachyus[b]	no	14	++	++	ND	ND	20	6
Erythrina spp.	yes	11	++	+	ND	fast	25	9
Flemingia macrophylla	yes	19–25	++++	+++	++	slow		7, 10
Gliricidia sepium	yes	10–20	+	+	+	fast	35–60	1, 2, 3, 7, 9, 11
Grevillia robusta[b]	no	32	+++	++	ND	slow	immobilizes	6
Inga edulis	yes	14–19	+++	+++	ND	slow	0–10	1, 8, 9
Leucaena leucocephala	yes	10–16	++	+++	+	fast	30	2, 3, 7, 9, 11
Myroxylon balsamum	yes	16	+	ND	+	moderate		13
Peltophorum dasyrrachis	no	20	+++	ND	+++	slow	12–15	2, 3
Piptadenia buchtienii	yes	16	+	+++	++++	slow		13
Pithecellobium saman	yes	14	+	+	ND	fast		9
Schizolobium amazonicum	yes	15	+	++	+	fast		13
Senna reticulata	no	18	+	+	ND	ND	0–20	8, 9
Senna spectabilis	no	14	+	+	+	moderate		4, 10
Senna siamea	no	16–18	+	++	+	moderate	immobilizes	1, 8, 12
Sesbania sesban	yes	15–16	+	+ to ++	+	fast	55	1, 7
Tephrosia candida	yes	13–20	+	+	ND	fast		5
Tephrosia vogelii	yes	15	+	+	ND	ND		10
Tithonia diversifolia[b]	no	14	ND	+	ND	ND	30	6

[a] 1, Constantinides and Fownes (1994b); 2, Handayanto et al. (1997b); 3, Handayanto et al. (1994; 1995); 4, Itimu OA, Cadisch G and Giller KE (unpublished results); 5, Joachim and Kandiah (1936); 6, Kwabiah and Palm C (unpublished results); 7, Mafongoya (1995); 8, Oglesby and Fownes (1992); 9, Palm and Sanchez (1991); 10, Palm C and Wangari N (unpublished results); 11, Tian et al. (1992a,b); 12, Tian et al. (1995); 13, Vargas E and Giller KE (unpublished results).
[b] Non-legumes.

and slowly decomposing leaves that release a much smaller proportion of their nitrogen.

The various types of decomposition patterns and the associated leaf-quality parameters are summarized in Table 2. Some decomposition patterns appear somewhat anomalous. For example, although *Senna* spp. have quite a narrow C-to-N ratio and little lignin or polyphenols, they decompose relatively slowly (O. A. Itimu, G. Cadisch and K. E. Giller, unpublished results). The reasons for this are at present unclear: the slow decomposition does not appear to be related to large amounts of soluble C but may be due to large amounts of the N being bound to condensed tannins in the fiber fraction.

Management of decomposition for efficient nutrient use

Why manage?

Efficiency of nutrient use depends on both the efficiency of acquisition of nutrients (nutrient capture) and the efficiency with which nutrients are then utilized by a growing plant (conversion efficiency). Two strategies can be used to manipulate litter decomposition in order to improve nutrient use efficiency: (1) regulate the rates of release of nutrients to improve the synchrony of nutrient supply with crop demand, and (2) provide a more favorable environment for plant growth. Improvements in the environment for plant growth may result in more extensive root development and thus enhance nutrient capture, or may improve the efficiency of nutrient use by enhancing plant growth in general. The two strategies relate to the short-term improvement of nutrient availability and the longer-term improvement of soil organic matter (Table 3). The long-term effects of building up SOM pools may be more important in some cases than short-term nutrient release patterns. The soil ecosystem takes years to show the effects of organic inputs. For example, it is commonly found that significant results from green manuring are obtained only after the practice has continued for several years.

The effect of organic inputs on both of these strategies is highly dependent on the quality of the inputs. Nutrient availability can be affected both directly by the contribution of nutrients released during decomposition, or indirectly through the effects of decomposition byproducts, such as organic ions that reduce the P fixation capacity of the soil (Hue, 1991). Soil organic matter content and the relative proportion of different organic matter fractions, which have different roles in nutrient cycling and in determining soil structure, also depend on the amount and quality of organic matter inputs. Thus the potential exists for manipulating soil fertility in various ways depending on the specific requirements of the soil and cropping system being managed.

Table 3. Beneficial roles of leaf prunings and litter from agroforestry trees.

Benefit	Timescale	Resulting from	Mediated by
Improved nutrient uptake	Short-term	Increased N availability	N release during decomposition
		Increased P availability	Reduced P fixation due to organic acids
		Enhanced root proliferation	Reduced aluminum toxicity due to complexation
Reduced nutrient losses	Long-term	Smaller leaching losses	Increased CEC
			Improved nutrient uptake
Improved moisture availability	Long-term	Increased moisture retention	Increased SOM
Reduced soil erosion	Long-term	Increased infiltration	Increased SOM
		Increased soil cover	Surface litter
		Increased soil stability	Increased SOM
		Reduced runoff	Increased SOM

Sources: Adapted from Giller and Wilson (1991); Young (1989).

Approaches to manipulating decomposition

Production of prunings of varied quality

Species selection is the most obvious method of selecting prunings of a particular property, although other considerations will often dictate the choice of an agroforestry tree. It is, of course, possible to obtain leaf litter of vastly different quality from a single species, for example by regulating pruning age, allowing leaves to senesce naturally, mixing different plant parts (leaves, pinnules, pinnae, woody branches), or selecting different provenances of the same species (Jackson et al., 1996).

The conditions prevailing during tree growth can also result in production of prunings with markedly differing quality. Nitrogen limitation in particular has been demonstrated to enhance the concentration of polyphenols in leaves of both legumes and non-legumes (Davies et al., 1964; Stitt et al., 1946), and N fertilization reduced polyphenol concentrations in quaking aspen (*Populus tremuloides*) (Bryant et al., 1987). By growing trees under axenic conditions so they could not nodulate and fix N_2, and then increasing N supply in the feeding solutions, Handayanto et al. (1995) produced litters of *Calliandra calothyrsus* and *Gliricidia sepium* of widely varying quality. The parameters altered most strongly by increasing N supply were increased N concentration in the leaves, reduced concentration of soluble polyphenols, and reduced protein-binding capacity of the polyphenols; these translate into predictable effects on decomposition and recovery of released-N by maize (Handayanto et al., 1995; Handayanto et al., 1997a). Thus, ensuring effective nodulation and N_2-fixation in multipurpose trees will promote high N contents and relatively low polyphenol contents, as compared with trees of the same species grown under N-limited conditions.

Managing prunings to alter their quality
Application of fresh prunings directly to soil, without drying, results in faster decomposition and may result in greater N recovery by crops (Wilson et al., 1986) although this will obviously depend on synchrony with crop growth. Mafongoya (unpublished results) found greater maize yields when fresh, rather than sun-dried, prunings of *Flemingia macrophylla* were applied in the field. Indeed, the sun-dried prunings immobilized N for more than eight weeks.

Pruning quality can be strongly affected by the temperature at which prunings are dried (Mafongoya et al., 1997) because elevated temperatures reduce the recovery of soluble polyphenols (Constantinides and Fownes, 1994b; Hagerman, 1988). Elevated temperatures may also cause an apparent increase in lignin contents because non-enzymatic 'browning' reactions between polyphenols, carbohydrates and protein degradation products form insoluble polymers which artificially raise lignin contents (Goering and van Soest, 1970).

The degree to which the prunings are chopped and mixed will determine the immediate accessibility of components to microbial attack, and may reduce the amounts of nutrients readily lost by leaching or available for plant uptake. Fine grinding of plant residues has been shown to enhance decomposition of materials with a high C:N ratio (Bremer et al., 1991; Jensen, 1994) and to stimulate immobilization (Jensen, 1994). Fine grinding is unlikely to be a practice adopted to manipulate decomposition in agriculture, although it may to some extent mimic effects of rapid comminution by microorganisms on decomposition. Incorporation of prunings into the soil also results in faster decomposition (Wilson et al., 1986) because there is more intimate contact between the prunings and the soil particles, and the prunings are maintained in a moist environment that is more favorable for decomposition. Incorporation of organic additions results in a bacterial dominated community, while fungi dominate with surface placement (Holland and Coleman, 1987). Such differences in the microbial community can influence nutrient dynamics and soil organic matter formation because of differences in carbon assimilation efficiencies and carbon-to-nutrient ratios of bacteria and fungi.

Field management of prunings
Perhaps the simplest way to regulate the rate of decomposition of prunings or litter is to vary the way in which they are incorporated into the soil. Wilson et al. (1986) showed much more rapid decomposition and greater N recovery of *Leucaena leucocephala* and *Gliricidia sepium* prunings when they were incorporated into soil, and similar results have been found with a range of species with both high- and low-quality prunings (Mafongoya and Nair, 1997). The timing of pruning application may also be varied to assist synchronization of nutrient release and crop demand. Mulongoy et al. (1993) showed

that it was advantageous to apply low-quality prunings two to four weeks before planting to allow sufficient time for decomposition.

Mixing prunings of different quality

Mixing prunings of different qualities is one obvious approach to regulating the rates of decomposition and nutrient (especially N) release from prunings. Mixtures can be comprised of different plant parts from a single species or of materials from species with widely differing quality. Regardless, the same rules govern decomposition interactions between the different quality materials. The effects of mixing different types of plant tissues in prunings was investigated by Constantinides and Fownes (1994a). They attributed the complex pattern of N accumulation or release to the immobilization that occurred in the twig fraction, as was hypothesized by Oglesby and Fownes (1992). Incubations of leaf-and-twig mixtures gave mineralization results that deviated from those expected from leaves and twigs acting independently, thus providing some evidence for interaction effects in the mixtures.

The decomposition pattern of a mixture may be the direct mean of the decomposition patterns of the component litters. In this case, there would be no interaction between the components, each behaving as it would if it were not mixed. Alternatively, there may be significant interactions between the component litters, such that the rates of decomposition and N release are altered and/or the amount of N released is changed in an unpredictable way.

If strong decomposition interactions are to occur between prunings of different types, the constituents of the different prunings must either be available to microorganisms at the same time or be able to interact directly. Thus, unless the component litters are finely ground and thoroughly mixed, such interactions can result only from the movement of soluble constituents. There are three types of soluble constituents that can move, resulting in interactions between litters: (1) compounds that contain readily available carbon as a substrate, (2) compounds that contain readily available N, and (3) soluble polyphenols which can complex with proteins, rendering them resistant to immediate microbial attack (Figure 4). Lignin degrades slowly, so protects other cell wall constituents, such as cellulose, from microbial attack (Chesson, 1997). Strong interactions between lignin or other recalcitrant substrates are thus unlikely to occur until the later stages of decomposition, when residues are more intimately mixed and fungi hyphal networks can bridge between resources (Rayner, 1994).

Interactions in decomposition have been demonstrated in mixtures of stems and leaves of cereals, wherein the stems contained a large amount of soluble C which stimulated immobilization of N from the leaves (Quemada and Cabrera, 1995). No decomposition interactions, however, were found between leaves and stems of clover.

Experiments mixing residues of *Gliricidia sepium* with those of *Peltophorum dasyrrachis*, at varying proportions, revealed a linear trend, with the polyphenol content increasing as the proportion of *Peltophorum* increased.

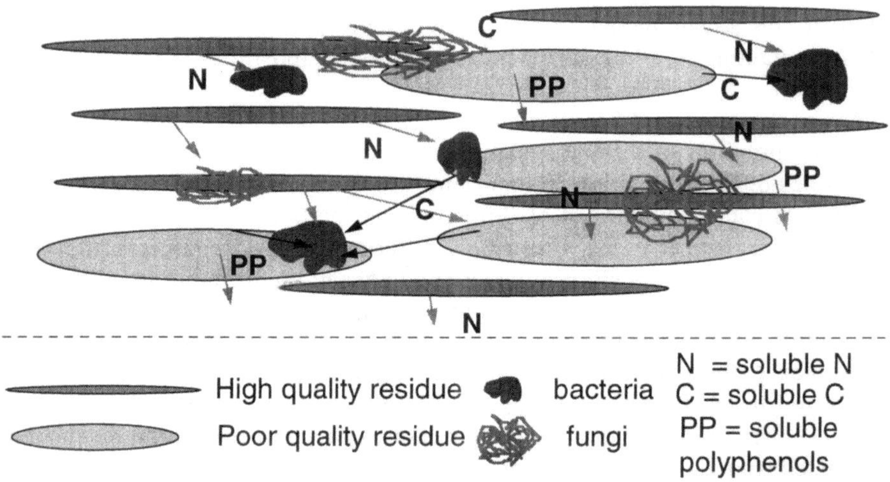

Fig. 4. Schematic diagram illustrating possible interactions in decomposition when leaf prunings or litters of differing quality are mixed. For interactions to occur, soluble components must come into contact with microorganisms or other constituents with which they can react. Microorganisms colonizing a resource will be able to compete readily for soluble C and N compounds as substrates for growth. Soluble reactive polyphenols may be leached from one residue and bind with proteins from another residue type. Alternatively, soluble proteins from one residue may be bound to insoluble reactive polyphenols in another type of residue.

However, the mixtures had a very low protein-binding capacity (PBC), similar to pure *Gliricidia* residues, unless more than 50% of the material was derived from *Peltophorum*. When *Peltophorum* prunings composed more than 50% of the mixture, the PBC rose sharply indicating that large quantities of soluble protein from the *Gliricidia* residues were able to bind to and saturate the reactive polyphenols of *Peltophorum* until the proportion of *Peltophorum* in the mixture reached 50% (Handayanto et al., 1994; Handayanto et al., 1997b). Decomposition and N release, and subsequent N recovery by maize, were inversely related to the protein-binding capacities of the mixtures, indicating that PBC is a robust predictor of N mineralization (Handayanto et al., 1997b). Within intact tissues, reactive polyphenols are localized either in the vacuoles or the cell walls, thus protecting enzymes and other proteins from complexation. Indeed, in some species, polyphenols are localized into particular cells such as trichomes (Li et al., 1996), which may limit their influence on decomposition rates.

A secondary mechanism for decomposition interactions between residues occurs when one residue alters the physical environment of another, for instance when a recalcitrant litter creates a surface mulch that increases moisture availability in the underlying litter and soil. Tian et al. (1993) found that when there was sufficient moisture, a low-quality mulch of *D. barteri* produced lower maize yields than did a high-quality mulch of *Gliricidia* which released more N. By contrast, during the dry season the lower-quality mulch produced higher maize yields because of the improved microclimate for decomposition and mineralization below the mulch. Another explanation for the increased moisture effect is that higher SOM leads to increased water retention.

Conclusions and future research needs

Many studies have shown that prunings or litter of agroforestry trees can supply sufficient nutrients to meet crop demand, with the notable exception of phosphorus. The potential of these organic inputs to supply nutrients immediately or through residual effects to subsequent crops depends on their resource quality. Various indices have been developed to predict decomposition and nutrient-release patterns of tree prunings. To date, the (polyphenol + lignin):N ratio seems to be the most robust indicator to predict mass loss and nutrient release, but no critical value can be given because of the different methods used to analyze polyphenols. Substantial gaps in our knowledge remain. To describe resource quality so it is a useful predictor for decomposition in a wide range of residue types, there must be a means of incorporating input from all of the various parameters. We suggest that development of a hierarchical approach, or decision tree, will be useful in this regard as it may eliminate the need to measure all parameters for all species or types of residues. More detailed characterization of the different types of polyphenols and how they interfere with the decomposition process will be necessary before a robust method of predicting decomposition of polyphenol-rich residues can be developed.

To date, almost all research has concentrated on the aboveground inputs in agroforestry systems, despite the substantial belowground contribution of root biomass. Root resource quality has been shown to influence root decomposition and nutrient-release patterns (Lehmann et al., 1995) but little is known of the relative importance of below- and above-ground organic matter inputs on nutrient cycling.

The residual effects on nutrient release and long-term changes in soil fertility resulting from application of prunings of different quality is a subject which has received minor attention to date. A series of functional pools of soil organic matter may be distinguished conceptually (Parton et al., 1994) or physically (e.g., Barrios et al., 1995; Vanlauwe et al., 1996; Vanlauwe et al., 1997c) which play different roles in soil fertility. It is likely that the relative sizes of such pools will be strongly influenced by the amounts and the quality of plant residues added to soil. Recent attempts to relate large, light fractions of SOM to the 'active' SOM pool that is important in supplying nutrients has been partially successful (Barrios et al., 1995), although it appears unlikely that such fractions will be physically isolatable (Magid et al., 1996). The long-term effects of agroforestry on soil fertility is an exciting area worthy of greater experimental investigation.

Experiments must be designed that will allow for evaluation of the short- and long-term effects of organic inputs of constrasting qualities on nutrient availability and soil organic matter content and fractions. Such experiments must be based either on applying organic inputs with equal amounts of the limiting nutrient, or organic inputs of equal mass. The former design may be better for nutrient studies and the latter for soil organic matter studies. If the latter design is used, then results must be presented according to the unit of nutrient added. It must be noted that organic inputs contain several nutrients, in addition to the one of interest for the experiment, and this may confound interpretation of the results.

Overall the greatest challenge is to develop ways of managing the decomposition rates of organic matter to optimize rates of short- and long-term nutrient release, and to maintain soil organic matter. Manipulation of residue quality by mixing resources from different tree species is one way in which this might be achieved, although as we have described, interactions in the decomposition of residue mixtures may be complex. Field methods for estimation of plant litter quality parameters are also required.

References

Aber JD and Melillo JM (1991) Terrestrial Ecosystems. Rinehart and Winston, Inc., Philadelphia, USA

Amato M, Jackson RB, Butler JHA and Ladd JN (1984) Decomposition of plant material in Australian soils. II. Residual organic ^{14}C and ^{15}N from legume plant parts decomposing under field and laboratory conditions. Australian Journal of Soils Research 22: 331–341

Anderson JM and Ingram JSI (1993) Tropical Soil Biology and Fertility: A Handbook of Methods. 2nd edition. CAB International, Wallingford, UK

Barrios E, Buresh RJ and Sprent JI (1995) Organic matter in soil particle size and density fractions from maize and legume cropping systems. Soil Biol Biochem. 28: 185–193

Bremer E, van Houtum W and van Kessel C (1991) Carbon dioxide evolution from wheat as affected by grinding, added nitrogen and the absence of soil. Biol Fert Soil 11: 221–227

Bryant JP, Clausen TP, Reichardt PB, McCarthky MC and Werner RA (1987) Effect of nitrogen fertilization upon the secondary chemistry and nutritional value of the quaking aspen (*Populus tremuloides* Michx.) leaves for the large aspen tortrix (*Choristoneura conflictana* (Walker)). Oecologia 73: 513–517

Budelman A (1989) Nutrient composition of the leaf biomass of three selected woody leguminous species. Agroforestry Systems 8: 39–51

Cadisch G and Giller KE (eds) (1997) Driven by Nature: Plant Litter Quality and Decomposition. CAB International, Wallingford, UK

Chesson A (1997) Plant degradation by ruminants: parallels with litter decomposition in soils. In: Cadisch G and Giller KE (eds) Driven by Nature: Plant Litter Quality and Decomposition, pp 47–66. CAB International, Wallingford, UK

Collins HP, Elliott LF, Rickman RW, Bezdicek DF and Papendick RI (1990) Decomposition and interactions among wheat residue components. Soil Science Society of America Journal 54: 780–785

Constantinides M and Fownes JH (1994a) Nitrogen mineralization from leaves and litter of tropical plants: relationship to nitrogen, lignin and soluble polyphenol concentrations. Soil Biol Biochem 26: 49–55

Constantinides M and Fownes JH (1994b) Tissue-to-solvent ratio and other factors affecting

determination of soluble polyphenols in tropical leaves. Comm. Soil Sci Pl Anal 25: 3221–3227
Davies RI, Coulson CB and Lewis DA (1964) Polyphenols in plant, humus and soil. IV. Factors leading to increase in biosynthesis of polyphenols in leaves and their relation to mull and mor formation. J Soil Sci 15: 310–318
Dawra RK, Makkar HSP and Singh B (1988) Protein-binding capacity of microquantities of tannins. Analytical Biochemistry 170: 50–53
Dubois M, Gilles KA, Hamilton JK, Rebers PA and Smith F (1956) Colorimetric method for determination of sugars and related substances. Analytical Chemistry 28: 350–356
Fox RH, Myers RJK and Vallis I (1990) The nitrogen mineralization rate of legume residues in soil as influenced by their polyphenol, lignin and nitrogen contents. Plant Soil 129: 251–259
Frankenberger WT and Abdelmagid HM (1985) Kinetic parameters of nitrogen mineralization rates of leguminous corps incorporated into soil. Plant Soil 87: 257–271
Giller KE and Cadisch G (1995) Future benefits from biological nitrogen fixation. An ecological approach to agriculture. Plant Soil 174: 255–277
Giller KE and Wilson KJ (1991) Nitrogen Fixation in Tropical Cropping Systems. CAB International, Wallingford, UK
Goering HK and van Soest PJ (1970) Forage Analyses, p. 20. USDA Agriculture Handbook. Vol. No. 379. Agricultural Research Service, Washington, DC
Hagerman AE (1988) Extraction of tannins from fresh and preserved leaves. Journal of Chemical Ecology 14: 453–461
Haggar JP, Tanner EU, Beer JW and Kass DCL (1993) Nitrogen dynamics of tropical agroforestry and annual cropping systems. Soil Biol Biochem 25: 1363–1378
Handayanto E, Cadisch G and Giller KE (1994) Nitrogen release from prunings of legume hedgerow trees in relation to quality of the prunings and incubation method. Plant Soil 160: 237–248
Handayanto E, Cadisch G and Giller KE (1995) Manipulation of quality and mineralization of tropical legume tree prunings by varying nitrogen supply. Plant Soil 176: 149–160
Handayanto E, Cadisch G and Giller KE (1997a) Regulating N mineralization from plant residues by manipulation of quality. In: Cadisch G and Giller KE (eds) Driven by Nature: Plant Litter Quality and Decomposition, pp 175–185. CAB International, Wallingford, UK
Handayanto E, Giller KE and Cadisch G (1997b) Nitrogen mineralization from mixtures of legume tree prunings of different quality and recovery of nitrogen by maize. Soil Biol Biochem 29: 1417–1426
Holland EA and Coleman DC (1987) Litter placement effects on microbial and organic matter dynamics in an ecosystem. Ecology 68: 425–433
Hue NV (1991) Effects of organic acids/anions on P sorption and phytoavailability in soils with different mineralogies. Soil Science 152: 463–471
Jackson FS, Barry TN and Lascano C (1996) The extractable and bound condensed tannin content of leaves from tropical trees, shrubs and forage legumes. J Sci Food Agric 71: 103–110
Jensen ES (1994) Mineralization-immobilization of nitrogen in soil amended with low C:N ratio plant residues with different particle sizes. Soil Biol Biochem 26: 519–521
Joachim AWR (1926) The analyses of some green manures. The Tropical Agriculturalist 67: 4
Joachim AWR and Kandiah S (1936) The change in composition and decomposability of typical Ceylon green manures with age. The Tropical Agriculturalist 82: 3–20
Lehmann J, Schroth G and Zech W (1995) Decomposition and nutrient release from leaves, twigs and roots of three alley-cropped tree legumes in central Togo. Agroforestry Systems 29: 21–36
Li Y-G, Tanner G and Larkin P (1996) The DCMA-HCl protocol and the threshold proanthocyanidin content for bloat safety in forage legumes. J Sci Food Agric 70: 89–101
Mafongoya PL (1995) Multipurpose tree prunings as a source of nitrogen to maize (*Zea mays* L.) under semi-arid conditions in Zimbabwe. PhD Diss, Univ of Florida, Gainesville, FL, USA (Diss. Abstr. 95-07412)

Mafongoya PL, Dzowela BH and Nair PKR (1997) Effect of multipurpose trees, age of cutting and drying method on pruning quality. In: Cadisch G and Giller KE (eds) Driven by Nature: Plant Litter Quality and Decomposition, pp 167–174. CAB International, Wallingford, UK

Magid J, Gorissen A and Giller KE (1996) In search of the elusive 'active' fraction of soil organic matter: three size-density fractionation methods for tracing the fate of homogeneously ^{14}C labelled plant materials in soil. Soil Biol Biochem 28: 89–99

Martin JP and Haider K (1980) Microbial degradation and stabilization of ^{14}C-labelled lignins, phenolics, and phenolic polymers in relation to soil humus formation. In: Kirk TK, Higuchi T and Chang HM (eds) Lignin Biodegradation: Microbiology, Chemistry, and Potential Applications, pp 78–100, Vol 2. CRC Press, West Palm Beach, FL, USA

Meentemeyer V (1978) Macroclimate and lignin control of litter decomposition rates. Ecology 59: 465–472

Melillo JM, Aber JD, Linkins AE, Ricca A, Fry B and Nadelhoffer KJ (1989) Carbon and nitrogen dynamics along a decay continuum: plant litter to soil organic matter. In: Clarholm M and Bergstrom L (eds) Ecology of Arable Land, pp 53–62. Kluwer, Dordrecht, The Netherlands

Mulongoy K, Ibewiro EB, Oseni O, Kilumba N, Opara-Nadi OA and Osonubi O (1993) Effect of management practices on alley cropped maize utilization of nitrogen derived from prunings on a degraded Alfisol in south-western Nigeria. In: Mulongoy K and Merckx R (eds) Soil Organic Matter and Sustainability of Tropical Agriculture, pp. 223–230. Wiley/Sayce, Chichester, UK

Mtambanengwe F and Kirchmann H (1995) Litter from tropical savanna woodland (miombo) chemical composition and C and N mineralization. Soil Biol Biochem. 27: 1639–1651

Nair PKR (1993) Introduction to Agroforestry. Kluwer, Dordrecht, The Netherlands

Oglesby KA and Fownes JH (1992) Effects of chemical-composition on nitrogen mineralization from green manures of 7 tropical leguminous trees. Plant Soil 143: 127–132

Palm CA (1995) Contribution of agroforestry trees to nutrient requirements in intercropped plants. Agrofor Sys 30: 105–124

Palm CA and Rowland AP (1997) A minimum dataset for characterization of plant quality for decomposition. In: Cadisch G and Giller KE (eds) Driven by Nature: Plant Litter Quality and Decomposition, pp 379–392. CAB International, Wallingford, UK

Palm CA and Sanchez PA (1990) Decomposition and nutrient release patterns of the leaves of three tropical legumes. Biotropica 22: 330–338

Palm CA and Sanchez PA (1991) Nitrogen release from the leaves of some tropical legumes as affected by their lignin and polyphenolic contents. Soil Biol Biochem 23: 83–88

Parton WJ, Woomer PL and Martin A (1994) Modelling soil organic matter dynamics and plant productivity in tropical ecosystems. In: Woomer PL and Swift MJ (eds) The Biological Management of Tropical Soil Fertility, pp 171–188. Wiley, Chichester, UK

Quemada M and Cabrera ML (1995) Carbon and nitrogen mineralized from leaves and stems of four cover crops. Soil Sci Soc Amer. J 59: 471–477

Rayner ADM (1994) Pattern-generating processes in fungal communities. In: Ritz K, Dighton J and Giller KE (eds) Beyond the Biomass. Compositional and Functional Analysis of Soil Microbial Communities, pp 247–258. John Wiley & Sons, Chichester, UK

Reed JD (1986) Relationships among soluble phenolics, insoluble proanthocyanidins and fiber in East African browse species. Journal of Range Management 39: 5–7

Reed JD (1995) Nutritional toxicology of tannins and related polyphenols in forage legumes. J Anim Sci 73: 1516–1528

Rowland AP and Roberts JD (1994) Lignin and cellulose fraction in decomposition studies using acid-detergent fibre methods. Communications in Soil Science and Plant Analysis 25: 269–277

Sanchez PA (1995) The science of agroforestry. Agrofor Sys 30: 1–55

Sanchez PA and Logan TJ (1992) Myth and science about the chemistry and fertility of soils in the tropics. In: Lal R and Sanchez PA (eds) Myths and Science of Soils of the Tropics, pp 34–46. SSSA Spec Publ No. 29. ASA, CSSA, and SSSA, Madison, WI, USA

Sandhu JM, Simha M and Ambasht RS (1990) Nitrogen release from decomposing litter of Leucaena leucocephala in the dry tropics. Soil Biol Biochem 7: 171–177

Schlesinger WH and Hasey MM (1981) Decomposition of chaparral shrub foliage: losses of organic and inorganic constituents from deciduous and evergreen leaves. Ecology 62: 762–774

Smith JL (1994) Cycling of nitrogen through microbial activity. In: Hatfield JL and Stewart BA (eds) Soil Biology: Effects on Soil Quality, pp 91–120. Advances in Soil Science. CRC Press, Boca Raton, FL, USA

Stitt RE, Hyland HL and McKee RL (1946) Tannin and growth variation of a *Sericea lespedeza* clone in relation to soil type. Journal of the American Society of Agronomy 38: 1003–1009

Swift MJ, Heal OW and Anderson JM (1979) Decomposition in Terrestrial Ecosystems. Studies in Ecology, Vol 5. University of California Press, Berkeley, CA, USA

TAPPI (1988) Water solubility of wood and pulp. T 207 OM-88, Technical Association of the Pulp and Paper Industry, Atlanta, GA, USA

Theander O and Westerlund E (1993) Quantitative analysis of cell wall components. In: Forage Cell Wall Structure and Digestibility, pp 83–104. ASA-CSSA-SSSA, Madison, WI, USA

Tian G, Kang BT and Brussaard L (1992a) Effects of chemical composition on N, Ca, and Mg release during incubation of leaves from selected agroforestry and fallow plant species. Biogeochemistry 16: 103–119

Tian G, Kang BT and Brussaard L (1992b) Biological effects of plant residues with contrasting chemical composition under humid tropical conditions: decomposition and nutrient release. Soil Biol Biochem 24: 1051–1060

Tian G, Brussaard L and Kang BT (1993) Biological effects of plant residues with contrasting chemical compositions under humid tropical conditions: effects on soil fauna. Soil Biol Biochem 25: 731–373

Tian G, Brussaard L and Kang BT (1995) An index for assessing the quality of plant residues and evaluating their effects on soil and crop in the (sub-)humid tropics. Applied Soil Ecology 2: 25–32

Trofymow JA, Preston CM and Prescott CE (1995) Litter quality and its potential effects on decay rates of materials from Canadian forests. Water, Air and Soil Pollution 82: 215–226

Vallis I and Jones RJ (1973) Net mineralization of nitrogen in leaves and litter of *Desmodium intortum* and *Phaseolus artropurpureus* mixed with soil. Soil Biol Biochem 5: 391–398

Vanlauwe B, Swift M and Merckx R (1996) Soil litter dynamics and N use in a leucaena (*Leucaena leucocephala* Lam de Wit.) alley cropping system in south-western Nigeria. Soil Biol Biochem 28: 739–749

Vanlauwe B, Sanginga N and Merckx R (1997a) Decomposition of four *Leucaena* and *Senna* prunings in alley cropping systems under sub-humid tropical conditions: the process and its modifiers. Soil Biol Biochem (in press)

Vanlauwe B, Sanginga N and Merckx R (1997b) Impact of residue particle size on its quality and C and N mineralization for leaf residues of five agroforestry species. Plant and Soil (in press)

Vanlauwe B, Sanginga N and Merckx R (1997c) Soil organic matter dynamics as affected by the fresh additions of *Leucaena* and *Dactyladenia* residues in alley cropping systems in Southwestern Nigeria. Plant and Soil (in press)

Van Soest PJ (1963) Use of detergents in analysis of fibrous feeds. II. A rapid method for the determination of fibre and lignin. Association of Official Agricultural Chemists Journal 46: 829–835

Vitousek PM (1984) Litterfall, nutrient cycling, and nutrient limitation in tropical forests. Ecology 65: 285–298

Waterman PG and Mole S (1994) Analysis of Phenolic Plant Metabolites. Methods in Ecology. Blackwell Scientific Publications, London, UK

Wilson GF, Kang BT and Mulongoy K (1986) Alley cropping: trees as sources of green-manure and mulch in the tropics. Biological Agriculture and Horticulture 3: 251–267

Young A (1989) Agroforestry for Soil Conservation. CAB International, Wallingford, UK

Agroforestry Systems **38**: 99–120, 1998.

Nutrient cycling under mixed-species tree systems in southeast Asia

P. K. KHANNA
CSIRO, Division of Forestry and Forest Products, P.O. Box E4008, Kingston-Canberra, ACT 2604, Australia; E-mail: partap.khanna@ffp.csiro.au

Key words: acacia, eucalyptus, N_2-fixing trees, N-transfer, P-dynamics

Abstract. *Eucalyptus* and *Acacia* are two tree genera that are commonly used in industrial plantations and as components of agroforestry systems in southeast Asia. These fast-growing trees are mostly grown in monocultures. However, questions are now being raised about the long-term sustainability of their growth as well as their effects on site quality. Losses of N and P from the site through biomass harvest and during site preparation constitute a major nutrient drain. As an alternative to monocultures, mixed-species cultures which include at least one N_2-fixing tree species can improve plant productivity and soil N dynamics. Among the various aspects of N dynamics in such stands, export of N during biomass harvest and inputs by the N_2-fixing component are the most important.

Reported estimates of the amount of N fixed by acacia and other N_2-fixing trees are highly variable, depending on inherited plant characteristics, tree age, site factors (e.g., drought), soil fertility (e.g., available P, metal toxicities) and unreliable methods of measuring N_2-fixation. Of the available techniques for assessing N_2-fixation by trees, the total N difference method (TND) is the simplest. The contribution of roots to assessments of N_2-fixation is recognized but rarely measured.

For short-rotation mixed-species plantations, the amount and time of N transfer from N_2-fixing trees to non-N_2-fixing trees are important issues to consider when attempting to develop productive nutrient management strategies. Based on limited information from trials in southeast Asia, it appears that acacia fixes substantial amounts of N during the first few years of establishment and a significant amount of that N is transferred to adjacent eucalypts, thereby improving the growth and nutrition of the eucalypts. The presumed transfer of N from acacias to eucalypts during the early stages of plantation development probably results from belowground turnover of roots and nodules because aboveground litter decomposition is slight at this stage, and contributes little to the overall N dynamics.

The available information on P cycling in mixed-species stands, during the early stages of stand growth, provides inconclusive evidence as to whether the availability of soil P increases, despite indications of higher levels of phosphatase activity in the fine roots of nitrogen-fixing trees. This would imply that additional inputs of P as fertilizer are required to remove any P deficiency in mixed-species stands. Long-term observations are required for better understanding of the nutritional and growth benefits of including N_2-fixing trees in mixed-species stands.

Introduction

Of the 32.3 million ha under forest plantations in 1990 in the Tropical Asia and Pacific (TAP) region, 19.8 million ha are in south Asia (Bangladesh, Bhutan, India, Nepal, Pakistan and Sri Lanka), 3.2 million ha in continental southeast Asia (Laos, Myanmar, Thailand and Vietnam), 9.16 million ha in insular southeast Asia (Indonesia, Malaysia and the Philippines) and 0.19 million ha in the Pacific Islands (Fiji, New Caledonia, Papua New Guinea, Samoa, Solomon Islands, and Vanuatu) (Pandey 1995). Eucalypts (*Eucalyptus* spp., 5.2 million ha), acacias (*Acacia* spp., 3.15 million ha) and teak (*Tectona grandis*, 2.2 million ha) are the most favored plantation species in the region. Estimates of eucalypt and acacia plantations in 1990 in the tropical region of the world are 10.06 million ha and 3.4 million ha respectively. Annual forest plantings in the TAP region during 1981-90 are estimated at 1.29 million ha (Pandey, 1995).

Eucalypts and acacias are the preferred plantation species, for both industrial and nonindustrial uses, because of their fast growth rates for wood production in short rotations. They are grown in single-species plantations or as components of agroforestry systems. However, questions are being raised about the long-term sustainability of their fast growth as well as their effects on site quality (FAO, 1995). Repeated loss of nutrients from a site in the harvested wood and during site preparation may adversely affect soil fertility, especially as related to N and P dynamics, and thus reduce long-term productivity. Mixed-species systems which include N_2-fixing species could therefore be a useful alternative to monocultures.

Mixed-species plantations are established for a variety of objectives. Ball et al. (1995) identified the following:

1) To provide stand protection – In a mixed-species system, one of the tree species may act as a nurse crop, providing protection against climatic factors, improving the stem form of another species, and improving the physical stability of stands through deeper rooting of some species. A typical example is the use of mixed-species stands as wind breaks (El-Lakany and Mohamed, 1993).
2) To enhance ecological stability – The ecological stability of a stand refers to its ability to resist external stress (resistance) and to return to its former dynamic state after perturbation (resilience). Mixing species may improve both resistance and resilience by improving nutrient cycling, enhancing resistance to pests or disease, and providing resilience to climate change. Montagnini et al. (1995) monitored pest damage in mixed and pure stands in Costa Rica for two years after planting and observed no damage, or no difference between mixed and pure stands, for nine of the 12 species tested; for three species, pest damage was less severe in mixed than in pure stands.
3) To exploit natural associations and synergy – Synergy refers to interactions in which one species enhances the biological performance of another, e.g., *Albizia* sp. and acacia increase the productivity of eucalypt in Hawaii (DeBell et al., 1985, 1989), and red alder (*Alnus rubra*) increases the productivity of Douglas-fir (*Pseudotsuga menziesii*) in North America (Binkley et al., 1992b). Species in mixed stands can be selected to cover a range of roles of succession on a site, e.g., pioneer species for a rapid

soil cover, climax species for a stand's final stages, and intermediate species which need neighbors to improve their growth and form.

4) To conserve biological diversity – This is characterized in terms of species richness, species rarity and species vulnerability. Mixed-species stands can be usefully developed to conserve biological diversity.
5) To obtain marketing advantages – Careful selection of species mixtures can insure against the risk of biological failure (caused by species characteristics or ecological conditions) and against economic failure (caused by excessive wood supply of a species and lack of markets for lesser-known species).
6) To provide a range of goods and services – When selecting species for mixed stands, it is important to have clear objectives about products. For example, a mixed-species stand may provide wood products such as fuel and poles, fodder for animals, resin or gum for industry, and nuts or other products for human consumption.

The low growth rates of trees under monocultural industrial plantations, especially in the TAP region, provide an additional incentive for considering mixed-species plantations. Pandey (1995) estimated that the annual wood production of teak was 2.4 to 3.0 m^3 ha^{-1} yr^{-1} in India and Bangladesh and < 2 m^3 ha^{-1} yr^{-1} in Indonesia. In India, eucalypt plantations at 8 to 10 years of age produced about 6 m^3 ha^{-1} yr^{-1} and *Acacia nilotica* between 15 and 20 years of age produced around 3 m^3 ha^{-1} yr^{-1}, whereas *A. mangium* in Malaysia and Indonesia produced > 20 m^3 ha^{-1} yr^{-1}. Potential productivities of most plantation species are high, as shown in well-managed field trials. For example, data reviewed by Lugo et al. (1990) showed that stemwood biomass production of tropical tree plantations averaged 7.9 Mg ha^{-1} yr^{-1}. Assuming an average wood density of 0.5, this value would translate into about 16 m^3 ha^{-1} yr^{-1}. It is not known if the growth rates of plantation species could be improved by growing them in mixed stands, but this subject should be one of the main research issues in the future.

Kelty (1992) considered the conditions under which the total yield of mixed-species stands was greater than that of monocultures of the component species, and concluded that yield of the mixture would exceed that of lower-yielding monocultures but not necessarily that of higher-yielding monocultures. Based on the degree of niche separation among species for resource utilization (aboveground and belowground components), the yield patterns in mixed stands can be explained by using the principles of competitive reduction and facilitation. According to Vandermeer (1989), the competitive reduction principle explains the yield increase in mixtures that is related to the efficient utilization of available resources by the mixtures. The facilitative production principle refers to a direct positive effect on growth. A good ecological mixture results when differences in the species' growth characteristics reduce competition and foster facilitation. Furthermore, species interactions must increase efficiency of use of those resources that limit productivity. Based on these considerations, the use of nitrogen-fixing trees (NFT) in association with non-N_2-fixing species may be a good ecological mixture on nitrogen-limiting sites (Kelty, 1992). A FAO review on mixed-species plantations in the tropics and subtropics, however, concluded that suitable studies and data are lacking for evaluation of mixed-species stands as an ecologically useful and economically viable option for land management (FAO, 1992).

The emphasis of the present review will be to compare mixed-species, fast-growing plantations with monocultures, in respect to the cycling of N and P. Here, a mixed-species stand is defined as a site which contains at least two tree components, of which one is a NFT, and in which direct interaction between trees is achieved through an appropriate planting design. The main objective of establishing such stands is to enhance the N status of low N sites under minimum nutrient inputs. Only NFTs are considered, rather than shrubs, because trees can provide the economic advantage of additional wood, e.g., fuelwood, within a short period after planting. As extensive data on mixed-species stands which cover these objectives are not available, this review will be developed at a conceptual level, using data from single-species stands and other systems to substantiate the concepts.

Nutrient cycling model

Nutrient cycling in forest ecosystems is driven primarily by transfer processes through which nutrients and carbon move into (inputs) and out of (outputs) the various ecosystem components or compartments. Therefore an ecosystem model provides a useful conceptual background for understanding the flow of elements in forests and the effects of nutrient flows on productivity. Such a model visualizes a forest ecosystem as a complex of interacting compartments, for example, the various plant components – foliage and twigs, wood, bark, coarse roots, fine roots; the litter and humus layers of the forest floor; soil organic matter; the soil solution; and the sorptive surfaces of soil minerals.

Various processes link the different compartments by transferring nutrients between them. Nutrient fluxes of interest include inputs (precipitation and interception, fertilizer additions, atmospheric fixation), outputs (leaching, harvesting, fire, denitrification), nutrient release and retention in soil (mineralization, immobilization, cation exchange and anion retention, mineral dissolution and retention), plant uptake, and litterfall. Of these, the present review will focus mainly on inputs and outputs, particularly of N and P. Nutrient fluxes are also affected by factors such as changes in microclimatic conditions (temperature and moisture) and the rate of litter decomposition. Due to space restrictions, however, these topics will not be discussed in detail. Nevertheless, it is acknowledged that the time and rate of both aboveground and belowground decomposition are of special significance because (1) mixing litter from different species and of different qualities may enhance the decomposition rate, and (2) under pure stands of NFTs, excessive amounts of N may accumulate, causing leaching losses of N and cations (Nykvist, 1997). Adding non-NFTs to such stands may increase the uptake of excess N.

Nutrient cycling processes are expected to differ between stands of single species and stands of mixed species because the vegetation components of various species may have different turnover times. Thus, in mixed-species stands, the role of these components in nutrient cycling needs to be considered separately for each species. Expected changes in the fluxes of N, P, and other nutrient-related parameters in mixed-species forest stands that include NFTs, as compared to pure stands without NFTs, are given in Table 1. These are based on the rationale that the addition of NFTs to a pure stand of non-N_2-fixing trees may alter the system in a number of ways:

1) direct effects from the addition of N by NFTs (refer to 3, 4, 9, 10, 13 in Table 1);
2) indirect effects due to interactions that are affected by the addition of N by NFTs (refer to 6, 7, 8, 12, 14, 15, and 17 in Table 1);
3) enhanced competition for nutrients (refer to 2, 12 in Table 1); and
4) changes due to other factors, which will be considered in the review (refer to 1, 5, 11, 16 in Table 1).

Inputs

Inputs of chemical elements occur via precipitation/deposition, fertilizer additions, biological fixation of atmospheric N, and mineral weathering. In the tropics, annual inputs of N and P in precipitation and interception by the tree canopy are relatively low, less than 5 kg ha^{-1} yr^{-1} of N and a fraction of a kg ha^{-1} yr^{-1} of P, but N values of up to 20 kg ha^{-1} yr^{-1} are sometimes reported (Bruijnzeel, 1990; Szott et al., 1991). It is expected that for similar types of canopy the magnitude of inputs via precipitation and interception will be similar in mixed-species and pure stands. When compared with other inputs (e.g., fertilizer additions and atmospheric fixation of N) and outputs (e.g., via biomass harvest in short-rotation, fast-growing plantations), inputs of N and P via precipitation represent a relatively minor component of nutrient cycling in tropical situations.

Two components of N_2 fixation need special consideration when assessing the amount of N fixed by NFTs: (1) the total amount of N in the aboveground and belowground components of NFTs, and (2) the fraction of total N in the plant that was obtained through N_2 fixation; this can vary from 10% to 76% of the total amount (Table 2). Because of the size and perennial nature of trees, there are special problems in assessing their N_2-fixation rates (Danso et al., 1992). Moreover high variability may exist in nodulation among isolines and provenances (Sanginga et al., 1989). A number of excellent reviews are available on the techniques of measuring atmospheric N_2 fixation (Knowles, 1980;

Table 1. Expected changes in fluxes of N, P, and related parameters when N_2-fixing trees (NFT) are added to pure stands of non-N_2-fixing trees, under fast-growth short-rotation conditions.

Nutrient flux/related parameter	Expected change in mixed stand	Remarks
1. Precipitation/interception	No change	For similar type canopy
2. Fertilizer inputs, except N	No change/higher	Additional P required for NFT and fast growth
3. N fertilizer inputs	Lower	Part of N supplied by NFT
4. Atmospheric N fixation	Higher	Site and soil conditions important
5. Mineral weathering and release of P	Higher	Production of acidity and enzymes
6. Mineralization	Higher	Climatic factors important
7. Nitrification	Higher	Site history and change in substrate quality
8. Denitrification	(?)	Higher if nitrates are excessive
9. Leaching losses, mainly for N	Moderately higher	Depends on site preparation
10. Harvest losses of N and P	Higher	For similar amount of biomass removed
11. Losses by fire	Minimal	Reduced slash burning
12. Plant uptake	Higher	Change in P uptake not certain
13. Foliage and root chemistry	Higher N	Positive effects on P seldom observed
14. Aboveground litter:		
– amount	Higher	Depends on the growth rates of mixtures
– decomposibility	Higher	Better C:N ratio
15. Turnover of roots and release of N	Higher	Better N status
16. Faunal activity	Higher (?)	Depends on change in food quality
17. Water use	Higher	Depends on climatic conditions

Rennie and Rennie, 1983; Shearer and Kohl, 1986; Giller and Wilson, 1991; Sanginga et al., 1995).

Data reported in the literature on N_2 fixation are often of limited usefulness because of methodological shortcomings. Most studies are based on short-term observations of a few days to a few months and are confined to discrete periods, lacking continuous measurements. Some of the data are obtained under laboratory conditions and others are site specific, often without information on the site variables (Table 2). The main weaknesses associated with data on NFTs are due to (1) lack of suitable field-based methodologies, (2) lack of field-based experiments to measure N in the biomass, and (3) lack of information on the biological and soil factors determining N_2 fixation by trees.

Under N-limiting conditions, the contribution of soil N to plant uptake may be small when compared to the contribution of N_2 fixation. In such situa-

Table 2. Amount of N fixed and the proportion of a plant's total N derived from biological nitrogen fixation (BNF).

Species	N fixed (kg N/ha/yr)	N from BNF (% of total N)	Method[a]	Authors
Acacia sp.	0.1–34	ND	ARA	[8], [9], [10]
Acacia sp.	< 1	ND	ARA	[13]
A. dealbata	12–32	ND	ARA	[18]
A. mearnsii	200	ND	ND	[17]
Acacia pellita	< 12	ND	ARA	[14]
A. pulchella	1.5	ND	ARA	[15]
A. verniciflua	32	ND	Other	[16]
A. seyal	ND	63	Other	[19]
A. raddiana	ND	62	Other	[19]
Leucaena leucocephala	304–548	45–65	TND	[1], [2], [3]
L. leucocephala	110	ND	ARA	[4]
Casuarina equisetifolia	43–60	39–76	TND	[5], [6], [7]
Gliricidia sepium	108	72	TND	[2]
G. sepium	13	ND	ARA	[10]
Albizia lebbeck	94	60	TND	[2]
Sesbania rostrata	505–581	35–45	TND	[11]
S. sesban	43–102	11–18	TND	[11]
Inga jinicuil	35	ND	ARA	[12]

[a] TND refers to total N difference, ARA to acetylene reduction analysis, and Other to methods such as changes in soil N.
ND = no data.
[1] Sanginga et al. (1985); [2] Liya et al. (1990); [3] Sanginga et al. (1990); [4] Högberg and Kvarnstöm (1982); [5] Duhoux and Dommergues (1985); [6] Sougoufara et al. (1990); [7] Gauthier et al. (1985); [8] Hansen et al. (1987); [9] Langkamp et al. (1979); [10] Roskoski et al. (1982); [11] Ndoye and Dreyfus (1988); [12] Roskoski (1981); [13] Lawrie (1981); [14] Langkamp and Dalling (1981); [15] Grove and Malajczuk (1981); [16] Turvey et al. (1983); [17] Orcahrd and Darb (1956); [18] Adams and Attiwill (1984); [19] Ndoye et al. (1995).

tions, the use of simpler methods of estimating N_2 fixation – such as the total N difference method (TND) – should be considered. In TND, the proportion of total N in a NFT that has accumulated from the soil (Ns) is estimated, and the remaining N is attributed to nitrogen fixation. Ns is assessed by considering it to be equal to the total N in the control non-NFT (reference plant). For estimating N in plant components – e.g., stems, foliage, coarse roots – two types of information are required: the weight of the components and the concentration of N in the components. Weight of the components can be obtained from allometric equations or growth models, and N concentrations can be measured or obtained from the literature. N in the litterfall can be estimated (initially some measurements may be required). Although difficult to measure, the concentration and turnover of N in fine roots is the most important flux requiring attention. In most biological nitrogen fixation (BNF) studies, the contribution of roots has not been included. Ignoring roots may result in significant underestimates of N_2 fixation, for example, by 26–38%

in Acacia species (Ndoye et al., 1995) and by 50–60% in *Leucaena leucocephala* and *Gliricidia sepium* (Sanginga et al., 1992).

An optimum supply of P to the NFT has been recognized as an important factor in increasing N_2 fixation. According to Sanginga et al. (1991), the increase in N_2 fixation results from the beneficial effect of P on overall plant growth, particularly root growth and root infection. However Sun et al. (1992) observed that, in a pot experiment, the maximum growth of *Acacia mangium* occurred at 10 mg P/kg soil, whereas N_2 fixation continued to increase with a P application of up to 30 mg P/kg soil. Sanginga et al. (1989) also reported a direct effect of P addition on the amount of N fixed by *Allocasuarina* and *Casuarina* species. Usually, the natural atmospheric inputs of P and the supply of P in highly acid tropical soils are low. Therefore, some input of P fertilizer will be important to stimulate growth and enhance N_2 fixation by NFTs. This is particularly important in short-rotation stands, where changes in P cycling are unlikely until natural litterfall or tree prunings form a significant fraction of the P cycle, through decomposition processes.

Outputs

Nutrient cycling in tropical plantations is affected by output parameters such as nutrient removal during harvest, and the mobilization and loss of nutrients during site preparation (including the management of slash and leaching). After evaluating these parameters, Fölster and Khanna (1997) concluded that losses of N and P in leaching and denitrification, although not precisely known in most cases, were probably low – ranging from a few kg ha^{-1} to 50 kg ha^{-1} – in comparison with losses from biomass harvest and site preparation. Management of slash – whether it is retained on site and incorporated into the soil, or burned – greatly affects nutrient losses, and thus the productivity of the following stand. Sim and Nykvist (1991) showed that in 3.7 years *Acacia mangium* in Sabah, Malaysia accumulated twice as much aboveground biomass when slash from the clearfelled forest (logs extracted manually) remained unburned than when it was burned (after tractor extraction of logs), indicating the combined effects of slash burning and soil disturbance. Both N and P are lost during burning and their losses are very hard to quantify. However, losses from fire are expected to be minimal under short-rotation management in the tropics because the utilization of wood and other vegetation components is high. Based on simple budgets of input/output of nutrients (Fölster and Khanna, 1997), the following generalized conclusions can be drawn:

1) For sustained rapid growth in short-rotation plantations, fertilizer inputs – especially of P and cations – will often be required;
2) Higher nutrient losses (exports with harvest or losses during site preparation) are expected in mixed-species stands than in monocultures of non-

N_2-fixing trees because of the higher N and cation contents in NFTs, and potentially higher levels of biomass export.

Nitrogen transfer processes

Most soil N occurs in an organically bound form which should be mineralized to ammonium or nitrate before it is taken up by plants. Only a small fraction of the total N (less than 10% in most cases) may be mineralized during the short rotation of a fast-growing plantation. In most soils, mineralization is a rate-limiting step in the supply of N to plants due either to the small amount of N mineralized relative to the quantity required for growth or to the lack of synchronization in the demand and supply of N. The amount of N mineralized and the demand for N by trees varies during the life of a stand (Miller, 1995). During the initial two to three years of plantation establishment, the amount of mineralized N is usually large (especially if harvest slash is retained on the site) and the plant needs are small. But as the trees become older, their demand for N increases and N mineralization becomes a limiting step, i.e., all the mineralized N is utilized by the system. Depending on the desired rates of growth under optimum soil moisture conditions, there may be a need to supply additional N at this stage, either as fertilizer (Nambiar, 1990/91) or by the transfer of N from NFTs.

Mineralization of soil organic N depends on the quality of the organic matter (C:N ratio; lignin and phenolic content), environmental conditions, and microbial factors (Mafongoya et al., 1997). Other soil properties such as pH, texture, and mineralogy are also important factors because of their role in stabilizing soil organic matter. In a study on nitrogen mineralization in a number of humid tropical forest soils from Costa Rica, Colombia, Peru and Brazil, the mineralogy of soils was found to be a major factor, with allophanic soils showing higher N mineralization than soils with smectite, oxidic and kaolinite mineralogies (Motavalli et al., 1995). Bernhard-Reversat (1988) measured soil N mineralization in a *E. camaldulensis* plantation and a natural forest of *Acacia seyal* in Senegal. In vitro incubation for 20 days produced 40 to 50 mg kg^{-1} N in acacia and 11 to 14 mg kg^{-1} in eucalypt soil. 'In-situ' mineralization was related to precipitation: 18 to 40 mg kg^{-1} during four weeks of the rainy season in acacia and 10 mg kg^{-1} during three weeks in the beginning of the rainy season in eucalypt. Mineralization data (Figure 1) obtained 31 months after planting a mixed-species stand of eucalypt and acacia (Khanna, 1997) showed much higher values for both laboratory and 'in situ' incubations (method of Raison et al., 1987) for soils from the 50:50 mixture of species than for soils from the pure eucalypt stands. 'In situ' N mineralization was 4.43 kg ha^{-1} month (30 days)$^{-1}$ in the 50:50 mixture, about double that observed in the pure eucalypt treatment. N transfer from NFTs to non-NFTs in mixed-species stands may enhance N_2-fixation by removing excess

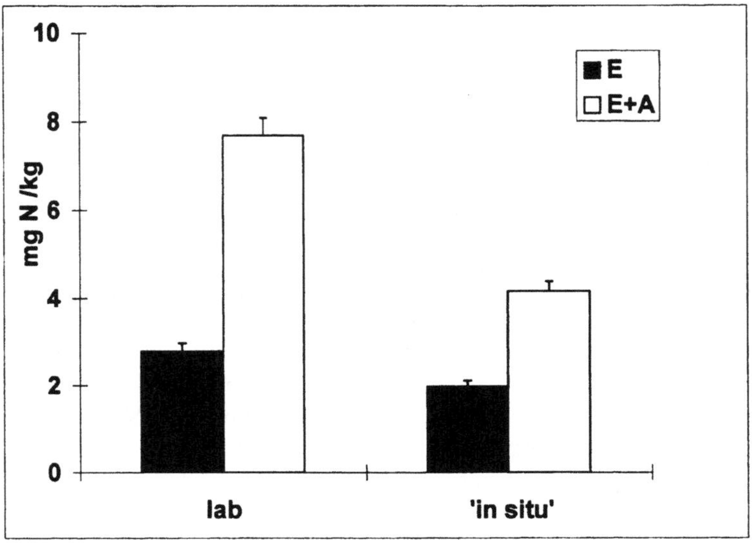

Source: Khanna (1997)

Fig. 1. Mineralization of soil nitrogen in pure eucalypt (E) and mixed eucalypt and acacia (50E:50A) plots at Cann River, Australia. Incubations were carried out under laboratory and 'in situ' conditions in undisturbed soils collected 31 months after tree planting. Mineralization values were calculated for a 30-day incubation period.

N (through uptake by non-fixers) and maintaining the soil at low mineral N levels (Tsai et al., 1993).

Transfer of N from a N_2-fixing plant to a non-N_2-fixing plant in a mixed stand can occur in various ways. Aboveground processes relate to the production of litterfall (or prunings), its nutrient content and decomposition rates. Following a review on agricultural ecosystems in the humid tropics, Ewel (1986) concluded that N_2 fixed in such systems would be used mostly by the N_2-fixing plants and would become available to other plants only through leaffall. In a comparison of six-year-old, mixed-species (*Albizia* sp. and eucalypt) and monoculture (eucalypt) stands, Binkley et al. (1992a) reported that the return of N in litterfall was 130 kg ha^{-1} yr^{-1} in the mixed plots and 40 kg ha^{-1} yr^{-1} in the pure eucalypt plots. However, the contribution of litterfall to nutrient cycling is expected to be small in mixed stands until the tree canopy has fully developed, which usually takes a few years after planting. Bargali et al. (1992) reported that in plantations of *Eucalyptus tereticornis* in northern India, the amount of N in leaf litterfall ranged from 6.3 kg ha^{-1} yr^{-1} at age 2 years, to 32.6 kg ha^{-1} yr^{-1} at age 8 years, whereas the amount of N retained in the vegetation (described as uptake) ranged from 22 kg ha^{-1} yr^{-1} to 88 kg ha^{-1} yr^{-1}, respectively. In another study on litterfall, in five-year-old plantations at Ratchaburi, Thailand, Bunyavejchewin (1989) found the following values for leaf litterfall (Mg ha^{-1} yr^{-1}) and N content in leaf litterfall (kg ha^{-1} yr^{-1}), respectively: *Eucalyptus camaldulensis*, 2.4 and 18.5; *Leucaena leucocephala*, 1.8 and 32.7; *Cassia siamea*, 2.7 and 29.8; *Azadirachta indica*, 3.4 and 46.6; and *Acacia auriculaeformis*, 3.0 and 38.6. However, considering the total amount of N in the aboveground biomass of trees (*E. camaldulensis*, 331 kg ha^{-1}; *L. leucocephala*, 688 kg ha^{-1}; *C. siamea*, 275 kg ha^{-1}; *A. indica*, 291 kg ha^{-1}; and *A. auriculaeformis*, 305 kg ha^{-1}) (Bunyavejchewin et al., 1989), it is obvious that the annual N requirement of these plantations was much higher than could be met by the decomposition of leaf litter. Moreover, the release of N from decomposing litter will require some time after the litter is deposited. Based on the above facts, it is possible that in tree plantations with short rotations (six to eight years), transfer of N through decomposition of litter will be unimportant until four or five years after planting.

Unlike aboveground litter, the turnover of fine roots and nodules may be a major source of N transfer from NFT to non-NFT during the first few years after plantation establishment, and is probably a major flux in short-rotation, mixed-species plantations. Hansen and Dawson (1982) reported that within two years after planting, *Populus* sp. growing in proximity to *Alnus* sp. had accumulated a significant amount of N. This N accretion was related to root turnover, and had a positive effect on the growth of *Populus*. In a climate where periodic drying and wetting phases are common, transfer of N from acacia to eucalypts was observed early in the plantation life, starting with the second year (Table 3) (Khanna, 1997). Nitrogen concentration in mature fully expanded eucalypt leaves, collected 25 months after planting, did not differ among treatment mixtures, although there was a tendency for a slight increase in N with an increasing proportion of acacia in the mixture. The N concentration of senescent eucalypt leaves increased with an increase in the acacia component of mixtures (linear response component significant at $P < 0.05$). Very low values of N in senescent eucalypt leaves indicated a major translocation of N prior to leaf shedding. N concentrations in acacia foliage pointed to similar translocation, although the concentrations (higher in acacia) and proportions of translocation (lower in acacia) were different. Acacia leaves (mature and senescent) tended to have slightly lower values of N when growing in monocultures than when growing in mixtures. The concentration of N in the fine roots (< 0.5 mm in diameter) of eucalypts, collected 31 months after planting, was higher (1.69 times) for the 50:50 treatment than for the pure eucalypts (Table 4). Similar results were obtained for medium roots (0.5–1.0 mm in diameter). Concentrations of N in fine roots of eucalypts collected from different soil depths did not vary, and the values were significantly lower than those in the leaves. The concentration of N in acacia roots was high, as expected – about 15 mg g^{-1} in fine roots and about 47 mg g^{-1} in nodules.

There is no direct method of assessing annual uptake of N by trees, but it is possible to make an estimate of uptake by indirect methods, for example, by using the 'in situ' sequential incubation technique (Raison et al., 1987), which involves estimation of the amount of N mineralized, change in mineral

Table 3. N and P concentrations in the leaves of eucalypt (E) and acacia (A) collected from pure and mixed-species stands at Cann River, Australia, 25 months after planting.

Treatment[a]	Eucalypt		Acacia	
	Fully expanded	Senescent	Fully expanded	Senescent
Nitrogen (mg g^{-1})				
100E	10.04	3.58	–	–
75E:25A	10.18	3.88	22.30	10.73
50E:50A	10.69	4.25	22.90	11.61
25E:75A	11.01	4.36	21.72	11.11
100A	–	–	19.16	9.60
Sed[b]	*1.111*	*0.362*	*2.513*	*0.924*
Signif.	*ns*	*lin < 5%*	*ns*	*ns*
Phosphorus (mg g^{-1})				
100E	0.67	0.18	–	–
75E:25A	0.59	0.16	0.79	0.20
50E:50A	0.55	0.15	0.71	0.18
25E:75A	0.57	0.15	0.69	0.16
100A	–	–	0.66	0.16
Sed	*0.043*	*0.017*	*0.061*	*0.0187*
Signif.	*lin < 5%*	*ns*	*lin < 8%*	*lin < 5%*

[a] Treatment refers to the proportions of eucalypt and acacia in a stand. For example, 100E and 100A are pure stands of eucalypt and acacia respectively; 75E:25A is 75% eucalypt and 25% acacia.

[b] Sed is the standard error of difference for comparing mean values. Lin refers to linear component of ANOVA (Snedecor and Cochran, 1967).

Source: Khanna (1997).

Table 4. Concentrations of N and P in the fine roots (< 0.5 mm diameter) of eucalypts (E), and phosphatase activity of eucalypt and acacia (A) fine roots, collected from pure and mixed-species stands at Cann River, Australia, 31 months after planting.

	N (mg/g)		P (mg/g)		Phosphatase (µg/g)	
Depth (cm)	100E[a]	50E	100E	50E	50E	100A
0–5	4.16 (1.23)	7.04 (2.21)	0.27 (0.07)	0.39 (0.23)	2674 (1376)	4673 (1399)
5–10	4.10 (1.36)	6.19 (2.16)	0.26 (0.05)	0.31 (0.13)	2612 (1525)	4599 (1765)
10–20	3.89 (1.22)	6.58 (1.73)	0.28 (0.06)	0.35 (0.13)	3325 (1786)	4407 (1214)

[a] Proportions of eucalypts and acacia in a stand: 100E and 100A are pure stands of eucalypt and acacia respectively; 50E = 50% eucalypt and 50% acacia. Mean values are given, with standard deviations in brackets (n = 10).

Source: Khanna (1997).

N in the rooting zone, and the amount of N leached from the rooting zone. Amount of N retained in the vegetation (plus the amount turned over in above- and below-ground litter), if measured periodically, can also give approximate values of N uptake during the measuring period. Most data reported in the literature are, however, too incomplete (usually lacking estimates of the amount of N retained by roots, and turned over in aboveground and belowground litter) for making reliable estimates of N uptake.

Phosphorus dynamics

Although P is the most limiting nutrient for plant productivity in soils of the humid tropics (Denslow et al., 1987), very little information on changes in P dynamics in forest stands is available for southeast Asia. Dynamics of P differ from those of N because P occurs in both organic and inorganic forms in the soil. Also, P is relatively immobile in the soil, so the activity of plant roots is a major factor in P uptake, and may result in differences among species. When comparing P dynamics in stands of mixed species with those of monocultures, the following information is useful for making some generalizations.

1) *Changes in the forms and availability of P in soils.* Giardina et al. (1995) reported that when compared with pure stands of Douglas-fir (740 stems/ha), a mixed stand of red alder and Douglas-fir (190 stems RA + 740 stems DF) had a higher amount of inorganic P fractions (anion exchange + NaOH + HCl). Zou et al. (1995) conducted a detailed study in two mixed stands (> 60-year-old stand of red alder plus different conifers, and an eight-year-old stand of *Paraserianthes* (syn. *Albizia*) *falcataria* and *Eucalyptus saligna*) on the mechanisms by which N_2-fixing trees can affect P forms by changing geochemical reactions in soils. Resin extractable inorganic P (Pi) and $NaHCO_3$ extractable Pi and organic P (Po) were measured in soils from mixed stands and compared with those from pure stands. All three fractions of P were higher in soils from mixed stands of red alder plus conifers. In the eucalypt and albizia stands only Po was higher. It has been postulated that a change in the solubility of soil P under N_2-fixing trees is related to rhizosphere acidification (Gillespie and Pope, 1990), which is associated with the balance in cation and anion uptake and the production of protons in nitrification. Another possible explanation for the increase in P solubility is that the excessive amount of organic acids produced by N_2-fixing plants may cause the direct dissolution of P or the complexation of Fe and Al.
2) *P acquisition processes and the role of plants in the dissolution and mineralization processes.* In addition to the role of plants in mobilizing inorganic forms of P in soils, the roots of N_2-fixing plants may exude excessive amounts of phosphatases to mineralize organic P in soils. Giardina et al. (1995) observed three times higher phosphatase activity in soils collected

under Douglas-fir mixed with red alder than under pure plots of Douglas-fir. However, high levels of phosphatase activity and possible increases in P supply will not necessarily remove P limitations to productivity. Khanna (1997) found that phosphatase activity was higher on the fine roots of acacia than on eucalypt roots (Table 4), suggesting the possibility that acacia may be able to mobilize greater amounts of organic P from soils, one of the strategies exhibited by plants under highly P-deficient conditions. The phosphorus nutrition of eucalypts was not affected when grown in mixture with acacia, as was evident from concentrations in root and foliage tissue (Tables 3 and 4). Binkley (1992) came to the same conclusion after reviewing information on mixed stands with NFTs. DeBell et al. (1989), however, reported a contrasting conclusion for mixed-species stands of eucalypt and albizia. The compiled data in Table 5 show higher P uptake values for red alder stands – either pure or mixed-species stands – than for pure stands of Douglas-fir or a mixture of conifers.

3) *Amount of P in aboveground and belowground components, and turnover processes.* Witkowski (1991) studied the effect of invasive alien acacia on nutrient cycling in the coastal lowlands of Cape Fynbos, South Africa. *Acacia* spp. (*A. saligna* and *A. cyclops*) produced more litter, with N content three times higher than that of the indigenous species, but no difference in P return was observed. However, in their study on mixed stands of albizia and eucalypt, Binkley et al. (1992a) reported a return in the litterfall of 6 kg P ha^{-1} yr^{-1} in the mixed plot as compared to 3 kg P ha^{-1} yr^{-1} in the pure eucalypt plot. Data compiled in Table 5 indicate higher values for P returned in litterfall in mixed or pure stands of red alder, but not in all cases for P retained in the vegetation and forest floor. In his review on the effect of N_2-fixing trees on the amount and chemistry of litterfall in mixed stands, Binkley (1992) observed two- to seven-fold greater amounts of phosphorus, sulfur, calcium, magnesium and potassium in the litterfall of mixed stands with NFTs than in non-NFT pure or mixed

Table 5. Phosphorus parameters in pure red alder (*Alnus rubra*) stands and mixed stands of red alder with Douglas-fir (*Pseudotsuga menziesii*) or various conifers (all values are relative to pure stands of Douglas-fir or conifer).

Species in mixture[a]	P content in vegetation	P content in forest floor	P in litterfall	Stand P uptake	Bray2 – P	Reference
RA	1.2	3.3	1.8	1.5	0.2	[1]
RA+DF	1.4	1.9	1.9	2.4	0.9	[2]
RA+C	0.7	0.7	2.5	1.6	14.0	[3]
RA+DF	2.8	4.4	3.2	2.7	1.6	[4]

[1] Cole et al. (1990), Johnson and Lindberg (1992); [2] Binkley et al. (1992b); [3] Binkley et al. (1992b), Zou et al. (1993); [4] Binkley (1983)
[a] RA = red alder; DF = Douglas-fir; C = various conifers.
Source: Giardina et al. (1995).

stands. He assigned the increase in cycling of P and S more to the turnover of soil organic pools than to inorganic pools in the soil.

In a study on different mixtures of eucalypt and acacia (Khanna, 1997), P concentration in all components of eucalypt foliage tended to decrease with an increase in the proportion of acacia in the stand (Table 3 shows values for fully expanded and senescent leaves; data on young leaves and twigs are not included). This decrease in P concentration may be the result of competition between acacias and eucalypts for a given amount of available P. A very significant withdrawal of P was observed during senescence of both eucalypt and acacia leaves. The concentrations in senescent leaves decreased to very low values of between 150 and 200 mg P kg^{-1} of dry matter. The concentration of P in the fine roots of eucalypts in the 50:50 mixtures did not differ from the concentration in monocultures (Table 4). Acacia roots did not show any differences in P (or N) concentrations among treatments.

4) *Effect of mixing species on the volume of soil exploited by fine roots, the role of mycorrhiza, the role of soil fauna, and changes in the requirements of P for growth.* An improvement in N and P levels in aboveground litter in mixed-species stands can improve the quality of litter as a food source for earthworms, changing their population density (Zou, 1993). Under P-limiting conditions, plants may exhibit special strategies, such as improved P uptake (including the role of mycorrhiza) and more efficient utilization of P. Data in Table 3 show very low P concentrations in the senescent leaves of both eucalypt and acacia, suggesting that the latter strategy may be important in this mixed stand. No changes in P cycling would be expected until litterfall, and P release during its decomposition, becomes significant. Even then, for short-rotation stands, some inputs of P fertilizer will be needed to stimulate growth and enhance N_2 fixation by N_2-fixing species. If the principal purpose for incorporating a legume in a stand is to provide N, it may be most efficient to use a small proportion of legumes but maximize their effectiveness by ensuring that P is not limiting.

Synthesis

Eucalyptus plantations have become a very important wood resource in countries of Mediterranean, subtropical and tropical climates, including Australia. Most soils on which eucalypts are grown are deficient in nitrogen. While nitrogen can be added in the form of mineral fertilizer, it may be too expensive in some areas where eucalypts are planted, in particular, in developing countries. Also, mineral fertilizer addition often causes negative effects both 'on site' and 'off-site', through soil acidification, leaching of nitrate into the groundwater, volatilization of ammonium, and nitrous-oxide (a 'greenhouse' gas) emissions.

Another way of enriching the forest stand with nitrogen is the introduction of N_2-fixing plants such as acacia, albizia, leucaena, casuarina and others. A few years after plantation establishment, when the system has been enriched with nitrogen, N_2-fixing trees can provide the additional benefit of wood harvest. This practice may be particularly promising in developing countries or for farm forestry, where the wood can be used as fuel and the leaves as fodder.

Species growing in mixtures often compete for site resources. Aboveground competition is related to physical resources (interception of light, rainfall and air-borne nutrients, space), whereas belowground competition pertains to available nutrients (especially P) and water. Some interactions among species may reduce competition (N supply by NFTs) or even improve plant growth (facilitation), for example, by increasing photosynthesis and improving microclimate (temperature, moisture and wind speed). Examples of such interactions in agroforestry systems are given by Ong and Huxley (1996).

In their study of eucalypt and albizia mixtures, Binkley et al. (1992a) observed the greatest biomass, aboveground net primary production, and annual increment in the 34% *Eucalyptus*/66% *Albizia* treatment at age 6 years. They attributed this to a combination of factors: greater nutrient-use efficiency of *Eucalyptus* and greater nutrient cycling under *Albizia*, resulted in greater light capture and high light-use efficiency in the mixed stands. Growth of eucalypts has also been stimulated in mixed stands with acacia. Despite a lack of detailed studies that describe the processes leading to enhanced growth, facilitation – through N additions to the system – was identified as one of the main processes (Khanna, 1997; Samraj et al., 1977). It is expected that this type of facilitation would be higher on N-limited sites than on those rich in N, as Binkley (1992) concluded after reviewing data from a number of studies with tree mixtures. Turvey et al. (1983) found no positive effect of N added by naturally regenerated *Acacia* sp. in *Pinus radiata* and *P. elliottii* plantations. Rather, they observed a negative effect, which was associated with competition for other resources (P, water), as indicated by a growth increase of pines after herbicide application controlled the acacia.

Three additional factors that affect interactions between species in mixed stands are the proportions of different species, the density of planting, and the age of the trees. Growth data from four-year-old, mixed stands containing *Eucalyptus globulus* and *Acacia mearnsii* in different proportions (Khanna, 1997) are shown in Figure 2 for mean tree basal area and in Figure 3 for basal area increments (between 3 and 4 years). The data are reported for two planting densities (1000 and 1500 trees ha^{-1}). Higher values were observed for both species when they were grown in mixtures rather than in pure stands. Acacia trees had higher growth rates than eucalypts, so were probably competing strongly for both water and phosphorus (e.g., Table 3 shows that acacia had higher P values in green and senescent foliage than did eucalypts in the same mixture). Despite strong competition from acacias, eucalypts attained higher total basal area and basal area increment per tree in a 50:50 combina-

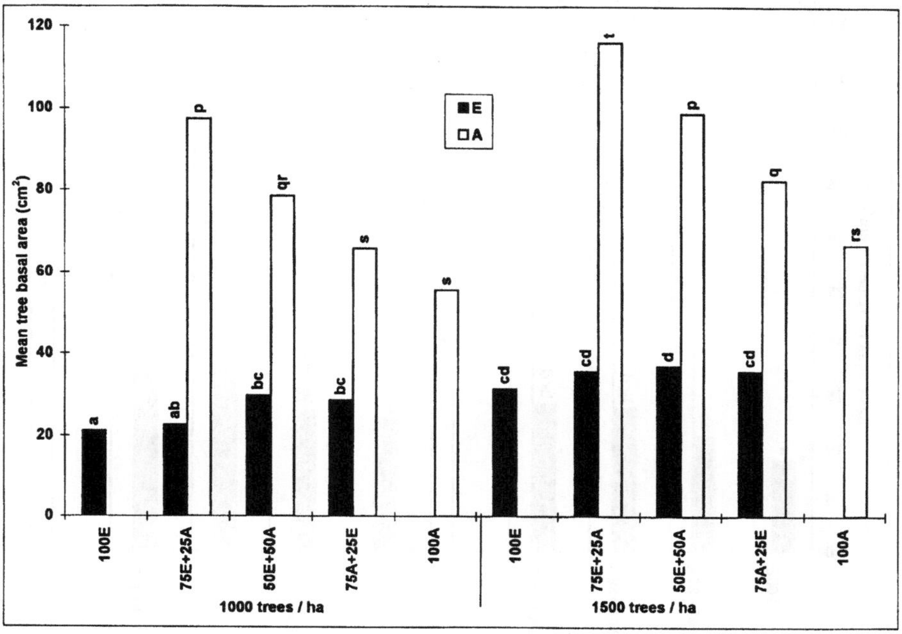

Source: Khanna (1997).

Fig 2. Mean tree basal areas in pure and mixed-species stands of *Eucalyptus globulus* (E) and *Acacia mearnsii* (A) at Cann River, Australia, 45 months after planting. Measurements are reported for various proportions of E and A, and two planting densities: 1000 trees ha^{-1} and 1500 trees ha^{-1}. Values with different letters are statistically different at $P < 0.05$.

tion with acacia than in a pure stand. This suggests that the advantage of additional N from N_2-fixation by acacia outweighed the detrimental effects of competition for light and water. The additional question of whether enhanced growth in mixtures is related to interactions between N supply and plant water use cannot be addressed because data on water use are lacking.

Little is known about belowground factors which might influence facilitation or competition between species in mixed stands. Relevant belowground interactions include the sharing of the soil volume (and the time of full site occupancy), spatial separation of roots (deep rooting versus shallow rooting), sharing of belowground resources (water and nutrients), transfer of nitrogen from N_2-fixing species to non-N_2-fixing species, changes in decomposition processes and nutrient release, and improvement of soil physical characteristics. The different growth rates for eucalypts (Figures 2 and 3), despite similar spacing of trees, suggest that the major competition between acacias and eucalypts takes place belowground. In addition, in plantation trials where acacias clearly promoted eucalypt growth, stimulation occurred before a substantial amount of acacia leaves had been shed, so only a small amount of N had been released via aboveground litter decomposition and mineralization

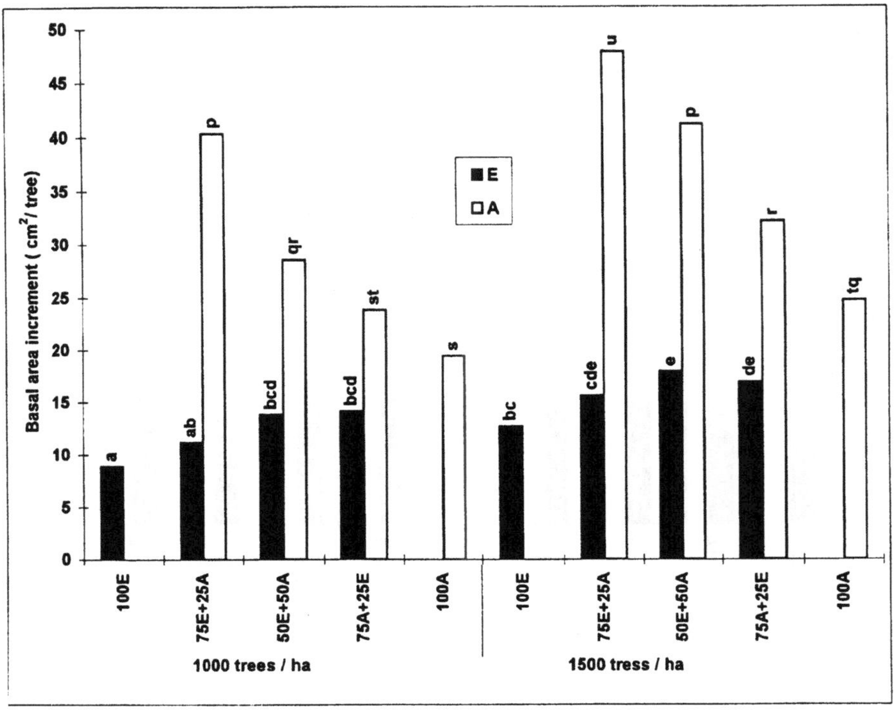

Source: Khanna (1997).

Fig. 3. Basal area increments in pure and mixed-species stands of *Eucalyptus globulus* (E) and *Acacia mearnsii* (A) at Cann River, Australia. Measurements represent increments between 33 and 45 months after planting, and are reported for various proportions of E and A and two planting densities: 1000 trees ha^{-1} and 1500 trees ha^{-1}. Values with different letters are statistically different at $P < 0.05$.

(Khanna, 1997). This suggests that eucalypts have access to soil nitrogen originating from the turnover of acacia fine roots, and that these roots must be short lived, so turn over rapidly. Distribution of roots, especially of fine roots, is an important factor in the issue of species compatibility. Yadav et al. (1993) observed the root distribution of two isolated *Acacia nilotica* trees growing in combination with a 126-day-old mustard crop in Haryana, India. At 6 m from the base of the trees, crop and tree roots were well intermixed, but the roots of *A. nilotica* tended to remain in the upper soil layers, so affected moisture relations of crops under rainfed conditions.

A number of nutrient interactions in mixed-species stands could result from decomposition of aboveground litter, changed microbial populations and their activities in the soil and rhizosphere (mycorrhizal effects, root diseases), and changed soil fauna (earthworms) and micro-environmental conditions (soil temperature and moisture levels). These have not been adequately addressed in the present paper but have been reported by others (Brown, 1992).

Most of the studies referred to here were confined to results obtained during the first few years of plantation establishment; the results may change during the subsequent phases of the plantation rotation. For example, competition for water (and light) will become more important as the plantations grow older. Moreover, the usefulness of growing mixed-species stands, rather than monocultures, will depend on the mixed stand's positive effects on the productivity of the target species and site fertility, as well as on the silvicultural options of managing mixtures. It is essential to consider the impact of harvesting on site nutrient budgets, and the productivity of subsequent rotations. Are greater quantities of critical nutrients lost or removed during the harvest of mixtures, as compared to monocultures? If so, how readily can these losses be offset? Incentives for the adoption of mixed-species plantations as alternatives to monocultures include economic considerations (higher productivity), improved plantation health (reduced losses due to disease and insect attacks), sustainability, and diversification of wood products. In a recent study in the humid neotropics, Montagnini et al. (1995) showed that mixed-species plantations gave favorable results with regard to plantation health and economic considerations, when compared with single-species plantations.

Acknowledgments

I wish to acknowledge the financial support by ACIAR, the technical support by S. J. Smith, K. Jacobsen and M. Connell, the help with the statistical analysis by P. Snowdon and professional support by R. J. Raison and A. Brown.

References

Adams MA and Attiwill PM (1984) Role of *Acacia* spp. in nutrient balance and cycling in regenerating *Eucalyptus regnans* F. Muell. forests. II. Field studies of acetylene reduction. Aust J Bot 32: 217–223

Ball JB, Wormald TJ and Russo L (1995) Experience with mixed and single species plantations. Commonwealth For Rev 74: 301–304

Bargali SS, Singh RP and Singh SP (1992) Structure and function of an age series of eucalypt plantations in central Himalaya. II. Nutrient dynamics. Annals of Bot 69: 413–421

Bernhard-Reversat F (1988) Soil nitrogen mineralization under a Eucalyptus plantation and a natural Acacia forest in Senegal. For Ecol Manage 23: 233–244

Binkley D (1983) Ecosystem production in Douglas-fir plantations: interactions of red alder and site fertility. For Ecol Manage 5: 215–227

Binkley D (1992) Mixtures of N_2-fixing and non-N_2-fixing species. In: Cannell MGR, Malcolm DC and Robertson PA (eds) The Ecology of Mixed-Species Stands of Trees, pp 99–123. Blackwell Sci Publi, London

Binkley D, Dunkin KA, DeBell D and Ryan MG (1992a) Production and nutrient cycling in mixed plantations of *Eucalyptus* and *Albizia* in Hawaii. Forest Sci 38: 393–408

Binkley D, Sollins P, Bell R, Sachs D and Myrold D (1992b) Biogeochemistry of adjacent conifer and alder-conifer stands. Ecology 73: 2022–2033

Bruijnzeel LA (1990) Hydrology of moist tropical forests and effects of conversion: a state of knowledge review. UNESCO – IHP, Humid Tropics Program, Paris

Brown AHF (1992) Functioning of mixed-species stands at Gisburn, NW England. In: Cannell MGR, Malcolm DC and Robertson PA (eds) The Ecology of Mixed-Species Stands of Trees, pp 125–150. Blackwell Sci Publi, London

Bunyavejchewin S (1989) Primary production of plots of five young close-spaced fast-growing tree species. III. Dry matter and nutrient content of litterfall. Nat Hist Bull Siam Soc 37: 65–73

Bunyavejchewin S, Kiratiprayoon S and Kumpun T (1989) Primary production of plots of five young close-spaced fast-growing tree species. II. Above-ground biomass, nutrient and energy content. Nat Hist Bull Siam Soc 37: 57–63

Cole D, Compton J, Van Miegroet H and Homann P (1990) Changes in soil properties and site productivity caused by red alder. Water Air Soil Pollut 54: 231–246

Danso SKA, Bowen GD and Sanginga N (1992) Biological fixation in trees in agro-ecosystems. Plant Soil 141: 177–196

DeBell DS, Whitesell CD and Schubert TH (1985) Mixed plantations of *Eucalyptus* and leguminous trees enhance biomass production. Research Paper, Pacific-Southwest Forest and Range Experiment Station, USDA Forest Service. No. PSW-175

DeBell DS, Whitesell CD and Schubert TH (1989) Using N_2-fixing *Albizia* to increase growth of *Eucalyptus* plantations in Hawaii. For Sci 35: 64–75

Denslow JS, Vitousek PM and Schultz JC (1987) Bioassay of nutrient limitations in a tropical rain forest soil. Oecologia 74: 370–376

Duhoux E and Dommergues YR (1985) The use of nitrogen fixing trees in forestry and soil restoration in the tropics. In: Ssali H and Keya SO (eds) Biological Nitrogen Fixation in Africa, pp 384–400. Nairobi MIRCEN, Nairobi, Kenya

El-Lakany MH and Mohamed SY (1993) Effects of species combination on the root characteristics of young *Acacia saligna*, *Casuarina cunninghamiana* and *Eucalyptus camaldulensis* trees. Alexandria J Agricultural Res 38: 211–227

Ewel JJ (1986) Designing agricultural ecosystems for the humid tropics. Annual Review of Ecology and Systematics 17: 245–271

Fölster H and Khanna PK (1997) Dynamics of nutrient supply in plantation soils. In: Nambiar EKS and Brown A (eds) Better Management of Soil, Nutrients and Water in Tropical Plantation Forests. ACIAR, Canberra, Australia (in press)

FAO (1992) Mixed and pure forest plantations in the tropics and sub-tropics. FAO Forestry paper 103 (based on the work of T.J. Wormald). FAO of the UN, Rome, Italy

FAO (1995) Plantations in tropical and subtropical regions – mixed and pure. FAO of the UN, Rome, Italy

Gauthier D, Diem HG, Dommergues YR and Ganry F (1985) Assessment of N_2 fixation by *Casuarina equisetifolia* inoculated with Frankia ORS021001 using ^{15}N methods. Soil Bio Biochem 17: 375–379

Giardina CP, Huffman S, Binkley D and Caldwell BA (1995) Alders increase soil phosphorus availability in a Douglas-fir plantation. Can J For Res 25: 1652–1657

Giller KE and Wilson KJ (1991) Nitrogen fixation in tropical cropping systems. CAB Int, Wallingford, United Kingdom

Gillespie AR and Pope PE (1990) Rhizosphere acidification increases phosphorus recovery of black locust: I. Model predictions and measured recovery. Soil Sci Soc Am J 54: 538–541

Grove TS and Malajczuk N (1981) Nitrogen inputs to *Eucalyptus marginata* and *E. diversicolor* forests. In: Rummery RA and Hingston FJ (eds) Managing Nitrogen Economies of Natural and Man Made Ecosystems, pp 199–204. CSIRO, Perth, WA

Hansen AP, Pate JS, Hansen A and Bell DT (1987) Nitrogen economy of post-fire stands of shrub legumes in jarrah (*Eucalyptus marginata*) forests of SW Australia. J Exp Bot 38: 26–41

Hansen EA and Dawson JO (1982) Effect of *Alnus glutinosa* on hybrid *Populus* height growth in a short rotation intensively cultured plantation. Forest Sci 28: 49–59

Högberg P and Kvarnström M (1982) Nitrogen fixation by the woody legume *Leucaena leucocephala* in Tanzania. Plant Soil 66: 21–28

Johnson DW and Lindberg SE (eds) (1992) Atmospheric Deposition and Forest Nutrient Cycling. Springer Verlag, New York

Kelty MJ (1992) Comparative productivity of monocultures and mixed-species stands. In: Kelty MJ (ed) The Ecology and Silviculture of Mixed-Species Forests, pp 125–141. Kluwer Academic Publ., The Netherlands

Khanna PK (1997) Comparison of growth and nutrition of young monocultures and mixed stands of *Eucalyptus globulus* and *Acacia mearnsii*. Forest Ecol Manage 94: 105–113

Knowles R (1980) Nitrogen fixation in natural plant communities and soil. In: Bergersen FJ (ed) Methods for Evaluating Biological Nitrogen Fixation, pp 557–582. Wiley, New York

Langkamp PJ, Swinden LB and Dalling MJ (1979) Nitrogen fixation (acetylene reduction) by *Acacia pellita* on areas restored after mining at Groote Eylandt, Northern Territory. Aust J Botany 27: 353–361

Langkamp PJ and Dalling MH (1981) Managing the nitrogen economies of rehabilitated sites on Groote Eylandt. In: Rummery RA and Hingston FJ (eds) Managing Nitrogen Economies of Natural and Man Made Ecosystems, pp 170–176. CSIRO, Perth, WA

Lawrie AC (1981) Nitrogen fixation by native Australian legumes. Aust J Bot 29: 143–157

Liya SM, Odu CTI, Agboola AA and Mulongoy K (1990) Estimation of N_2 fixation by nitrogen fixing trees in the subhumid tropics using 15-N dilution and difference methods. Paper presented at the fourth African Association for Biological Nitrogen Fixation. IITA. Ibadan, 24–28 Sept 1990

Lugo AE, Wang D and Bormann FH (1990) A comparative analysis of biomass production in five tropical tree species. For Ecol Manage 31: 153–166

Mafongoya PL, Giller KE and Palm CA (1997) Decomposition and nutrient release patterns of tree prunings and litter. Agroforestry Systems 38: 77–97

Miller HG (1995) The influence of stand development on nutrient demand, growth and allocation. Plant Soil 168/169: 225–232

Montagnini F, Gonzalez E, Porras C and Rheingans R (1995) Mixed and pure forest plantations in the humid neotropics: a comparison of early growth, pest damage and establishment costs. Commonwealth For Rev 74: 306–314

Motavalli PP, Palm CA, Elliot ET, Frey SD and Smithson PC (1995) Nitrogen mineralization in humid tropical soils: mineralogy, texture, and measured nitrogen fractions. Soil Sci Soc Am J 59: 1168–1175

Nambiar EKS (1990/91) Management of forests under nutrient and water stress. Water, Air and Soil Pollution 54: 209–230

Ndoye I and Dreyfus B (1988) N_2 fixation by *Sesbania rostrata* and *Sesbania sesban* estimated using ^{15}N and total N difference methods. Soil Bio Biochem 20: 209–213

Ndoye I, Gueye M, Danso SKA and Dreyfus B (1995) Nitrogen fixation in *Faidherbia albida*, *Acacia raddiana*, *Acacia senegal* and *Acacia seyal* estimated using the ^{15}N isotope dilution technique. Plant Soil 172: 175–180

Nykvist N (1997) Uptake of nutrients in a plantation of *Acacia mangium* in relation to decrease in soil amounts. J Sustainable Forestry 4(1/2): 131–139

Ong CK and Huxley P (eds) (1996) Tree-Crop Interactions: a physiological approach. CAB International, Wallingford, UK

Orcahrd ER and Darb GD (1956) Fertility changes under continual wattle culture with special reference to nitrogen and base status of the soil. Trans 6th Intern Congr Soil Sci, Paris, 1956, Vol. D, pp 305–310

Pandey D (1995) Forest resources assessment 1990. Tropical forest plantation resources. FAO Forestry paper 128. FAO, Swedish University of Agricultural Sciences, Swedish International Development Cooperation Agency

Raison RJ, Connell MJ and Khanna PK (1987) Methodology for studying fluxes of soil mineral-N in situ. Soil Biol Biochem 19: 521–530

Rennie RJ and Rennie AD (1983) Techniques for quantifying N_2 fixation in association with non-legumes under field and greenhouse conditions. Can J Microbiol 29: 1022–1035

Roskoski JP (1981) Nodulation and N_2-fixation by *Inga jinicuil*, a woody legume in coffee

plantations. I. Measurements of nodule biomass and field C_2H_2 reduction rates. Plant Soil 59: 201–206
Roskoski JP, Montano J, van Kessel C and Catillejo G (1982) Nitrogen fixation by tropical woody legumes: potential source of soil enrichment. In: Graham PH and Harris SC (eds) BNF Technology for Tropical Agriculture, pp 447–454. CIAT, Cali, Colombia
Samraj P, Chinnamani S and Haldorai B (1977) Natural versus man-made forests in Nilgiris with special reference to run-off, soil loss and productivity. Indian Forester 103: 460–465
Sanginga N, Mulongoy K and Ayanaba A (1985) Effect of nodulation and mineral nutrients on nodulation and growth of *Leucaena leucocephala*. In: Ssali H and Keya SO (eds) Biological Nitrogen Fixation in Africa, pp 419–427. Nairobi MIRCEN, Nairobi, Kenya
Sanginga N, Danso SKA and Bowen GD (1989) Nodulation and growth of *Allocasuarina* and *Casuarina* species to phosphorus fertilization. Plant Soil 118: 125–132
Sanginga N, Bowen GD and Danso SKA (1991) Intra-specific variation in growth and P accumulation of *Leucaena leucocephala* and *Gliricidia sepium* as influenced by soil phosphate status. Plant Soil 133: 201–208
Sanginga N, Vanlauwe B and Danso SKA (1995) Management of biological N_2 fixation in alley cropping systems: estimation and contribution to N balance. Plant and Soil 174: 119–141
Sanginga N, Zapata F, Danso SKA and Bowen GD (1990) Effect of successive cutting on nodulation and nitrogen fixation of *Leucaena leucocephala* using ^{15}N dilution and the difference methods. In: van Beusichem ML (ed) Plant Nutrition-Physiology and Application, pp 667–674. Kluwer Academic Publ., Dordrecht, The Netherlands
Sanginga N, Zapata F, Danso SKA and Bowen GD (1992) Estimating nitrogen fixation in *Leucaena* and *Gliricidia* using different ^{15}N labelling methods. In: Mulongoy K, Gueye M and Spencer DC (eds) Biological Nitrogen Fixation and Sustainability of Tropical Agriculture, pp 265–275. Wiley-Sayce and AABNF co-publication, Chichester, UK
Shearer G and Kohl H (1986) N_2-fixation in field settings: estimations based on natural ^{15}N abundance. Aust J Plant Physiol 13: 699–757
Sim BL and Nykvist N (1991) Impact of forest harvesting and replanting. J Tropical Forest Sci 3: 251–284
Snedecor GW and Cochran WG (1967) Statistical Methods, 6th ed. Iowa State Univ Press, Ames, Iowa, USA
Sougoufara B, Danso SKA, Diem HG and Dommergues YR (1990) Estimating N_2 fixation and N derived from soil by *Casuarina equisetifolia* using labelled ^{15}N fertilizer: some problems and solutions. Soil Biol Biochem 22: 695–701
Sun JS, Simpson RJ and Sands R (1992) Nitrogenase activity of two genotypes of *Acacia mangium* as affected by phosphorus nutrition. Plant Soil 144: 51–58
Szott LT, Fernandes ECM and Sanchez PA (1991) Soil-plant interactions in agroforestry systems. For Ecol Manage 45: 127–152
Tsai SM, Da Silva PM, Cabezas WL and Bonetti R (1993) Variability in nitrogen fixation of common bean (*Phaseolus vulgaris* L.) intercropped with maize. Plant Soil 152: 93–101
Turvey ND, Attiwill PM, Cameron JN and Smethurst PJ (1983) Growth of planted pine trees in response to variation in the densities of naturally regenerated acacias. For Ecol Manage 7: 103–117
Vandermeer J (1989) The Ecology of Intercropping. Cambridge Univ Press, Cambridge, UK
Witkowski ETF (1991) Effects of invasive alien acacias on nutrient cycling in the coastal lowlands of the Cape Fynbos. J Applied Ecol 28: 1–15
Yadav JP, Sharma KK and Khanna P (1993) Effect of *Acacia nilotica* on mustard crop. Agroforestry Systems 21: 91–98
Zou X (1993) Species effects on earthworm density in tropical tree plantations in Hawaii. Biol and Ferti Soils 15: 35–38
Zou X, Binkley D and Doxtader KG (1993) A new method for estimating gross phosphorus mineralization and immobilization rates in soils. Plant and Soil 147(1): 243–250
Zou X, Binkley D and Caldwell BA (1995) Effects of dinitrogen-fixing trees on phosphorus biogeochemical cycling in contrasting forests. Soil Sci Soc Am J 59: 1452–1458

Agroforestry Systems **38**: 121–137, 1998.

Agroforestry in the management of sloping lands in Asia and the Pacific

E. T. CRASWELL*, A. SAJJAPONGSE, D. J. B. HOWLETT and A. J. DOWLING
International Board for Soil Research and Management, P.O. Box 9-109, Bangkhen, Bangkok, 10900, Thailand (Address for correspondence: E-mail: e.craswell@cgnet.com)*

Key words: hedgerow systems, Pacific, soil conservation, Southeast Asia, steeplands, tropical soils

Abstract. Steeply sloping lands are widespread in the tropics. An estimated 500 million people practice subsistence agriculture in these marginal areas. Continued population growth has led to the intensified cultivation of large areas of the sloping lands, exacerbating the problem of soil erosion. Although research shows that alley cropping and other contour agroforestry systems can stabilize the sloping lands, these systems have not been widely adopted by farmers.

The Framework for Evaluating Sustainable Land Management (FESLM) has been tested in sloping land areas in the Philippines. Sustainable land management must be productive, stable, viable, and acceptable to farmers, while protecting soil and water resources. Farms on which contour hedgerow intercropping has been adopted meet the multifaceted requirements of FESLM, whereas the farmers' current practice does not. Appropriate land management measures for particular locations depend on a complex suite of social, economic, and biophysical factors, and need to be developed in participation with farmers.

The role of agroforestry in sustainable management of sloping lands is the subject of networks coordinated by the International Board for Soil Research and Management (IBSRAM) in seven countries in Asia (ASIALAND) and four countries in the Pacific (PACIFICLAND). We review selected outcomes from a wealth of network data. From these results the following conclusions about the sustainability of various agroforestry systems for sloping lands can be drawn:

- In the Pacific, soil loss from sloping lands due to water erosion under farmers' current practices is episodic, unpredictable, and possibly not severe;
- Agroforestry systems that utilize legume shrubs, fruit trees, coffee (*Coffea* spp.) or rubber (*Hevea brasiliensis*) provide useful economic returns, but are not an essential component in terms of soil protection because grass or pineapple (*Ananas comosus*) planted on the contour are equally effective in reducing erosion;
- Agricultural intensification will lead to nutrient mining, reduction of aboveground biomass, declining yields, and less soil protection unless external sources of nutrients are used; nitrogen can be effectively supplied using legumes;
- Cash derived from hedgerow trees and/or shrubs may provide an incentive for their adoption by farmers, as well as funds to purchase external inputs such as fertilizers;
- Labor may be a major constraint to the adoption of complex agroforestry systems.

We also discuss the information management systems required to effectively manage and utilize the extensive sets of experimental and indigenous data being accumulated. We believe such information systems can facilitate technology transfer across and between regions, and improve the efficiency of research into agroforestry and other land-management approaches.

Introduction

Large areas of the warm, humid tropics in Southeast Asia, the Pacific, Latin America, the Caribbean, and Africa are hilly or mountainous. Jackson and Scherr (1995) estimate that these tropical hillside areas are inhabited by 500 million people, or one-tenth of the current world population, many of whom practice subsistence agriculture. The region most affected is Asia which has the lowest area of arable land per capita. Aside from limited areas of irrigated terraces, most of the sloping land, which constitutes 60% to 90% of the land resources in many Southeast Asian countries, has been by-passed in the economic development of the region (Maglinao and Hashim, 1993). Poverty in these areas is often high, in contrast to the relative wealth of irrigated rice farms in lowland areas that benefited from the green revolution. Rapid population growth in some countries is also exacerbating the problems of hillside areas. Increasingly, people are migrating from high-potential lowland areas where land is scarce to more remote hillside areas. Such migration, together with inherent high population growth, is forcing a transformation in land use from subsistence to permanent agriculture on fragile slopes, and is creating a new suite of social, economic, and environmental problems (Garrity, 1993; Maglinao and Hashim, 1993).

Soils of the tropical sloping lands vary from highly leached Ultisols to recent Andisols. However, steep slopes cause most of these areas to be classified as unsuitable for continuous production of arable crops unless conservation measures are introduced to stabilize the landscape. The combination of steep slopes and highly erosive rainfall in many areas of the humid tropics makes on-site soil erosion and off-site sedimentation major concerns. Consequent changes in the regularity of stream flows due to increased runoff in the wet season are also costly (de Graaf, 1996).

This paper reviews some aspects of the role agroforestry can play in the sustainable management of sloping land areas in Southeast Asia and the Pacific. The paper draws on recent research by scientists in seven Southeast Asian and four Pacific countries collaborating in the ASIALAND and PACIFICLAND sloping land networks. Both networks are coordinated by the International Board for Soil Research and Management (IBSRAM). Core experiments in each country use standardized designs and techniques to collect data on climate, soil loss, runoff and productivity, and aim to compare farmers' current practices with alternative land management practices. Alternative practices range from contour hedgerow intercropping with shrub legumes to contour grass strips, and utilize cash trees such as coffee, bananas or rubber, or crops such as pineapple or ginger. Network participants meet regularly to compare results. Although both networks emphasize on-farm participatory research, long-term soil erosion data continue to be collected at many sites.

Agroforestry and soil erosion

Agroforestry has recently been defined as a dynamic, ecologically based, natural resource management system that, through the integration of trees into farm- and rangeland, diversifies and sustains smallholder production for increased social, economic, and environmental benefits (Leakey, 1996). Clearly, such increased benefits are an objective of any successful land-use system. Farmers in developed and developing countries have devised literally thousands of agroforestry practices and systems that are in active use (Nair, 1989). In the tropics, the most studied agroforestry technology is the alley cropping system, based on the cultivation of annual crops between hedgerows of perennial shrubs that provide fuelwood, poles and other benefits (Kang and Wilson, 1987; Howlett et al., 1996). If leguminous shrubs are used, the foliage can provide a high-protein forage or a nutrient-rich mulch. The scientific literature on alley cropping and other agroforestry systems has expanded considerably in recent years as the field of agroforestry evolves from a descriptive to a strategic, process-oriented research phase (Sanchez, 1995). Nevertheless, experimental data on the long-term effects of contour hedgerow systems on the sustainability of sloping land agriculture are still relatively rare.

We have observed that many alley cropping experiments have been conducted on relatively flat land where soil erosion is not a serious threat. In such areas, the extra labor and intensified competition created by planting hedgerows are offset by fewer benefits than on steeplands. The participatory research by Rogers (1992) shows that in flat land areas in the Pacific, shrubs planted on a grid rather than in rows may have significant benefits through reduced labor and competition. Shrubs can be used to shade out weeds and provide nutrient-rich mulch by lopping branches as crops grow and demand more light and nutrients.

For sloping lands susceptible to erosion, agroforestry systems based on contour hedgerow intercropping have been advocated as a means of biological control of erosion, without the high capital costs of mechanical structures (Young, 1989). In these systems, an A-frame or some other leveling device is commonly used to ensure that hedgerows of trees or shrubs, such as *Leucaena* spp. (leucaena), *Gliricidia sepium* (gliricidia) or *Erythrina* spp. (erythrina), are planted on the contour. Stems and cut branches of hedgerow plants are placed on the ground to slow runoff, soil particles are deposited and accumulate to create terraces, and water infiltration increases behind the barrier. According to Kiepe (1995), this approach was advocated by Kerkhoven for soil conservation in Indonesia as long ago as 1913. However, the slow rate of farmer adoption of alley cropping on sloping lands has been disappointing to many advocates of this system. Major social and economic constraints appear to be the main problem (Fujisaka, 1994); some studies (e.g., Nelson et al., 1996) raise serious questions about the high labor costs of establishing hedgerows. Future research on alley cropping would clearly benefit from a modern holistic approach that evaluates the social, economic, and cultural dimensions of the technology as well as the biophysical aspects.

Sustainable land management

International agencies, such as the Food and Agriculture Organization of the United Nations (FAO), have emphasized the need for greater attention to the sustainability of agriculture and rural development. Consequently, the framework and approach for assessing technologies or land-use systems, such as agroforestry, are currently undergoing major changes designed to increase the impact that research has on farmers' practices. The new paradigm for research requires a broad-based approach that considers social and economic dimensions as well as biological and physical dimensions. To achieve sustainable land management, research must address the needs of land users, use a participatory mode, and take into account policy and institutional dimensions that may affect outcomes (Greenland et al., 1994). This new approach requires strong interdisciplinary cooperation. The broad foundations of the new approach are encapsulated in the Framework for Evaluating Sustainable Land Management (FESLM) (FAO, 1993) that was developed by an international working group convened by IBSRAM. Sustainable land management was defined as land management systems that combine technologies, policies, and activities aimed at integrating socioeconomic principles with environmental concerns so as to simultaneously satisfy the five pillars of FESLM:

Protection	to protect the potential of natural resources and prevent degradation of soil and water quality;
Viability	to be economically viable;
Acceptability	to be socially acceptable;
Productivity	to maintain or enhance production/services; and
Security	to reduce the level of production risk.

The FESLM provides a comprehensive basis for evaluating land management practices such as agroforestry systems. A recent preliminary report of its use in sloping land areas of the Philippines (Gomez et al., 1996) illustrates the difficulty of assessing in the field the sustainability of particular land-use practices. In particular, the selection of indicators and thresholds presents practical and theoretical problems, particularly in terms of weighting parameters so that the different pillars of the FESLM can be assessed evenhandedly. Nevertheless, an interdisciplinary team convened by IBSRAM undertook a series of field visits aimed at deciding on practical indicators that could be used to assess the sustainability of agriculture in the field. The indicators and thresholds selected from a survey of adopters and non-adopters of contour hedgerow intercropping in sloping land areas of Cebu, the Philippines, are shown in Table 1. The approach taken in defining parameters based on community average values would not be suitable in an area

Table 1. Selected field indicators and thresholds for sustainability assessment of land-use practices, from Cebu in the Philippines.

Indicator	Threshold level
Yield (crop)	20% more than average community yield
Net income	20% better than community average
Frequency of crop failure	20%, or average frequency in community, whichever is lower
Soil depth	50 cm or average of similar soil types in the community
Organic carbon	1% or community average, whichever is higher
Permanent ground cover	15% or community average, whichever is higher

Source: Gomez et al. (1996).

where agriculture is uniformly sustainable or unsustainable. Gomez et al. (1996) normalized values for these indicators by expressing data from a land-use system as a proportion of their threshold levels, as defined in Table 1. They contend that a system would not be sustainable if the average of the indicators is less than one, i.e., when the average value for the system's indicators is less than the average threshold value calculated according to Table 1. The field data collected can be represented in a radar polygon that clearly displays the sustainability of the hedgerow cropping system in relative terms (Figure 1). Data from the non-adopter (e.g., sample farm 5) exceeds the threshold value only for the frequency of the crop failure indicator, showing that the farmer's current practice of clean cultivation and continuous cropping without conservation measures has lower yield and profit; reduced soil cover, organic carbon, and soil depth due to nutrient mining and erosion; and a high production risk. For the adopter farm with contour alley cropping (e.g., sample farm 10), Figure 1 indicates lower risk and values exceeding the threshold for all of the other indicators.

The work of Gomez et al. (1996) illustrates a novel and effective way to assess the multiple dimensions of sustainability in a farmer's field. Their work also shows that hedgerow intercropping with leguminous shrubs planted on the contour is sustainable in the sloping lands of Cebu that were surveyed in this study. Adoption of contour alley cropping is widespread in some parts of Cebu, a fact that owes much to the educational efforts of nongovernmental organizations (NGOs) such as the Mag-uugmad Foundation (Cerna et al., 1995). The work of Gomez et al. (1996) also indicates that FESLM provides a promising multifaceted framework for assessing sustainability in agroforestry and other land management systems in the field. Information below, drawn from the ASIALAND and PACIFICLAND networks, illustrates particular facets of the sustainability complex.

The ASIALAND network

The ASIALAND management of sloping lands network was established in 1988 to assist participating national agricultural centers in identifying and

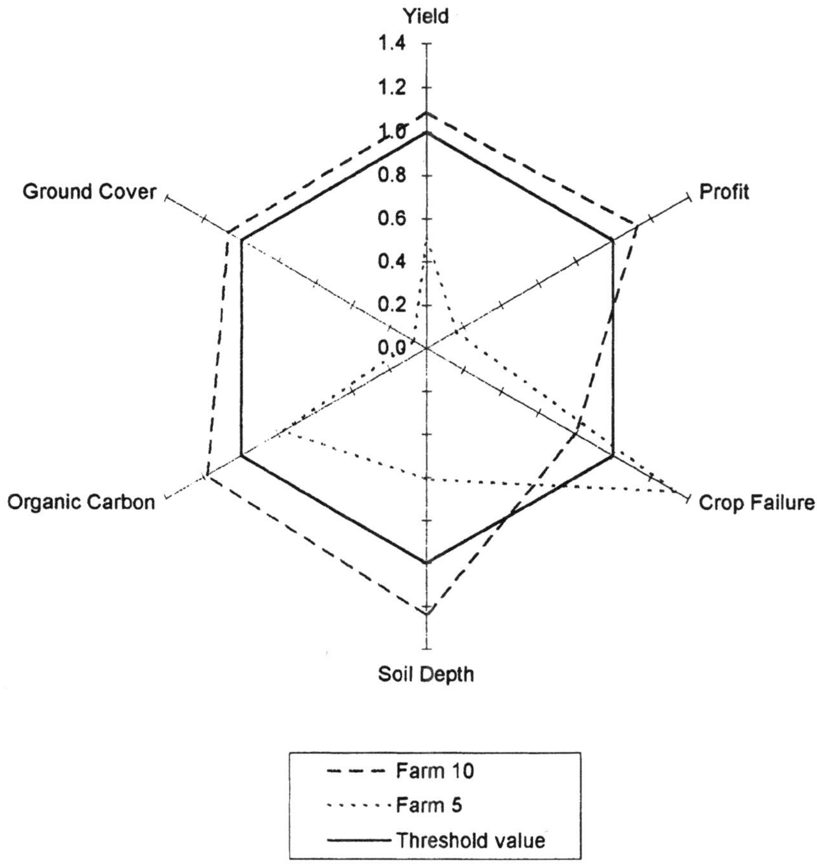

Fig. 1. Radar polygon of sustainability indicators on two farms in Cebu, Philippines. Farm 5 followed current farmer practice of maize (*Zea mays*) production without conservation measures, and farm 10 used an agroforestry system with maize planted between hedgerows of leucaena. The threshold value line represents the normalized value of 1 for each parameter (see Table 1).

promoting appropriate and sustainable technologies for the management of their sloping agricultural lands. Seven countries – China, Indonesia, Laos, Malaysia, Philippines, Thailand, and Vietnam – are participating in the network. Technologies being assessed against farmers' current practices include alley cropping and various agroforestry systems that were selected collaboratively by researchers and farmers. Selected results from network research in the Philippines and Thailand are used to illustrate the sustainability of particular land management treatments within the FESLM context.

Protection

Measurements of soil loss and runoff at a core network experiment in Thailand in 1994 are shown in Table 2 (IBSRAM, 1995b). Simulated farmers' practice

Table 2. Annual soil loss and runoff as affected by fertilizer application and treatment at Doi Tung, Thailand in 1994 (all treatments cropped with upland rice).

Treatment[a]	Soil loss (t/ha)		Runoff (mm)	
	–F[b]	+F	–F	+F
Farmers' practice	460	193	332	165
Alley cropping	33	21	103	78
Grass strip	28	14	98	64
Hillside ditches	19	9	96	52
Coffee	185	40	190	67

[a] Farmers' practice = rice monocrop planted up and down the slope; alley cropping = *Leucaena leucocephala* and *Cajanus cajun*, every 5 m on the contour; Grass strip = *Paspalum notatum* every 5 m on the contour; Hillside ditches = every 5 m on the contour; Coffee = shrubs planted on 2-m grid.
[b] –F = no fertilizer application; +F = with fertilizer application.
Source: IBSRAM (1995b).

– a monoculture of upland rice planted up and down the slope – led to extensive soil loss and high rates of runoff. Conservation measures every 5 m on the contour, such as hillside ditches, 1-m-wide strips of Bahia grass (*Paspalum notatum*), and hedgerows of leucaena (*L. leucocephela*) and pigeon pea (*Cajanus cajan*), reduced soil loss twenty-fold and runoff three-fold relative to farmers' practice. An agroforestry treatment that included upland rice and coffee shrubs planted on a 2-m grid also reduced the amount of soil loss and runoff, but the reduction was less than under the other treatments. Fertilizers were applied to replace nutrients lost due to erosion and continuous cultivation. Fertilizer applications also increased biomass production and markedly enhanced the soil-conserving effects of all treatments, including farmers' practice. The data suggest that trees such as coffee and legume shrubs in the agroforestry systems used in this study are perhaps less effective than other conservation measures at reducing runoff and soil loss, but their importance in meeting economic criteria for sustainability are shown below.

Over a period of four years at the network site in the Philippines, an agroforestry treatment utilizing bananas in contour hedgerows with vetiver grass (*Vetiveria zizanioides*) significantly reduced soil loss compared with farmers' practice (Table 3). The vetiver grass effectively controlled soil erosion whereas the banana was a source of income (see below). While rates of soil erosion at the Philippines site were high, annual losses varied greatly. Similar fluctuations in annual rates of soil loss were observed at other network sites.

Economic viability

Coffee in the agroforestry treatment at the site in Thailand markedly improved the economic return to labor and materials (Table 4). At this site, the return to labor from upland rice in the farmers' practice treatment was poor, and declined further when fertilizer inputs were used, due to the low value of the

Table 3. Soil loss under different treatments over four years at the Mabini site in the Philippines.

Treatment	Soil loss (Mg/ha)			
	1989	1990	1991	1992
Farmers' practice[a]	20	97	18	41
Agroforestry[b]	17	2	< 1	< 1

[a] Farmers' practice = corn and peanut cultivated up and down the slope.
[b] Agroforestry = banana and vetiver grass as contour hedgerows with corn and peanut intercrops.
Source: IBSRAM (1995b).

Table 4. Economic returns (annual) as affected by fertilizer application under different treatments at Doi Tung, Thailand in 1994 (all treatments cropped with upland rice).

Treatment[a]	Return to labour ($/ha)		Return to materials ($/ha)	
	–F*	+F	–F	+F
Farmers' practice	56	–46	–233	–206
Alley cropping	62	–66	–169	–163
Grass strip	21	–6	–170	–92
Hillside ditches	278	263	51	150
Coffee	831	1163	490	852

[a] Farmers' practice = rice monocrop planted up and down the slope; alley cropping = *Leucaena leucocephala* and *Cajanus cajun*, every 5 m on the contour; Grass strip = *Paspalum notatum* every 5 m on the contour; Hillside ditches = every 5 m on the contour; Coffee = shrubs planted on 2-m grid.
[b] –F = no fertilizer application; +F = with fertilizer application.
Source: IBSRAM (1995b).

harvested rice. These economic benefits must be weighed against the high soil loss in the coffee agroforestry treatment.

Social acceptability

Surveys and case studies in northern Thailand (Renaud, 1997) suggest that the availability of labor may significantly affect the adoption of agroforestry systems by farmers. Out-migration of farm family members for employment overseas or in Thai cities is causing considerable social disruption and reducing labor availability for farm production. In many cases, women are left to run the farms in addition to caring for the family. Clearly, this will be a major factor influencing the sustainability of land management practices that require high labor inputs.

Productivity

Soil productivity can be assessed using a Productivity Index (PI) calculated from data on soil pH, organic carbon, and rates of soil loss and soil formation (Sajjapongse, 1995). Data from network experiments showed that PI declined with time in plots simulating farmers' practice. This decline correlated well with the decline in yield at sites in Thailand and the Philippines, but not in Indonesia.

Although crop yield integrates a range of soil productivity factors, its utility as an indicator of productivity is reduced by variations in seasonal weather patterns and the incidence of pests and diseases. Network results show that yield commonly declines in the establishment years of hedgerow systems because some land area is lost to the hedgerow. Further data analysis is necessary to define better indices of soil productivity change.

Security

Additional research is also needed to better assess the security pillar of FESLM. Variations in annual rainfall affect crop production and income, and require further assessment. Also, cyclones or typhoons are common in Southeast Asia and the Pacific. The high winds and torrential rainfall that are associated with these storms can cause extensive wind damage, landslides and/or surface erosion and off-site flooding. Conservation technologies should be able to survive natural disasters without increasing the level of on-farm risk to the farmer. This risk must be factored into the design of appropriate soil conservation and agroforestry technologies.

The PACIFICLAND network

A great variety of traditional, or indigenous, smallholder agricultural systems have evolved in the South Pacific to meet the diverse range of biophysical and social environments that characterize the region (Brookfield, 1989). The vast majority of these systems are based on subsistence root crops [e.g., sweet potato (*Ipomoea batatas*), taro (*Colocasia* spp.), yam (*Dioscorea* spp.), cassava (*Manihot esculenta*)] with grass, brush, or forest fallows. Traditional cash crops include coconut (*Cocos nucifera*), coffee, cacao (*Theobroma cacao*), kava (*Piper methysticum*) and betel nut (*Areca catechu*). Post-colonial times have seen an increase in the development of cash and commercial crops [e.g., ginger (*Zingiber officinale*), taro, oil palm (*Elaeis guineensis*), sugar cane (*Saccharum officinarum*)] and an increase in population (SPREP, 1992). As a result, fallow periods are declining. For example, on the Vanuatu island of Pentecost, the duration of fallow periods has declined from 4–20 years to 4–5 years (Liebbrecht, 1993; Craswell and Morrison, 1994; Billonneau, 1995). Flatter areas of fertile arable land tend to be used for commercial plantations and, in some cases, urban development. In the smallholder sector, tree cash crops have also been introduced. As a result, the use of sloping lands for both subsistence and cash cropping is increasing across the Pacific (Howlett, 1995). These changes have often been accompanied by an increase in logging activity over the last decade, particularly in Melanesian countries.

Agricultural systems in the Pacific are thus under increasing pressure from a number of different social, economic, and physical forces (Howlett, 1995). The majority of farmers in the Pacific are resource-poor, use few or no external inputs such as fertilizer (Craswell, 1989), and rely on subsistence agriculture for their food and cash. With an intensification in the use of sloping lands in the Pacific, problems associated with declining productivity, soil loss, and land degradation are increasing. In addition, adverse downstream impacts on inland waterways and fisheries, beaches, and near-shore reefs and fisheries are of increasing concern.

The PACIFICLAND network was set up in 1992 to evaluate sloping land management systems and, during its first phase (1992–96), six sites were established: Vanuatu (two sites), Fiji (one), Papua New Guinea (PNG) (two), and Samoa (one) (D.Howlett, unpublished). Additional sites will be established in Tonga and Solomon Islands during the second phase (1996–99). These researcher-based network experiments aim to compare different land management systems to farmers' practice. Climate, soil loss, runoff, crop productivity, and a range of soil chemical and physical properties are being measured and related to land management practices. Currently, sites in Vanuatu, PNG and Samoa are under fallow, having been cropped for up to four years. Land management practices are in place. Detailed country reports are given elsewhere (IBSRAM, 1993; 1994; 1995a; 1996).

In this review, we focus on first phase activities and discuss results related to the protection pillar of FESLM, i.e., the measurement of soil erosion. Farming systems aspects of sustainable land management in sloping lands of the Pacific are also discussed briefly.

Average soil erosion rates from PACIFICLAND network sites planted to root crops are given in Table 5. Except for PNG, these results indicate that rates of soil loss are less than those suggested in Howlett (1995) where, using data adapted from Humphreys (1994), rates of 20 to 90 Mg/ha/y in Fiji and < 10 to 38 Mg/ha/y in PNG are indicated. In keeping with the low rates of soil loss, infiltration rates are high. For example, in Samoa, infiltration rates are often over 95% on slopes of 25° with over 4000 mm of rainfall (Pratap et al., 1996). Similar high infiltration rates have been recorded at other sites. The on-farm impact of these low rates of erosion appears to be negligible from a productivity point of view. However, as fertilizers are not commonly used, continuous cultivation and cropping may have accelerated the rate of nutrient decline causing a reduction in vegetative cover and greater rates of soil erosion.

Observed low rates of soil loss can be attributed to a combination of four factors: (1) low soil erodibility, (2) low rainfall erositivity, (3) good tillage practice, and (4) the presence of vegetative cover. The soils at the PACIFICLAND sites are generally well structured clay loams and are not highly erodible. Although total annual rainfall is high (Table 5), rainfall events with intensities above 50 mm/hr are short (typically < 12 min) and hence rainfall erosivity is relatively low. Furthermore, farmers' tillage practices mostly entail little ground disturbance, with digging sticks used to make holes at planting;

Table 5. Soil erosion rates from the PACIFICLAND Network sites.[a]

Country, site and treatment	Slope range (degrees)	Mean annual rainfall (mm)	Rate of soil loss (Mg/ha)			
			1992	1993	1994	1995
Fiji – Waibau	24–29	4000				[1]
Farmers' practice			0.3	1.4	1.2	7.1
Pineapple strips on contour			0.1	0.2	0.9	0.2
Vetiver strips on contour			0.2	0.1	0.1	0
PNG – Aiyura, southern slope	10–20	2500				
Farmers' practice			na	na	28.0	nd
Vetiver strips			na	na	25.2	nd
Leucaena hedgerow			na	na	26.3	nd
Coffee (1m spacing)			na	na	36.2	nd
PNG – Aiyura, northern slope	10–20	2500		[2]		
Farmers' practice			na	na	16.3	nd
Vetiver strips			na	na	7.0	nd
Leucaena			na	na	13.9	nd
Coffee			na	na	9.0	nd
Vanuatu – Lakura	10–19	1962	[3]			
Farmers' practice			12.0	24.0	nd	
Flemingia hedgerow			0.1	0.6	2.5	nd
Vanuatu – Sara	18–25	3486	[3]			
Farmers' practice			2.5	4.7	nd	nd
Flemingia hedgerow			2.2	3.3	nd	nd
W. Samoa – Tapatapao	20–25	4500		[4]	[5]	
Farmers' practice			na	0.1	0.1	nd
Cover crop (lablab)/mulch			na	0.1	0.1	nd
Bananas and pineapple hedgerow			na	0.8	0.1	nd
Flemingia/erythrina hedgerow			na	0	0.1	nd

na = not applicable as experimentation not commenced; nd = not determined. 1 – data for 5 months only; 2 – data for 9 months only; 3 – data for 8 months only, an exceptionally dry period; 4 – data for 7-month period; 5 – data for 6-month period.

[a] All treatments were cropped to root crops, such as sweet potato, taro, and yam; farmer practice varied from site to site. Most 1995 data was not collated at time of reporting.

Source: IBSRAM (1996).

weeds are removed by hand as required and retained as a surface mulch. Bare soil is exposed only rarely.

The often-held belief that the combination of sloping land, high rainfall and traditional smallholder agriculture leads to high rates of soil erosion cannot therefore be uniformly applied to the South Pacific. Nevertheless, contemporary examples of degraded lands can be found in, for example, Gumine in the Highlands of PNG (M.B. Kanua, pers. comm.) and in the sugar cane lands of Fiji (Gawander, 1993). The potential for erosion (and nutrient mining) thus remains very real, particularly with an intensification in the use of sloping lands for both subsistence and cash cropping. Sustainable sloping land management in the region will require a more holistic approach if it is to address the interrelated challenges of declining yield, shortening fallow periods, and maintaining soil fertility levels.

Given the social and cultural diversity within the South Pacific, the active involvement of farmers as resource users is essential in the development and promotion of appropriate sloping land management technologies. Most PACIFICLAND trials are on-farm, and socioeconomic surveys of the surrounding areas were undertaken before trials commenced (Billonneau, 1995; Humphrey and Igua, 1995; Konabe and Powae, 1995; Limalevu, 1995; Nagatalevu et al., 1996; Pratap, 1995). These surveys identified key issues on the use of sloping lands by farmers. Gender was seen as a particularly important issue given the different roles of men and women in the farming systems and household. These issues are being incorporated into the design of new PACIFICLAND activities.

Cost-benefit analysis (labor and chemicals versus marketable yield) indicates that for the early years after establishment, the benefits of the hedgerow treatments do not outweigh the costs to the farmer, unless the species planted on the contour is a cash crop such as pineapple (Pratap et al., 1996). Where farmers do not have long-term security of tenure, such as in Fiji (Nagatalevu and Ratukalou, 1996), there is even less likelihood that they will invest in the long-term sustainability of their land. This points to a need for more research on economic and policy issues – including land tenure, land use, market accessibility, public investment and rural organizations – that are likely to affect the development and adoption of more sustainable farming systems. Such systems will have to provide both short- and longer-term benefits to the farmer, and/or be supported by appropriate government policies and incentives to promote their adoption. In addition, these technologies should strive to either reduce the amount of labor required or change the type of work; this could be facilitated by providing shade or reducing time spent on weeding.

PACIFICLAND research is assessing the effectiveness of hedgerows of leucaena and coffee in Aiyura PNG, flemingia (*Flemingia macrophylla*) and gliricidia in Keravat PNG, flemingia in Vanuatu, and flemingia and erythrina in Samoa (Howlett, 1995). Positive effects from a legume shrub or tree hedgerow – i.e., nitrogen fixation, addition of organic matter, and conservation of soil and fertility – are likely to develop over time before being expressed as either increased or more sustained crop yields. Short-duration fallows and cropping cycles typical of the Pacific may therefore augur against hedgerow systems. Although fuelwood and building poles are in demand, tree hedgerows may not be the most appropriate method for their production. Demand for other hedgerow products, such as fodder for livestock, is commonly limited. In summary, the chances of this type of technology being adopted are slim, unless the farmer receives external incentives in recognition of the off-site benefits from reduced soil erosion.

Network research is also assessing the effectiveness of other types of vegetative barriers, including common alternatives to tree hedgerows such as vetiver grass and pineapple barriers (Howlett, 1995). As well as these traditional barriers, locally important plants such as pitpit (*Saccharum edule*); valangur (*Polyscias grandiflora*), a stem vegetable; gaga (*Heliconia bihai*), a low-lying broadleaved plant used for wrapping betel nut; and banana have been included as possible alternatives. The aim of these treatments is to provide a benefit to the farmer from the hedgerow in return for his labor required to establish it. In this case, the conservation value of the hedgerow becomes an added benefit over the mid- to longer term. Other alternative hedges or vegetative barriers can also be considered. For example, in Fiji, on-farm work with the participation of farmers is currently testing the effectiveness of kava, bele (*Abelmoschus manihot*), and voivoi (*Pandanus thurstoni*) planted in strips or contour bunds as erosion control measures capable of providing vegetative barriers of economic value.

To date, vetiver has proved the most effective barrier to soil erosion. However, farmers in Fiji express reluctance to adopt vetiver as a conservation measure. This is particularly interesting as vetiver has a long history in Fiji. The principal concerns of the farmers are the following: (1) vetiver will become a weed and spread beyond the farmer's field, though there are no scientific grounds for this; (2) the hedgerows will spread within the field and take up too much land; and (3) there are no perceived economic returns from the grass itself, although it could be used as a thatch or for handicrafts. In the sugar cane lands of Fiji, farmers are actually removing old vetiver hedgerows established in the 1950s and 1960s.

Many traditional farming systems in the South Pacific have a lengthy fallow cycle. In many parts of the region, the shift to a more commercially based agricultural system is leading to greater intensification, shorter fallows, and less time for soil fertility to regenerate. Management of the fallow to maintain or enhance soil fertility is becoming more important in the development of sustainable land management systems. Farmers in some countries already practice some degree of fallow management and in PNG the planting of casuarina (*Casuarina* spp.) prior to a fallow is a common practice. Similarly, the use of other nitrogen-fixing trees is common across the region. In grassland areas, fires, accidental or deliberate, are common. These can present serious degradation and erosion risks in themselves and increase the risk of using trees or other legumes that are not tolerant to fire in the fallow cycle.

Conclusions and research needs

The ASIALAND and PACIFICLAND networks are producing a wealth of data on systems for the management of sloping lands in their regions (see IBSRAM, 1993; 1994; 1995a; 1995b; 1996). This paper presents only a few examples that highlight some of the major findings. Based on network results, we draw the following specific conclusions about the sustainability of various agroforestry systems for sloping lands.

- In sloping land areas of the South Pacific, soil loss due to water erosion under farmers' current practices is episodic and unpredictable, and may not be severe.
- Agroforestry systems that utilize legume shrubs, fruit trees, coffee, or rubber provide useful economic returns, but are not an essential component in terms of soil protection because contour strips of grasses or pineapple are equally effective in reducing erosion.
- Agricultural intensification is leading to nutrient mining. This will reduce yields, aboveground biomass and soil protection unless external sources of nutrients are used; nitrogen can be effectively supplied using legumes.
- Cash derived from hedgerow trees and/or shrubs may provide an incentive for their adoption by farmers, as well as funds to purchase external nutrients such as fertilizers.
- Labor may be a major constraint to the adoption of complex agroforestry systems.
- FESLM provides a useful framework for evaluating the multiple facets of sustainability.

Past research has been driven too much by technology testing on-station. Future research programs will need to involve farmers and other land users if they are to move from a process of consultation to one of interactive participation in the process of technology development and testing (Bechstedt, 1997). Research methods and tools are needed that can deal with the complexity of sustainability assessment and with the extensive sets of experimental data from field trials, such as those discussed in this paper. We believe that tools such as advanced computer and information technology can help deal with this complexity and assist with the effective utilization of the large knowledge and data bases. For example, IBSRAM is using the ASIALAND network data to develop a Decision Support System (DSS) for sustainable land management at farm, village, watershed, and regional scales (Rais, 1997). This will allow improved and effective decision-making through the application of information technologies to access and manage the large and complex data bases that have accumulated. This approach will also allow the integration and synthesis of socioeconomic and biophysical information with indigenous practices and knowledge systems. New information management systems can also help through the development of Resource Management Domains; these can facilitate technology transfer across regions and improve the efficiency of research into agroforestry and other land management issues (Craswell et al., 1997) .

Acknowledgments

The authors thank our colleagues in the ASIALAND and PACIFICLAND networks whose data we have reviewed. We are also grateful to the Swiss Agency for Development and Cooperation, the Asian Development Bank, the Australian Agency for International Development, the Australian Center for International Agricultural Research, the European Union Pacific Regional Agricultural Programme, and the UK Overseas Development Administration (now the Department of International Development) for their support, without which the research reported would not have been possible.

References

Bechstedt HD (1997) Moving further from consultative to interactive participation. In: ASIALAND Management of Sloping lands Network Document No. 13. IBSRAM, Bangkok, Thailand (in press)

Billonneau O (1995) Farming systems and socio-economic survey, Pentecost Island, Vanuatu. In: The Management of Sloping Lands in the South Pacific Islands (IBSRAM/PACIFICLAND), pp 151–165. IBSRAM Network Document No. 15. IBSRAM, Bangkok, Thailand

Brookfield H (1989) The human context of sustainable smallholder development in the Pacific. In: Soil Management and Smallholder Development in the Pacific Islands, pp 189–204. IBSRAM Proceedings No. 8. IBSRAM, Bangkok, Thailand

Cerna L, Moneva L and Geradino EC (1995) Mag-uugmad research and extension for soil, water, and nutrient management. In: The Zschortau Plan for the Implementation of Soil, Water, and Nutrient Management Research, pp 191–198. DSE: Zschortau/IBSRAM, Bangkok, Thailand

Craswell ET (1989) The role of fertilizers in smallholder agriculture. In: Soil Management and Smallholder Development in the Pacific Islands, pp 99–109. IBSRAM Proceedings No. 8. IBSRAM, Bangkok, Thailand

Craswell ET and Morrison RJ (1994) Resource-poor farmers in small nations – the Pacific perspective. Proceedings 15th International Congress of Soil Science 9: 397–415

Craswell ET, Rais M and Dumanski J (1997) Resource management domains as a vehicle for sustainable development. International Journal on Failures and Lessons Learned in Information Technology Management: Special Issue on Information Systems for Sustainable Development (in press)

de Graaf J (1996) The price of soil erosion: An economic evaluation of soil conservation and watershed development. Mansholt studies 3, p 299. Wageningen Agricultural University, Wageningen, The Netherlands

FAO (1993) FESLM: An International Framework for Evaluating Sustainable Land Management. World Soil Resources Report 73. FAO, Rome, Italy

Fujisaka S (1994) Learning from six reasons why farmers do not adopt innovations intended to improve the sustainability of upland agriculture. Agricultural Systems 46: 409–425

Garrity DP (1993) Sustainable land use systems for sloping uplands of Southeast Asia In: Technologies for Sustainable Agriculture in the Tropics, pp 41–64. ASA Special Publication 56. American Society of Agronomy, Madison, USA

Gawander JS (1993) Sugarcane and erosion. In: The Management of Sloping Lands in the South Pacific Islands (IBSRAM / PACIFICLAND), pp 25–31. Network Document No. 7. IBSRAM, Bangkok, Thailand

Gomez AA, Swete-Kelly DE, Syers JK, and Coughlan KJ (1996) Measuring sustainability of agricultural systems at the farm level. In: Methods for Assessing Soil Quality, pp 401–410. SSSA Special Publication 49. Soil Science Society of America, Madison, WI, USA

Greenland DJ, Bowen GD, Eswaran H, Rhoades R and Valentin C (1994) Soil, Water and Nutrient Management Research – A New Agenda. IBSRAM, Bangkok, Thailand

Howlett DJB (1995) Sustainable smallholder agriculture and the role of fertilizers in Fiji and the South Pacific. Fiji Agricultural Journal 51: 59–72

Howlett DJB, Singh A and Nakalevu T (1996) Institutional development of agroforestry in Fiji: 1980 to 1996. Fiji Agricultural Journal 52: 1–17

Humphrey B and Igua P (1995) PACIFICLAND trial: Keravat socioeconomic survey, Papua New Guinea. In: The Management of Sloping Lands in the South Pacific Islands (IBSRAM/ PACIFICLAND), pp 129–149. Network Document No. 15. IBSRAM, Bangkok, Thailand

Humphreys GS (1994) The interpretation of soil erosion measurements. In: The Management of Sloping Lands in the South Pacific Islands (IBSRAM / PACIFICLAND), pp 111–138. Network Document No. 10. IBSRAM, Bangkok, Thailand

IBSRAM (1993) The management of sloping lands in the South Pacific islands (IBSRAM/ PACIFICLAND). Network Document No. 7. IBSRAM, Bangkok, Thailand

IBSRAM (1994) The management of sloping lands in the South Pacific islands (IBSRAM/ PACIFICLAND). Network Document No. 10. IBSRAM, Bangkok, Thailand

IBSRAM (1995a) The management of sloping lands in the South Pacific islands (IBSRAM/ PACIFICLAND). Network Document No. 15. IBSRAM, Bangkok, Thailand

IBSRAM (1995b) ASIALAND: The management of sloping for sustainable agriculture in Asia (Phase 2, 1992–1994). Network Document No. 12. IBSRAM, Bangkok, Thailand

IBSRAM (1996) The management of sloping lands in the South Pacific islands (IBSRAM/ PACIFICLAND). Network Document No. 19. IBSRAM, Bangkok, Thailand

Jackson LA and Scherr S (1995) Nondegrading Land Use Strategies for Tropical Hillsides. 2020 Brief No. 27. International Food Policy Research Institute, Washington, DC, USA

Kang BT and Wilson GF (1987) The development of alley cropping as a promising agroforestry technology. In: Steppler HA and Nair PKR (eds) Agroforestry: A Decade of Development, pp 227–243. ICRAF, Nairobi, Kenya

Kiepe P (1995) No Runoff, No Soil Loss: Soil and Water Conservation in Hedgerow Barrier Systems. Tropical Resource Management Papers 10, Wageningen Agricultural University, Wageningen, The Netherlands

Konabe B and Powae W (1995) Exploratory socio-economic survey: Kainantu and Heganofi districts, Papua New Guinea. In: The Management of Sloping Lands in the South Pacific Islands (IBSRAM/PACIFICLAND), pp 101–112. Network Document No. 15. IBSRAM, Bangkok, Thailand

Leakey R (1996) Definition of agroforestry revisited. Agroforestry Today 8(1): 5–7

Liebbrecht S (1993) Rapport d'enquête socio-économique, Pentecôte. Department of Agriculture and Horticulture, Vanuatu

Limalevu L (1995) Preliminary socio-economic survey of Waibau, Fiji. In: The Management of Sloping Lands in the South Pacific Islands (IBSRAM/PACIFICLAND), pp 69–77. Network Document No. 15. IBSRAM, Bangkok, Thailand

Maglinao AR and Hashim GM (1993) Management of sloping lands for sustainable agriculture in Southeast Asia. Soil Management Abstracts 5(4): 111–127

Nagatalevu M and Ratukalou I (1996) Issues affecting sustainable land management in Fiji. In: The Management of Sloping Lands in the South Pacific Islands (IBSRAM/PACIFICLAND), pp 243–252. Network Document No. 19. IBSRAM, Bangkok, Thailand

Nagatalevu M, Foarete HM and Field SP (1996) The ginger industry in Fiji: A case study. Fiji Agriculture Journal 52: 19–34

Nair PKR (ed) (1989) Agroforestry Systems in the Tropics. Kluwer, London, UK

Nelson R, Grist P, Menz K, Paningbatan E and Mamicpic M (1996) A cost-benefit analysis of hedgerow intercropping in the Philippine uplands using SCUAF. Imperata Project Paper 1996/2. Australian National University, Canberra, Australia

Pratap V (1995) Western Samoa report. In: The Management of Sloping Lands in the South Pacific Islands (IBSRAM/PACIFICLAND), pp 167–170. Network Document No. 15. IBSRAM, Bangkok, Thailand

Pratap V, Yapa LGG, Rogers S, Tuivavalagi N and Lemalu S (1996) Management of sloping land in Western Samoa: the effect of various cropping systems on crop yield and soil loss. In: The Management of Sloping Lands in the South Pacific Islands (IBSRAM/PACIFICLAND), pp 63–84. Network Document No. 19. IBSRAM, Bangkok, Thailand

Rais M (1997) Decision support system for sustainable land management in sloping lands in Asia. In: Syers JK and Bouma J (eds) International Workshop on Resource Management Domains, Kuala Lumpur, 26–29 August, 1996. IBSRAM, Bangkok, Thailand (in press)

Renaud F (1997) The financial cost-benefit analysis of soil conservation practices in Northern Thailand. Mountain Research and Development 17: 11–18

Rogers S (1992) The potential for alley cropping as a sustainable land use system in the volcanic islands of the South Pacific. Journal of South Pacific Agriculture 1: 3–10

Sanchez P (1995) Science in agroforestry. Agroforestry Systems 30: 5–55

Sajjapongse A (1995) Productivity index: an assessment of the effect of long-term soil erosion. In: Maglinao A and Sajjapongse A (eds) International Workshop on Conservation Farming for Sloping Uplands in Southeast Asia: Challenges, Opportunities, and Prospects, pp 169–181. Proceedings No. 14. IBSRAM, Bangkok, Thailand

South Pacific Regional Environmental Programme (SPREP) (1992) Environment and Development: A Pacific Island Perspective. Asian Development Bank (ADP), Manila, Philippines

Young A (1989) Agroforestry for Soil Conservation. CAB International, Wallingford, UK

Agroforestry Systems **38**: 139–164, 1998.

Shade management in coffee and cacao plantations

J. BEER*, R. MUSCHLER, D. KASS and E. SOMARRIBA
Apdo. 44, Area of Watersheds and Agroforestry Systems, CATIE, Turrialba, Costa Rica
(Address for correspondence: E-mail: jbeer@catie.ac.cr)*

Key words: *Coffea arabica*, interactions, shaded perennials, silviculture, *Theobroma cacao*, wood production

Abstract. Shade trees reduce the stress of coffee (*Coffea* spp.) and cacao (*Theobroma cacao*) by ameliorating adverse climatic conditions and nutritional imbalances, but they may also compete for growth resources. For example, shade trees buffer high and low temperature extremes by as much as 5 °C and can produce up to 14 Mg ha^{-1} yr^{-1} of litterfall and pruning residues, containing up to 340 kg N ha^{-1} yr^{-1}. However, N_2 fixation by leguminous shade trees grown at a density of 100 to 300 trees ha^{-1} may not exceed 60 kg N ha^{-1} yr^{-1}. Shade tree selection and management are potentially important tools for integrated pest management because increased shade may increase the incidence of some commercially important pests and diseases (such as *Phythphora palmivora* and *Mycena citricolor*) and decrease the incidence of others (such as *Colletotrichum gloeosporioides* and *Cercospora coffeicola*). In Central America, merchantable timber production from commercially important shade tree species, such as *Cordia alliodora*, is in the range of 4–6 m^3 ha^{-1} yr^{-1}.

The relative importance and overall effect of the different interactions between shade trees and coffee/cacao are dependent upon site conditions (soil/climate), component selection (species/varieties/provenances), belowground and aboveground characteristics of the trees and crops, and management practices. On optimal sites, coffee can be grown without shade using high agrochemical inputs. However, economic evaluations, which include off-site impacts such as ground water contamination, are needed to judge the desirability of this approach. Moreover, standard silvicultural practices for closed plantations need to be adapted for open-grown trees within coffee/cacao plantations.

Introduction

In 1995, coffee (*Coffea* spp.) and cacao (*Theobroma cacao*) plantations worldwide totaled 16,700,000 ha (FAO, 1996). Plantations of these perennial crops are one of the most important forms of land use and are of enormous economic importance for developing countries in the humid tropics (Graaff, 1986; Wood and Lass, 1985). In addition to these crops, many other well-known perennials, such as black pepper (*Piper nigrum*) and vanilla (*Vanilla fragrans*), and lesser-known crops, such as Yerba mate (*Ilex paraguariensis*) (Evans and Rombold, 1984) and cupuazú (*Theobroma grandiflorum*) – a close relative to cacao (Duarte, 1992), are usually grown under shade trees (Purseglove, 1968; 1972); i.e., in agroforestry systems based on two or more perennial species (Nair, 1993). Nevertheless, unshaded intensively managed plantations have been promoted in some countries.

One reason for maintaining shade trees in perennial-crop plantations is the income provided by their fruit and/or timber; these products may supplement farmers' income when coffee/cacao prices are low. At a national level, increasing awareness of the environmental costs associated with high-input monocultures – for example, ground water pollution and soil erosion on sloping land in the case of coffee (Boyce et al., 1994; PROMECAFE, 1995) – has also led to renewed interest in the use of shade trees in areas where they had previously been eliminated. At the international level, attention has recently focused on the contribution of shaded coffee or cacao fields for maintaining biodiversity (Perfecto et al., 1996; Young, 1994) and for stop-over points for migrating birds (Wille, 1994). Low and variable prices for coffee and cacao, rising prices for inputs, and environmental problems have raised questions about the 'sustainability' of high-input, unshaded plantations and rekindled interest in the use of shade trees.

In this context, it is important to take stock of the accumulated information on shade trees. The purpose of this paper is to summarize the voluminous literature on ecological aspects of shaded coffee and cacao plantations, and on the productivity and management of the shade trees. The scope of the paper does not permit detailed discussion of socioeconomic aspects; however, a few salient features are mentioned. Finally, we discuss priority research topics relating to the ecology and management of shade tree systems. The emphasis of the review is on Central America, which is CATIE's mandate area, and where most of the reported work on the topic has been conducted.

The effects of maintaining shade trees over perennial crops were described as early as the late nineteenth century. For example, Lock (1888) provided a comprehensive description of the positive and negative effects of shade trees, based on studies of coffee management in Ceylon (now Sri Lanka) (Table 1). Sâenz (1895), Cook (1901), and authors from Colombia (FNC, 1932), Venezuela (INC, 1942) and Honduras (Ortega, 1951), also published early accounts of the potential role of shade trees over coffee. The relative importance of the different effects of shade trees, and hence the need for shade, is strongly affected by site conditions – a fact that has generated considerable

Table 1. Key aspects of shade and shade trees for coffee in Ceylon (now Sri Lanka).

- *Climatic range:* Shade is not universally beneficial. The need for shade is a function of climate (it is especially important in hot and dry climates).
- *Benefits of shade:* Diminished crop exhaustion, and increased longevity of coffee plants; reduced costs; maintenance/improvement of soil fertility; increased litter (and hence nutrient availability); value of timber.
- *Drawback:* Coffee yield reduction, but compensated by increased longevity.
- *Beneficial shade tree attributes:* Small foliage, provision of timber, fruit or other useful products, sub-soil feeder so that nutrients are recycled by fallen leaves.

Source: Lock (1888).

controversy concerning the use of shade (Willey, 1975; Beer, 1987). The increasing use of inorganic fertilizers and pesticides during the past fifty years, sometimes associated with partial or total elimination of the shade trees, has stimulated discussion on the advantages and disadvantages of shade trees, and their selection criteria (Willey, 1975; Beer, 1987; Wrigley, 1988; ANACAFE, 1991; Smith et al., 1992).

The major physiological benefits that coffee and cacao receive from shade trees can be placed into two main categories, both associated with reduced plant stress:

1) Amelioration of climatic and site conditions through (i) reduction of air and soil temperature extremes (heat at lower elevations and cold at higher elevations), (ii) reduction of wind speeds, (iii) buffering of humidity and soil moisture availability, and (iv) improvement or maintenance of soil fertility including erosion reduction; and
2) Reduction in the quantity and quality of transmitted light and hence avoidance of over-bearing (e.g., in coffee) and/or excessive vegetative growth (e.g., flushing in cacao). Shade reduces nutritional imbalances and dieback.

The main physiological drawback is competition, especially when excessive shade is used. If high agrochemical applications are feasible, for example when crop prices are high and environmental issues are not considered, maximum attainable yields per hectare are often higher without shade (Willey, 1975). In seasonally dry zones, root competition for water may limit the use of shade trees for coffee (Franco, 1951). Pest and disease problems may increase (Fonseca, 1939; Dakwa, 1980; Smith, 1981; Wrigley, 1988; ICAFE, 1989b) or decrease (Tapley, 1961; Ahenkorah et al., 1974; Nataraj and Subamanian, 1975; Smith, 1981; Eskes, 1982; Campbell, 1984) with increased shade. Yield potential, competition for water, and pest/disease incidence depend on site conditions, and are fundamental issues in the controversy over the use of shade trees in coffee and cacao plantations.

The plant-plant interactions associated with the above effects mostly occur indirectly via effects on the environment, i.e., the so-called 'response (of the plant) and effect (on the environment)' axiom (Goldberg and Werner, 1983). In this way, a shade tree species may increase (+), decrease (–), or have no effect (0) on the vigor and productivity of an associated crop species (Anderson and Sinclair, 1993). Most studies have emphasized the short-term effects of the shade species on the associated crop, for example, effects on crop yield, disease incidence and nutrient availability (Beer, 1987). Few studies, such as that of Ahenkorah et al. (1974), have evaluated the long-term effects of shade on crop vigor and productivity.

The characteristics of the shade stratum – as determined by the botanical composition, number of vertical layers (sub-strata), degree of canopy closure, tree crown characteristics, and shade tree management (pruning, thinning and/or replanting) – vary greatly between major ecological zones and between farms within each zone in response to both biophysical and socioeconomic factors (Somarriba et al., 1996). Highly diverse shade structures are typical of small farms in marginal or suboptimal sites (Jiménez, 1979; Escalante, 1985; Espinoza, 1985; Alcorn, 1989). Simple shade structures, and in the case of coffee, even high-input unshaded plantations, are often established by wealthy farmers on optimal sites or are maintained with external inputs on suboptimal sites, e.g., in Brazil, Colombia and Costa Rica.

Little socioeconomic research on shaded coffee or cacao systems has been published, and labor input, often a limiting factor in agriculture in the humid tropics, has received inadequate attention. An annotated bibliography of pre-1990 publications included only 17 references (mostly in 'grey literature') that presented financial and economic analysis of coffee- or cacao-shade systems (Swinkels and Scherr, 1991). However, this bibliography did not include all of the rich literature on intercropping cacao with coconuts, oil palm and rubber, systems now labelled as agroforestry (e.g., Lee and Kasbi, 1980; Pushparajah and Soon, 1986).

Agronomic consequences of using shade trees in coffee and cacao plantations

Soil organic matter and soil fauna

Soil organic matter (SOM) content may increase with time under agroforestry systems of coffee and cacao. For example, over a 10-year period following conversion of sugar cane fields to cacao plantations, SOM increased by 21 percent under pruned leguminous *Erythrina poeppigiana* and by nine percent under unpruned nonleguminous *Cordia alliodora* (Beer et al., 1990). Nair and Rao (1977) found that the association of cacao with coconuts increased the number of bacteria and fungi in the coconut rhizosphere. They attributed this positive effect to increased litter inputs. Maintenance of high SOM levels by shade trees might help to stabilize coffee nematode (*Meloidogyne* and *Pratylenchus* spp.) populations below critical levels (Araya, 1994). At the same time, the reduction of environmental stress provided by shade increases the tolerance of coffee plants to infestations of these nematodes (OFICAFE, 1978). However, an injudicious choice of the shade species could have the opposite effect. *Inga* spp., for example, have been identified as alternative hosts for coffee nematodes (Zamora and Soto, 1976) but it is not clear whether this would actually increase or decrease the nematode pressure on the coffee.

Biological nitrogen fixation

Management practices will affect N_2 fixation by leguminous shade trees in coffee and cacao plantations. Plantations may be heavily fertilized with N and other elements, or not fertilized at all (Wood and Lass, 1985; Wrigley, 1988).

Populations of leguminous shade trees may range from fewer than 100 trees ha^{-1} to more than 300 trees ha^{-1} (Ling, 1984; Barker, 1991). Shade trees in many plantations are allowed to grow freely while in more intensively managed areas several pollardings per year may be carried out. Pruning residues may be left around the trees, chopped and spread on the ground, or exported for fodder and firewood. All of these practices will affect levels of N_2 fixation and N availability in plantations. The choice of species, and even the choice of clones, will also affect N_2 fixation rates (Nygren and Ramírez, 1995).

Lindblad and Russo (1986) found levels of acetylene reduction in a heavily fertilized coffee plantation shaded with *E. poeppigiana* that were similar to those reported by Roskowski and van Kessel (1985) and Escalante et al. (1984) in unfertilized coffee and cacao plantations shaded with *Inga jinicuil*, *Gliricidia sepium* or *E. poeppigiana*. These latter authors reported N_2 fixation of 35–60 kg ha^{-1} yr^{-1}, but this may be an underestimate since the acetylene reduction method measures only short-term nitrogenase activity (Peoples and Herridge, 1990). However, Fassbender (1987), comparing the nutrient balances of leguminous and nonleguminous coffee-shade tree associations, also estimated that 60 kg N ha^{-1} yr^{-1} were fixed by *E. poeppigiana*. Nygren and Ramírez (1995) found that *E. poeppigiana* nodules disappeared almost completely for ten weeks after pruning, which suggests that there may be 20 weeks in the year during which these biannually pruned trees do not fix N_2 and compete with the associated crop for soil N. Herrera et al. (1987) reported that the nodules of unpruned *E. poeppigiana* shade trees in cacao plantations in Venezuela disappeared during the dry season.

The literature cited above reports relatively low contributions of N, through N_2 fixation, from the shade trees in coffee and cacao plantations. Studies in other agroforestry systems (Rao et al., this volume) have also shown the limited contribution of N_2 fixation by trees to the current associated crops, and suggest that the importance of this characteristic has been over-emphasized. The ability of a shade species to produce large quantities of organic material, as litter and pruning residues, can be more important than N_2 fixation because of the positive effects on soil chemical and physical properties, especially in plantations that are fertilized (Beer, 1988).

Nutrient cycling

Most aspects of nutrient cycling will be directly affected by the choice of shade species since these species differ significantly in aboveground biomass productivity, rate of biomass decomposition and fine root biomass productivity (Palm, 1995). They may also differ in rooting depth though in the humid tropics, where most coffee and cacao plantations are found, the majority of the roots will be found near the soil surface. Cuenca et al. (1983) reported that coffee roots were infected with vesicular arbuscular mycorrhizae and that these mycorrhizal-fine root associations invaded the litter layer, resulting in micro-stratification of coffee and tree roots (Herrera et al., 1987).

A comparison of Ca and Mg in the stems and branches of unpruned non-leguminous *C. alliodora* and pruned leguminous *E. poeppigiana* found greater accumulations of these nutrients in *C. alliodora*. However, transfer of N, P, K, Ca and Mg to the soil was greater from *E. poeppigiana* (Fassbender, 1993). Working in the same plots, Muñoz (1993) found that the combined fine roots in the *E. poeppigiana*-cacao association decomposed more quickly than the leaves of either species. Although total biomass and nutrients were greater in leaf litter, he pointed out that the more rapid turnover of smaller amounts of nutrients in fine roots within the rhizospere may be of considerable ecological significance. A higher total nutrient content was found in the *C. alliodora* associations (Alpizar et al., 1986), which had a greater biomass of fine roots than the *E. poeppigiana* associations (Fassbender, 1993; Muñoz, 1993). Greater leaching losses of Ca and Mg occurred in these *E. poeppigiana* associations (Imbach et al., 1989a; 1989b) possibly because the return of nutrients to the soil surface was concentrated in green (pruned) biomass which decomposed rapidly (Heuveldop et al., 1985). Despite the greater nutrient losses, cycling indices (relation of nutrient turnover to nutrients in the biomass) were higher in the pruned *E. poeppigiana* association (Beer et al., 1990; Nair et al., 1995).

Transfer of N_2 fixed or extracted from the soil by leguminous shade trees to the non-N_2-fixing associated crop has generally been assumed to occur largely through the aboveground pruning residues and litterfall (e.g., Fassbender, 1993). Studies carried out in Latin American coffee and cacao plantations, with 120–560 leguminous shade trees per hectare (0–3 pollardings per year), showed that these inputs can vary from 3–14 Mg ha^{-1} yr^{-1} of dry matter containing 60–340 kg N ha^{-1} yr^{-1} (Beer, 1988). Escalante et al. (1984) calculated that 57–66 kg N ha^{-1} yr^{-1} was released through nodule senescence and decomposition of unpruned *E. poeppigiana* with no difference in nodule N content (22–23 kg N ha^{-1}) between fertilized and unfertilized plots. Nygren and Ramírez (1995) found a turnover of 6.8 to 35.4 g N $tree^{-1}$ in a 23-week pruning cycle (9.6 to 50.0 kg N ha^{-1} yr^{-1}) through *E. poeppigiana* nodule senescence and decomposition. These two studies suggest that a significant proportion of N_2 fixed by shade trees may be transferred below ground to non-N_2-fixing plants.

Babbar and Zak (1994, 1995) found higher rates of N mineralization in Costa Rican coffee plantations shaded by *E. poeppigiana* (148 kg N ha^{-1} yr^{-1}) compared to unshaded plantations (111 kg N ha^{-1} yr^{-1}). Both were heavily fertilized with mineral N at rates up to 300 kg N ha^{-1} yr^{-1}. They concluded that N cycling was more efficient in shaded plantations because, despite the greater availability of mineralized N, less N was lost through leaching.

Shade management, especially pruning, has a critical influence on nutrient cycling and, hence, in addition to its use in managing the microclimate of the underlying crop, provides a tool to manipulate the timing and quantity of nutrient transfer from tree to soil. Although some information exists on the belowground processes of these systems (see above), this stratum is still a 'black box' whose internal biological and chemical mechanisms are poorly quantified and little understood.

Soil erosion

Runoff and soil loss are lower in shaded than in unshaded plantations (Bermudez, 1980; Wiersum, 1984; Leon, 1990). However, similar benefits can be obtained through mulching without shade (Willey, 1975). If shade trees are to contribute to erosion control, natural litterfall and/or pruning residues should maintain a mulch layer during the rainy season. In this respect, slow litter decomposition would be an advantage.

A dense shade canopy will provide better soil protection than an open canopy during high intensity rainfall. Trees, however, can also adversely redistribute precipitation. For example, during low- to moderate-intensity rainfall, coalescence and drip from the leaves of tall timber trees can loosen soil particles and increase soil surface erosion (Wiersum, 1984). Hence, a low crown with small leaves is preferable to reduce drip damage.

Greenhouse gases

In heavily fertilized (including 300 kg N ha^{-1} yr^{-1}) Andisols from shaded and unshaded Costa Rican coffee plantations, laboratory incubation showed higher rates of total denitrification in soil from the shaded plantations (732 vs 455 μg N_2O–N kg^{-1} d^{-1}, respectively) (Babbar and Zak, 1995). Thus, it appears that the use of shade trees in fertilized plantations may contribute to greenhouse gases. It is unlikely, however, that such high denitrification rates, the highest of which was equivalent to approximately 10 kg ha^{-1} day^{-1}, will occur in the field for any prolonged period.

Kursten and Burschel (1993) calculated that 14–52 Mg C ha^{-1} are stored in the aboveground woody biomass of shade trees in coffee and cacao plantations. Compared to annual crops, these agroforestry systems also stored an additional 10–50 Mg C ha^{-1} in the litter layer and soil organic matter. However, the most valuable potential contribution of shaded perennial systems, with respect to atmospheric CO_2 levels, lies in the protection of the remaining forest by offering farmers on the agricultural frontier a sustainable cash-crop alternative to slash-and-burn cultivation for annual crops. This could prevent the release of up to 1000 Mg C ha^{-1}.

Environmental contamination

Besides its undesirable contribution to greenhouse gases and global warming (Duxbury et al., 1993), fertilizer N can pollute groundwaters with nitrite and nitrate. Groundwater NO_3–N concentrations in the Central Valley of Costa Rica, where 50% of the groundwater recharge area is under intensive coffee management with little or no shade, occasionally exceed the 10 mg L^{-1} (Reynolds, 1991) considered to constitute a human health hazard (Frazer et al., 1980). Annual leaching losses at 60 or 100 cm depth were 5 to 9 kg N ha^{-1} in shaded and 24 kg N ha^{-1} in unshaded coffee plantations (Imbach et al., 1989a; Babbar and Zak, 1995), both of which were much lower than the 50 to 100 kg N ha^{-1} yr^{-1} that is typical for tropical agroecosystems (Imbach et al., 1989b). Thus, adding shade trees has the potential to reduce nitrate contamination of groundwater in areas of intensive coffee management. The need to reduce contamination of the environment by pesticides, which are heavily used in high-input coffee and cacao monocultures (Willey, 1975; Graaff, 1986), provides another argument for maintaining shade trees (Boyce et al., 1994).

Light availability

Quantitative data about the level of shading in coffee and cacao plantations are often lacking or are difficult to compare since they have been measured using different light sensors and methodologies (e.g., for cacao, Murray, 1954; Vernon, 1967; Bonaparte, 1975; Luján, 1992). In addition to modifying light availability, tree canopies also negatively affect light quality (Nair, 1979; Bainbridge et al., 1996). The upper limit of acceptable shade for coffee (a C3 plant) is considered to be between 40% and 70% (Kumar and Tieszen, 1980; ICAFE, 1989b; Muschler, 1995). A similar range may apply for cacao (Alvim, 1977). For both crops, nutrient and/or moisture limitations increase the degree of shading needed.

The effects of modifying light availability are determined by the specific requirements of each crop, which vary with age and site conditions, particularly with soil fertility (Willey, 1975). Photosynthetic rates of coffee are at a maximum at intermediate shade levels in many of the climatic conditions found in the tropics (Nutman, 1937a; 1937b). While saturating light levels were as low as 300 μmol m^{-2} s^{-1} for shaded coffee, the level for unshaded plants was as high as 600 μmol m^{-2} s^{-1} (Kumar and Tieszen, 1980). The reduced photosynthetic rate that occurred above these radiation levels was explained as a reaction to leaf temperatures above 25 °C, a condition to which coffee is very sensitive (Maestri and Barros, 1977). However, these measurements were made on individual leaves or isolated plants, which do not represent the conditions that affect most of the coffee foliage in an unshaded, high density plantation, where self-shading is considerable (Willey, 1975).

Temperature, wind speed and humidity

Temperatures of exposed coffee leaves can exceed 40 °C (Maestri and Barros, 1977). In a Mexican coffee plantation under the shade of *Inga jinicuil* (205 trees ha^{-1}; average tree height 14 m), average maximum temperature was reduced by 5.4 °C, average minimum temperature was increased by 1.5 °C, and the vapor pressure deficit was substantially reduced, compared to unshaded plantations (Barradas and Fanjul, 1986). In East African coffee plantations shaded by tall *Paraserianthes falcataria* (syn. *Albizia moluccana*, *A. falcataria*) trees, temperatures were 5–6 °C lower and humidity variations were greater than in unshaded sites (Kirkpatrick, 1935). Reduced heat-load of the coffee plants during the daytime and reduced heat losses at night explain this buffering effect. The inclusion of shade trees also reduces wind speed in the crop strata (Schroeder, 1951; Alvim, 1977). This is of particular importance for young cacao plants which are highly susceptible to desiccation (Leite et al., 1981). Nevertheless, since coffee is planted over a wider climatic range than cacao, the importance of these microclimatic effects may be greater for coffee.

Crop phenology, yields and quality

When nutrient availability is not limiting growth, a positive correlation between cacao yields and light availability has been reported (Murray, 1954; Bonaparte, 1975). Vernon (1967b) concluded that the relationship between cacao yields and available light was approximately linear in the range of 30% to 60% of full sunlight. However, when modelling the complete range from 0 to 100%, a quadratic model gave a better fit than the simple linear model, suggesting that some degree of shading for cacao is desirable. Note that photosynthetically active radiation (PAR) was not measured in any of the studies cited in this paragraph.

Cannell (1975) suggested that close spacing of coffee bushes results in mutual shading that may inhibit floral initiation at existing nodes on coffee branches. Castillo and López (1966) and Jaramillo and Valencia (1980) reported fewer coffee flowers under shade than in full sunlight. Using artificial shade treatments, Montoya et al. (1961) and Castillo and López (1966) found significant increases in the number of nodes per coffee branch and flower buds per node as sunlight levels increased. Montoya et al. (1961) also reported a significant positive correlation between the increases in the number of nodes per branch and yields per bush the following year. Cannell (1975) stated that the most important component of yield is the number of nodes formed. Therefore, it seems logical to conclude that, because the number of nodes formed and the number of fruit set at each node can both be affected by light levels, shading on good sites can directly reduce coffee yields even when all other growth factors are favorable. Using multiple regression analysis, Beer (1992) detected such a direct negative effect of increased shade tree density on coffee berry production, in addition to the indirect effect on yield associated with reduced bush growth.

Lagemann and Heuveldop, (1983) reported that higher shade density had a negative effect on coffee yield. Unshaded, high-input coffee can yield more than shaded plantations with the same management (Fournier, 1988; Chamorro et al., 1994). Nevertheless, in some trials in Costa Rica, production of the varieties Bourbon and Caturra under biannually pollarded *E. poeppigiana* was equal to or even greater than production from unshaded coffee that had been given the same management (ICAFE, 1989a; Ramírez, 1993). Machado (1959), working in Columbia, found yield advantages for *Coffea arabica* var. Típica shaded by *Inga* spp., as compared to unshaded coffee (20% higher yields in fertilized plots and over 60% higher in unfertilized plots). However, shading by *Calliandra* spp. reduced the yield by about 30% compared to the unshaded plots. In another experiment (Machado, 1959), production of the unshaded treatments for var. Bourbon, and also to a much lesser degree for var. Típica, exceeded the shade treatments, especially in fertilized plots, but the advantage of the unshaded plots diminished from year to year.

Thus, some studies have demonstrated that higher yields can be obtained from intensively managed unshaded coffee but the results are inconsistent, probably because of widely varying site conditions and management. When comparing shade versus no-shade, or comparing different shade species in on-farm applied research, a group of factors vary rather than just the factor 'species.' In each treatment, optimal management should be used. This implies different coffee/cacao and shade tree planting densities, pruning, fertilization and other agrochemical applications, in order to judge the potential of each shade species and/or management without shade (Somarriba et al., 1996).

The outcome of adding shade trees to coffee or cacao plantations depends on many factors, the most important of which are related to the following influences: (1) production objective(s), (2) inputs available, and (3) environmental characteristics (Muschler, 1997, 1997). While the factors in groups 1 and 2 can easily be altered, environmental characteristics, such as soil properties, water availability or elevation, permit limited or no changes at all. Hence these latter factors may be more critical in determining the shade response in different plantations. Figure 1 is an attempt to reconcile seemingly contradictory shade responses of coffee by considering them according to environments. This figure is hypothetical, but data from long-term experiments in different environments fit this model well (R. Muschler, 1997, unpublished data).

When coffee production is plotted against elevation, for soils with and without nutrient and/or moisture limitations, the highest production of unshaded coffee typically occurs within the range between 900 and 1200 m.a.s.l. Actual values for each region depend on the site-specific climatic conditions, including the effects of latitude on average temperatures. At lower elevations, unshaded coffee production decreases strongly in response to high temperatures, while at higher elevations it decreases in response to low temperatures, and possibly wind damage. Under these sub-optimal conditions, trees can buffer the microclimatic extremes and can increase coffee production over that of unshaded sites ('shade contribution'). However, shading of coffee in the optimal elevation range results in a yield depression. On soils with nutrient or water storage limitations at all elevations, the potential yield advantage of unshaded versus shaded coffee is reduced or even reversed; i.e., the relative proportion of the shade contribution increases on sites with these soil limitations (Figure 1B).

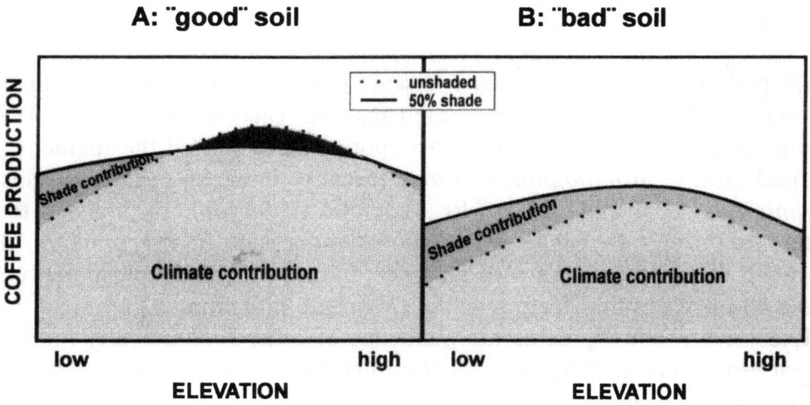

Fig. 1. Idealized hypothetical coffee production without shade and under trees giving 50% shade as a function of elevation for soils without (A) or with (B) limitations of rooting depth, nutrients or moisture. While production of unshaded coffee for given soil conditions and elevation depends primarily on climatic conditions ('Climatic contribution'), shade may improve production ('Shade contribution') in marginal conditions. At ideal coffee elevations and on soils without limitations, shading may reduce production (dark area) (modified from Muschler, 1997).

Despite the common belief that coffee produced under shade is of higher quality, Willey (1975) points out that although light attenuation may cause etiolation and changes in leaf morphology, it has little effect on the morphology and hence the chemical constituents of fruits. He therefore concludes that shade will affect the quality of a vegetative product, such as tea, but that quality effects on cocoa or coffee beans are unlikely to be significant. Hernández (1995) reported insignificant differences between cherry or bean sizes from shade and no-shade coffee plots. However, during two consecutive years, the conversion factor for dried beans per unit weight of green cherries was 0.6% higher under shade, which translated into an additional 44 kg ha^{-1} of processed coffee at average Costa Rican production levels. Guyot et al. (1996) demonstrated that shade and altitude have similar significant positive effects on coffee bean size and the chemical parameters which determine quality because they slow ripening.

The effects of shade on coffee and cocoa yields are difficult to characterize since the premise of 'proper shade management and design' is seldom assured. Furthermore, the relative yield advantage of unshaded coffee or cacao may be limited: (1) to 'ideal' soil and climatic conditions (Figure 1); (2) to one or two decades of production, after which environmental degradation, especially via soil erosion and pesticide residues, may seriously reduce productivity and/or environmental quality (Boyce et al., 1994); and (3) to frequently replanted and pruned plantations since unshaded coffee or cacao bushes have a shorter life expectancy than shaded bushes (Ahenkorah et al., 1974; Wrigley, 1988).

Pests and diseases

Black pod (*Phythphora palmivora*) disease of cacao is reported to be favored by increased humidity due to increased shading (Dakwa, 1980; Smith, 1981). Comparing microclimatic conditions, spore liberation and the incidence of the pod disease *Moniliophthora roreri* in cacao under the shade species *E. poeppigiana*, *G. sepium* or *I. edulis*, Meléndez (1993) did not find significant differences except for higher spore counts underneath *E. poeppigiana*. This, however, did not correlate with a higher disease incidence, presumably due to the absence of other factors which favor pod infection.

Excessive shade increases the incidence of other economically important fungal diseases (e.g., *Mycena citricolor* in coffee) (ICAFE, 1989b), especially in very moist situations such as river sides or valley bottoms. In contrast, damage caused by *Cercospora coffeicola*, a fungus disease which can completely defoliate coffee plants, is greater in unshaded plantations (Nataraj and Subramanian, 1975), possibly due to the higher susceptibility of water-stressed or nutrient-deficient plants (Wrigley, 1988). The provision of more or less shade to help control these diseases must seek a balance since they occur together in many plantations. While these generalizations are valid for *M. citricolor* and *C. coffeicola* on most if not all sites, the correlation between the level of shade and the incidence of *Hemileia vastatrix* seems much weaker (Eskes, 1982). This coffee leaf rust causes serious defoliation both in unshaded and shaded conditions. Shade trees can also act as alternative hosts, and hence as sources of inoculum, for crop diseases. For example, *Inga* spp. and many fruit trees – such as oranges and mangos – may be infected by *M. citricolor* (Wellman, 1961).

The coffee berry borer (*Hypothenemus hampei*) is reported to be favored by dense shade (Fonseca, 1939; Wrigley, 1988) but it has been suggested (B. Medina, 1997, pers. comm.) that some shade favors the persistence of the biological control agents *Beauveria bassiana*, an entompathogenic fungus, and *Cephalonomia stephanoderis*, a parasitic wasp. On the other hand, mealybug (Homoptera) infestations (Campbell, 1984) and Anthracnosis fungal disease (*Colletotrichum gloeosporioides*) of unshaded cacao (Porras and Sánchez, 1991) are two of the reasons why cacao monocultures are not economically justified despite the initial production advantage (Ahenkorah et al., 1974).

Weeds

The proper selection and management of permanent shade species can reduce labor input and weeding costs considerably (Silva et al., 1990). These can amount to 70% of all costs during the first two to three years of a cacao plantation (Corven, 1993). Shade can shift species composition towards less aggressive weeds (e.g., more broadleaf, less Gramineae) (Vernon, 1967a; Huxley, 1975), a change of considerable economic importance even if total weed biomass were unaffected. After removing shade trees and thinning coffee plants, weed biomass increased almost tenfold to > 12 Mg ha^{-1} (dry weight), with a higher incidence of the more light-demanding aggressive graminaceous weeds (Goldberg and Kigel, 1986). Weed growth in a coffee plantation was virtually eliminated and grassy weeds completely suppressed under a ≥ 40% homogeneous shade treatment (Muschler, 1997). The savings that resulted from not having to weed (2.5%–3% of total production costs in mature plantations) were double the costs of managing shade trees that are pollarded twice per year (Rojas Cubero, 1996). This is one of the reasons why Costa Rican coffee farmers plant more leguminous shade trees when coffee prices are low.

Buffer zone agroforestry and conservation of biodiversity

In terms of their architecture and ecology, many traditional shaded coffee and cacao plantations (which often have a diverse shade component that includes > 50 tree species (Espinoza, 1985)) resemble natural forest more than most other agricultural systems. Shaded coffee plantations in Mexico compare quite favorably to natural forest as refuges for migratory birds (Wille, 1994; Greenberg et al., in press). These agroforestry systems also have a high potential as refuges for biodiversity, particularly in deforested areas (Young, 1994; Perfecto et al., 1996), and for buffer zone management. Moreover, they can serve as pathways or stopovers for the migration of animal species between natural reserves. When native species are used as shade trees in a buffer zone, a larger gene pool of these species can be maintained than would be possible in the protected area alone.

Productivity of shade trees

Timber-producing shade trees have low management costs (Calvo and Platen, 1996) and are considered a 'saving account' that can be realized at times of low prices or failure of the underlying crop (Rodríguez, 1982; Mussak and Laarman, 1989; Barker, 1991; Somarriba, 1992). Income from fruit trees, timber or firewood, and other perennial crops used as shade for coffee or cacao is significant (Sabogal, 1983; Kajomulo-Tibaijuka, 1985; O'kting'ati and Mongi, 1986; Escalante et al., 1987) and may result in better financial performance than would occur in plantations using conventional, leguminous, 'service' shade trees (Glover, 1981; Platen, 1993; Hernández and Platen, 1995; Trejos and Platen, 1995).

Timber production

A conservative estimate of the merchantable timber increment from 100 trees ha^{-1} of *C. alliodora* in a coffee plantation is 4–6 m^3 ha^{-1} yr^{-1} (Table 2). Assuming high coffee prices and an average coffee production of 30 'fanegas'

Table 2. Timber production from Costa Rican and Colombian coffee plantations.

Species	Place	Elevation (m.a.s.l.)	Rainfall (mm/yr)	Tree density (#/ha)	MAI Vtot[a] (m^3 ha^{-1} yr^{-1})	MAI Vcom[b] (m^3 ha^{-1} yr^{-1})	Reference
Cordia alliodora	CATIE	550	2600	228	13		Rosero and Gewald (1979)
Cordia alliodora	Costa Rica	50–750	> 2600	68–290		6–13	Somarriba and Beer (1987)
Cordia alliodora	Colombia	1200–1800	–	170		4	Escobar (1979)
Cordia alliodora	CATIE	600	2600	139	13		Heuveldop et al. (1985)
Cordia alliodora	La Suiza, CR	600	2600	117–275	7–12		Beer et al. (1981)
Cordia alliodora	CATIE	600	2600	110–350	10–15		Hernández (1995)
Cordia alliodora	Turrialba	600–800	2600	50–150		1.5–2.5	Barker (1991)
Cordia alliodora	Colombia	1300	2475	100		4–5	Chamorro et al. (1994)
Cordia alliodora	Turrialba	600	2600	120–410		4–20	González (1980)
Cordia alliodora	Turrialba	600	2600	120–290		6–15	Somarriba (1990)
Eucalyptus grandis	Turrialba	500–1400	2600	1000	2–26		Salas (1994)
Eucalyptus grandis	Turrialba	500–1400	2600	950	1.5–32.5		Salas (1994)
Cedrela odorata	Puriscal, CR	800	2500	78+141		7–12	Ford (1979)
Cedrela odorata	Sn Carlos, CR	250	4100	84		8.3	Ford (1979)
Cedrela odorata	Puriscal, CR	800	2500	32–87		3–4.5	Sabogal (1983)
Cedrela odorata	Puriscal, CR	1000–1100	2200	150–290		0.9–3.1	Espinoza (1985)

[a] Mean annual increment (MAI) of total stem volume.
[b] Mean annual increment of commercial wood volume. For *C. alliodora* it is 64% of Vtot (Somarriba and Beer, 1987).
CR = Costa Rica.
#/ha = Number of trees per hectare.

ha^{-1} yr^{-1} (see Table 3 for explanation), the value of the timber increment could compensate for a reduction of coffee yield of up to 17%. For intermediate prices, timber value could compensate for yield reductions of up to 33%, while at low coffee prices – which generate a profit of less than US$5 per fanega – the value of the timber can compensate for the loss of all the coffee production. Of course, this crude comparison does not take into account discounting, tree management costs (including felling damage (Somarriba, 1992)) and other factors that affect the financial comparison of coffee production with and without the timber trees (Hernández, 1995). Moreover, the coffee revenue is generated immediately while revenue from timber, though substantial, is available only at the end of the rotation.

The above contrasts are based on conservative estimates of the relative values of coffee and timber since timber prices are likely to increase in the future (Gómez, 1995) while prolonged coffee price increases are unlikely. The recent interest of Central American coffee institutions in sustainable management (PROMECAFE, 1995), including the use of timber-producing shade trees, is partly due to these trends. Diversification and new markets, such as specialty coffees and organic coffee, may be the only options for farmers to increase their income from perennial crop plantations. The above studies of timber production in coffee plantations were mostly carried out at lower altitudes and in areas with relatively high rainfall (Table 2). These are the zones where this practice has the most potential since the duration of timber rotations and competition for water between these unpruned trees and underlying crops are minimized.

Damage of coffee/cacao during the harvest of intermixed timber trees is a major concern of many farmers. This is particularly true in the case of cacao, which forms a more permanent structure than coffee and is, therefore, slower to recover from physical damage. Felling of 29 mature *C. alliodora* in coffee plantations in the tropical humid zone of Turrialba, Costa Rica affected 498

Table 3. Comparison of the values[a] of coffee production and potential annual increment of *Cordia alliodora* (100 trees ha^{-1}) in a Costa Rican coffee plantation.

Price paid for fresh coffee before pulping (US-$/fan) (1)		Production cost per fanega[b] (US-$) (2)	Profit from sale of one fanega of coffee (US-$) [(1) – (2)] = (3)	Value of annual timber increment (US-$) (4)	Number of fanegas of coffee equivalent to timber value [(4) – (3)] = (5)
High	75	45[c]	30	150	5
Intermediate	60	45	15	150	10
Low	< 50	45	< 5	150	> 30

[a] Prices for December 1995 converted to US-$ @ 200 Colones/US-$. A mean annual increment of 4 m^3 ha^{-1} yr^{-1} of merchantable timber of *Cordia alliodora* is assumed (Table 2).
[b] 1 fanega = 256 kg of fresh coffee berries.
[c] 9000 Colones is a low estimate, considering that Rojas (1996) gives a cost of 15,500 Colones for producing one fanega.

coffee plants but most (91%) were only slightly damaged, principally by the tree crowns (Somarriba, 1992). Thus, tree crown dimensions and characteristics are important criteria when considering possible timber harvesting damage in perennial crop plantations. Log skidding was a minor source of damage. Damage to crops, and the consequences of this damage, can be reduced by felling during periods of low coffee prices or yields, felling immediately after the harvest but prior to annual coffee pruning when some of the damage can be repaired, and by planting the trees between the rows of coffee, along roads and on plantation borders. Despite possible crop damage, the low restoration costs of affected plants, the small reductions in coffee yields, and the large financial gain from timber sales strongly support the use of timber trees in coffee plantations.

Fruits and fuelwood production

In a comparison of unshaded coffee with mixed plantations, which included oranges and the timber tree *Cedrela odorata*, Espinoza and Hidalgo (1991) found that all economic indicators favored the diversified systems. However, market limitations greatly restrict the commercial potential of fruit trees in many regions. Bananas (*Musa* spp.), which are often used as temporary shade and are sometimes a permanent component of plantations, compete strongly with coffee (Robinson, 1961; Mitchell, 1963). Furthermore, during harvest or because of windfall, banana stems may damage underlying crops and newly established permanent shade trees. The priorities of the farmers and availability of markets need to be verified before promoting the use of fruit trees as shade for coffee or cacao.

Shade trees in coffee plantations can produce large quantities of high-quality fuelwood: e.g., 36.5, 8.3 and 8.8 m^3 ha^{-1} yr^{-1} from 635 trees ha^{-1} of *Mimosa scabrella* (Picado, 1985), 1250 trees ha^{-1} of *Inga densiflora* (Salazar, 1985) and 330 trees ha^{-1} of *Gliricidia sepium* (Salazar, 1984), respectively. The fuelwood obtained from coffee pruning and renewal is also an important resource for many rural families.

Future research priorities

So far, ecological research on shade trees in coffee and cacao plantations has emphasized: (1) effects of shade-soil fertility interactions on coffee/cacao production; (2) shade-microclimate-pest/pathogen interactions; and (3) nutrient cycling, soil fertility maintenance and sustainability issues. Applied research has focused principally on species selection, propagation methods, determination of optimal population densities, planting arrangements and pruning practices. The aim has been to intensify coffee/cacao management, in comparison with traditional shaded systems, for higher yields and economic returns. A wealth of site- and/or species-specific data exists on crop productivity under shade and on many of the underlying physiological changes and processes in both crop and tree, as well as the effects of shade trees on microclimate and soils. It is therefore valid to ask what further biophysical research on this system should be a priority. Before this question can be answered, it is necessary to examine the justifications for further research and to identify the people it should benefit.

A number of factors, both economic and environmental, support the claim that improved management of shade trees in perennial crop systems is still a priority topic. There is increasing demand and willingness of consumers to pay premium prices for export crops that are sustainably and/or organically produced, controlled in some cases by certification requirements (Boyce et al., 1994; Revista Forestal Centro-Americano, 1996). Carbon sequestration by tropical forests and agroforestry systems (Kursten and Burschel, 1993) is another option which can be 'sold' to developed countries (Dixon, 1995). These international influences, together with internal pressures in the producing countries to protect watersheds and biodiversity, reduce environmental contamination and diversify insecure monocropping economies, have changed the high-input 'green revolution' approach of the national coffee institutes of Central America (PROMECAFE, 1995). Nevertheless, the rate of adoption of diversified sustainable technologies is slowed by the conspicuous absence of clear guidelines on how to select and manage appropriate shade trees for different sites in order to augment the advantages of tree-crop associations, while minimizing the disadvantages. Shade tree selection criteria (Beer, 1987) and management recommendations (e.g., desirable light levels) are often based on empirical observation rather than on objective quantification with control of confounding factors (Sanchez, 1995). Future research should integrate and supplement the existing results in order to provide a comprehensive understanding of the processes occurring within and between the components in shaded perennial crop plantations, thereby permitting the extrapolation of the site- and/or species-specific results. Only then can generalized recommendations (e.g., using decision support tools such as expert systems) be provided as a framework for validation in the varying biophysical and socioeconomic circumstances where coffee and cacao are produced.

The following sections cover, in order of decreasing priority, three ecological research areas of particular concern to the national organizations. We do not attempt to specifically discuss the socioeconomic priorities which are equally important to farmers. Related applied research topics are then discussed in the final section.

Competition between trees and crops

Basic studies are still required on the relative importance of competition for light, water and nutrients along ecological gradients, e.g., to identify the limiting factor(s) for a range of biophysical sites. Studies are needed on the effects of tree crown architecture and dynamics (phenological and/or pruning cycle), for different tree species and sizes, on the availability of photosynthetically active radiation for the understory crop (Somarriba et al., 1996). The effects of shade on crop phenology, particularly the initiation and development of flowers and fruits (Beer, 1992; Young, 1994), on coffee bean size, on the conversion factor bean/cherry and on coffee quality (taste studies) are related topics that need additional study since they have economically important implications (Hernández, 1995; Muschler, 1995).

Greater emphasis should be given to studies of belowground interactions, in part because there is a dearth of relevant information (Anderson and Sinclair, 1993). The influence of shade trees on water availability for the understory crop in seasonally dry zones is as important a research topic now as when Franco (1951) identified it as the main justification for eliminating shade trees from coffee plantations in many areas of Brazil. Competition for water can occur, even in tropical rain forest areas, during short dry periods (Kozlowski, 1982).

Information is available on seasonal nutrient uptake by coffee and cacao (Rodríguez et al., 1963; Carvajal et al., 1969) but not for the shade trees. Although considerable information is available on annual nutrient cycling totals and nutrient export in timber, fuelwood, and crop products (Fassbender, 1993), seasonal nutrient cycling patterns and hence competition are little understood. Decomposition patterns of fine roots and litter, and nutrient release, will affect the relative importance for crops of maintaining/increasing SOM (a medium- to long-term nutrient reserve) versus providing a nutrient flush (a short-term nutrient source) (Haggar et al., 1993). The effect of site conditions on the functioning of shade tree root associations with mycorrhiza or rhizobium (e.g., on N_2 fixation) is another research area that would facilitate fine tuning of management and the extrapolation of site-specific results. Little is known about the pathways and efficiency of nutrient uptake from organic residues produced in these systems, e.g., the role of different SOM fractions and micro-flora/fauna. In part, this is due to the difficulty in using tracers such as ^{15}N to study systems including large trees (Clinton and Mead, 1994). The interactions between inorganic and organic fertilizers should be studied to improve the efficiency of use of the former. Research has tended to concentrate on N when the supply of other nutrients may be more important: e.g., P in the volcanic P-fixing soils of Central America, and cations in the acid soils of many humid tropical areas. Studies of fine root dynamics could permit matching fertilization timetables to periods of fine root growth of crops when the trees are less competitive (Muñoz, 1993). Fertilizer distribution and potential root barrier species, such as some Gramineae, should be tested as a means to manage the spatial distribution of coffee and tree roots (Schroth, 1995), at least during the juvenile phase.

Coffee pests and diseases

There is a lack of agreement among farmers and scientists as to whether shade trees reduce or increase diseases and pests of economic importance, such as leaf rust (*H. vastatrix*) and the coffee berry borer (*Hypothenemus hampei*). Further research is needed to understand how shade trees modify the microclimate to the detriment or benefit of these diseases and pests, and hence on their epidemiology. For example, many coffee farmers believe that the higher incidence of *M. citricolor* under shade is related to rain drip from upperstory tree leaves, particularly with timber species whose height can not be regulated by pruning; this is a topic that requires testing. The trade-off between competition with the crop and the value of shade to control weed biomass and to favor less competitive weed species merits further study. The effect of different shade management options (including no-shade) on soil pests of coffee and cacao, such as nematodes, is another area of research with immediate economically important applications. In general, the role of shade trees in integrated pest management for coffee is an increasingly important research area.

Sustainability and environmental benefits

Many coffee farmers and their organizations have demonstrated that they are concerned about medium- to long-term perspectives by implementing erosion control recommendations and ignoring advice to eliminate shade trees. However, they generally lack solid information about the factors which control sustainability and environmental effects, aspects which they are increasingly pressured to consider and which can offer them marketing advantages (e.g., certification). Subjects that require field study in commercial and experimental plantations include the medium- to long-term trends in total and available macronutrients, SOM (total and fractions), soil structure, the hydrological regime (infiltration vs run-off, internal drainage, water-holding capacity), leaching, ground water quality, and soil erosion (surface erosion and superficial soil slippage). Further work is needed to quantify the ability of these agroforestry systems to capture CO_2 (e.g., by increasing SOM), or to reduce the emissions by providing a sustainable agricultural alternative to slash and burn (probably more relevant in the case of cacao than coffee).

Applied research

Applied research, drawing on the process-oriented research discussed above, should emphasize silvicultural interventions to reduce competition, reduce crop pests and diseases and address environmental problems. Topics include regional selection trials and criteria for shade species/varieties/provenances based on root architecture as well as aboveground structural characteristics; the effects of tree establishment methods on rooting patterns (Schlönvoigt, 1993); and spacing, thinning and pruning regimes (Somarriba et al., 1996). Studies of the potential of tree mulch, produced on- or off-site, to substitute for inorganic fertilizer will have increasing importance due to the rapid growth in demand for "organic" products. Despite the above emphasis on crop production, the need for research to improve the productivity of the tree component should not be forgotten, e.g., adaptation of plantation silviculture to improve management of widely spaced timber trees in perennial crop plantations. Many of the principles that have been illustrated above could be applied to improve combinations of upperstory trees with other perennial crops, such as those mentioned in the introduction.

Acknowledgments

The authors wish to acknowledge the financial assistance of CATIE, GTZ (Deutsche Gesellschaft für Technische Zusammenarbeit, GmbH), and Danida (Danish International Assistance) who have supported the agroforestry research group in CATIE which is focused on shade perennial crop systems. We also wish to thank M. R. Rao, E. Torquebiau, P. Strømgaard and M. Kanninen for their comments on this review.

References

Ahenkorah Y, Akrofi GS and Adri AK (1974) The end of the first cacao shade and manurial experiment at the Cacao Research Institute of Ghana. J Hort Sci 49: 43–51

Alcorn JB (1989) An economic analysis of Huastec Mayan forest management. In: Browder JO (ed) Fragile Lands of Latin America: Strategies for Sustainable Development, pp 182–203. Westview Press Boulder, Colorado

Alpizar L, Fassbender HW, Heuveldop J, Folster H and Enriquez G (1986) Modelling agroforestry systems of cacao (*Theobroma cacao*) with laurel (*Cordia alliodora*) and poro (*Erythrina poeppigiana*) in Costa Rica. I. Inventory of organic matter and nutrients. Agrofor Syst 4: 3–15

Alvim P de T (1977) Cacao. In: Alvim P de T and Kozlowski TT (eds) Ecophysiology of Tropical Crops, pp 279–313 Academic Press, New York

ANACAFE (1991) Manual de Caficultura. Asociación Nacional del Café, Guatemala

Anderson LS and Sinclair FL (1993) Ecological interactions in agroforestry systems. Agrofor Abstr 6: 57–89

Araya M (1994) Distribucion y niveles poblacionales de Meloidogyne spp. Y Pratylenchus spp. En ocho cantones productores de café en Costa Rica. Agronomia Costarricense 18: 183–187

Babbar LI and Zak DR (1994) Nitrogen cycling in coffee agroecosystems: net nitrogen mineralization and nitrification in the presence and absence of shade trees. Agric Ecosystems Environ 48: 107–113

Babbar LI and Zak DR (1995) Nitrogen loss from coffee agroecosystems in Costa Rica. Leaching and denitrification in the presence and absence of shade trees. J Envir Qual 24: 227–233

Bainbridge R, Evans GC and Rackham O (eds) (1996). Light as an Ecological Factor. Blackwell, Oxford, England

Barker DJ (1991) An economic analysis of farming coffee and trees at Turrialba, Costa Rica: comparing small farms with poro (*Erythrina poeppigiana*) only to those with both laurel (*Cordia alliodora*) and poro. MSc Thesis, CATIE, Turrialba, Costa Rica

Barradas VL and Fanjul L (1986) Microclimatic characterization of shaded and open-grown coffee (*Coffea arabica* L.) plantations in Mexico. Agric For Meteor 38: 101–112

Beer JW (1987) Advantages, disadvantages and desirable characteristics of shade trees for coffee, cacao and tea. Agrofor Syst 5: 3–13

Beer JW (1988) Litter production and nutrient cycling in coffee (*Coffea arabica*) or cacao (*Theobroma cacao*) plantations with shade trees. Agrofor Syst 7: 103–114

Beer JW (1992) Production and competitive effects of the shade trees *Cordia alliodora* and *Erythrina poeppigiana* in an agroforestry system with *Coffea arabica*. Ph.D. thesis, University of Oxford, Oxford, England

Beer JW, Bonneman A, Chavez W, Fassbender HW, Imbach AC and Martel I (1990) Modeling agroforestry systems of cacao with *Cordia alliodora* and *Erythrina poeppigiana* in Costa Rica. V. Productivity indices, organic matter models and sustainability over ten years. Agrofor Syst 12: 229–249

Beer JW, Clarkin KL, De las Salas G and Glover NL (1981) A case study of traditional agroforestry practices in a wet tropical zone: the 'La Suiza' project. In: Chavarria M (ed) Las Ciencias Forestales y su Contribución al Desarrollo de la America Tropical, pp 191–209. CONICIT-INTERCIENCIA-SCITEC, San Jose, Costa Rica

Bermudez MM (1980) Erosión hídrica y escorrentía superficial en el sistema de café (*Coffea arabica* L.) poro (*Erythrina poeppigiana* (Walper) O.F. Cook) en Turrialba, Costa Rica. M.Sc. Thesis. CATIE-UCR, Turrialba, Costa Rica

Bonaparte EE (1975) Yield gradients in cocoa (*Theobroma cacao* L.) shade and fertilizer experiments. Acta Horticulturae 49: 251–257

Boyce JK, Fernández Gonzalez A, Fürst E and Segura Bonilla O (1994) Café y Desarrollo Sostenible: Del Cultivo Agroquímico a la Producción Orgánica en Costa Rica. EFUNA, Heredia, Costa Rica

Calvo G and von Platen H (1996) Cacao-laurel-plátano. Costos y beneficios financieros. Serie Técnica Informe Técnico No. 264. CATIE, Turrialba, Costa Rica

Campbell CAM (1984) The influence of overhead shade and fertilizers on the Homoptera of mature Upper-Amazon cocoa trees in Ghana. Bull Entomol Res 74: 163–174

Cannell MGR (1975) Crop physiological aspects of coffee bean yield: a review. J Coffee Res 5: 7–20

Carvajal JF, Acevedo A and Lopez CA (1969) Nutrient uptake by the coffee tree during a yearly cycle. Turrialba 19: 13–20

Castillo ZJ and López AR (1966) Nota sobre el efecto de la intensidad de la luz en la floración del cafeto. Cenicafé 17: 51–60

Chamorro G, Gallo A and Lopez R (1994) Evaluación económica del sistema agroforestal café asociado con nogal. Cenicafé 45(4): 164–170

Clinton PW and Mead DJ (1994) Competition for nitrogen between *Pinus radiata* and pasture. I. Recovery of ^{15}N after one growing season. Can J For Res 24: 882–888

Cook OF (1901) Shade in Coffee Culture. Department of Agriculture Bulletin No. 25, Washington, DC, USA

Corven JM (1993) Asociación de cultivos en cacao: aspectos económicos. In: Phillips-Mora W (ed) Sombras y Cultivos Asociados con Cacao, pp 11–18. Serie Técnica Informe Técnico No. 206. CATIE, Turrialba, Costa Rica

Cuenca G, Aranguren J and Herrera R (1983) Root growth and litter decomposition in a coffee plantation under shade trees. Plant and Soil 71: 477–486

Dakwa JT (1980) The effects of shade and NPK fertilizers on the incidence of cocoa black pod disease in Ghana. Horticultural Abstracts No. 2164. 50(3): 187 (Ghana Agr Sc 9(3): 179–184, 1976)

Dixon RK (1995) Agroforestry systems: sources or sinks of greenhouse gases. Agrofor Syst 31: 99–116

Duarte J (1992) A cultura do cupuacuzeiro em Rondonia. EMBRAPA-CPAF, Porto Velho, Rondonia, Brazil

Duxbury J, Harper LA and Mosier AR (1993) Contributions of agroecosystems to global climate change. In: Harper LA, Mosier AR, Duxbury JM and Rolston (eds) Agricultural Ecosystem Effects on Trace Gases and Global Climate Change, pp 1–18. Amer Soc Agron Spec. Pub. No. 55. Madison, USA

Escalante EE (1985) Promising agroforestry systems in Venezuela. Agrofor Syst 3: 209–221

Escalante EE, Aguilar A and Lugo R (1987) Identificación, evaluación y distribución espacial de especies utilizadas como sombra en sistemas tradicionales de café (*Coffea arabica*) en dos zonas del Estado de Trujillo, Venezuela. Venezuela Forestal 3(11): 50–62

Escalante G, Herrera R and Aranguren J (1984) Nitrogen fixation in shade trees (*Erythrina poeppigiana*) in cacao plantations in North Venezuela. Pes Agropec Bras 19 (s/n): 223–230

Escobar ML (1979) El crecimiento y el rendimiento del guácimo nogal *Cordia alliodora* (Ruiz y Pavon) Cham. asociado con café en el Suroeste de Antioquía. Tesis Ingeniero Forestal. Universidad Nacional, Medellín, Colombia

Eskes AB (1982) The effect of light intensity on incomplete resistance of coffee to *Hemileia vastatrix*. Neth J Plant Path 88: 191–202

Espinoza LP (1985) Untersuchungen über die Bedeutung der Baumkomponente bei agroforstwirtschaftlichem affeeanbau an Beispielen aus Costa Rica. Ph.D.Thesis. Göttinger Beitr ge zur Land- und Forstwirtschaft in den Tropen und Subtropen. Heft 10

Espinoza LP and Hidalgo T (1991) Cafetal arbolado. Proyecto Desarrollo Agrícola Forestal (PRODAF), Puriscal, Costa Rica

Evans PT and Rombold JS (1984) Paraíso (*Melia azedarach* var. gigante) woodlots: an agroforestry alternative for the small farmer in Paraguay. Agrofor Syst 2: 199–214.

FAO (1996). Production Yearbook, Vol. 49, pp. 171–173. FAO, Rome

Fassbender HW (1987) Nutrient cycling in agroforestry systems of coffee (*Coffea arabica*) with shade trees in the Central Experiment of CATIE. In: Beer J, Fassbender H and Heuveldop J (eds) Advances in Agroforestry Research, pp 155–165. CATIE, Turrialba, Costa Rica

Fassbender HW (1993) Modelos Edafológicos de Sistemas Agroforestales. 2nd Ed. CATIE, Turrialba, Costa Rica

FNC (1932) Manual del Cafetero Colombiano. Federación Nacional de Cafeteros, Bogotá, Colombia

Fonseca JP da (1939) A 'broca' e o sombreamento dos cafezais. Biológico 5: 133–136

Ford L (1979) An estimate of the yield of *Cedrela odorata* L. (syn. *C. mexicana* Roem.) grown in association with coffee. In: De las Salas G (ed) Workshop Agro-forestry Systems in Latin America, pp 177–183. CATIE, Turrialba, Costa Rica

Fournier LA (1988) El cultivo del cafeto (*Coffea arabica* L.) al sol o a la sombra: un enfoque agronómico y ecofisiológico. Agronomía Costarricense 12(1): 131–146

Franco CM (1951) A agua do solo e o sombreamento dos cafezais na America Central. Bragantia 11: 99–119

Frazer P, Chilvers C, Beral B and Hill MJ (1980) Nitrate and human cancer: a review of the evidence. Int J Epidemiology 9: 3–11

Glover N (1981) Coffee yields in a plantation of *Coffea arabica* var. Caturra shaded by *Erythrina poeppigiana* with and without *Cordia alliodora*. Informe Técnico No. 17. CATIE, Turrialba, Costa Rica

Goldberg AD and Kigel J (1986) Dynamics of the weed community in coffee plantations grown under shade trees: effect of clearing. Isr J Bot 35: 121–131

Goldberg DE and Werner PA (1983) Equivalence of competitors in plant communities: a null hypothesis and a field experiment approach. Amer J Bot 70: 1098–1104

Gómez M (1995) Precios de productos forestales en América Central. Revista Forestal Centroamericana 12: 25–30

González LE (1980) Efecto de la asociación de laurel (*Cordia alliodora* (Ruiz y Pav.) Oken) sobre producción de café (*Coffea arabica*) con y sin sombra de poró (*Erythrina poeppigiana* (Walpers) O.F. Cook). M.Sc. Thesis. CATIE, Turrialba, Costa Rica

Graaff J de (1986) The Economics of Coffee. Pudoc, Wageningen, The Netherlands

Greenberg R, Bichier P and Sterling J (In press). Bird populations and planted shade coffee plantations of Eastern Chiapas. Biotrópica

Guyot B, Gueule D, Manez JC, Perriot JJ, Giron J and Villain L (1996) Influence de l'altitude et de l'ombrage sur la qualite des cafes Arabica. Plantations, recherche, developpement (Juillet-Aout): 272–283

Haggar JP, Beer JW, Tanner EVJ and Rippin M (1993) Nitrogen dynamics of tropical agroforestry and annual cropping systems. Soil Biol Biochem 25: 1363–1378

Hernández I and von Platen HH (1995) Maderables como alternativa para la substitución de sombras en cacaotales establecidos: la economía. Serie Técnica Informe Técnico No. 259. CATIE, Turrialba, Costa Rica

Hernández OR (1995) Rendimiento y análisis financiero del sistema agroforestal café (*Coffea arabica* cv caturra) con poró (*Erythrina poeppigiana*) bajo diferentes densidades de laurel (*Cordia alliodora*). MSc Thesis, CATIE, Turrialba, Costa Rica

Herrera R, Accardi A, Aranguren J, Navidad E, Escalante G, Toro M and Cuenca G (1987) Coffee and cacao plantations under shade trees in Venezuela In: Beer J, Fassbender HW Heuveldop J (eds) Advances in Agroforestry Research, pp 173–181. CATIE, Turrialba, Costa Rica

Heuveldop J, Alpizar L, Fassbender HW, Enriquez G and Folster H (1985) Sistemas agroforestales de café (*Coffea arabica*) con laurel (*Cordia alliodora*) y café con poró (*Erythrina poeppigiana*) en Turrialba, Costa Rica. II. Producción agrícola, maderable y de residuos vegetales. Turrialba 35: 347–355

Huxley PA (1970) Some aspects of the physiology of *Arabica coffee*: the central problem and the need for synthesis. In: Luckwill LC and Cutting CV (eds) Physiology of Tree Crops, pp 255–268. Academic Press, London, UK

ICAFE (1989a) Informe Anual de Labores. Programa Cooperativo Instituto del Café de Costa Rica (ICAFE) – Ministerio de Agricultura y Ganaderia (MAG). San José, Costa Rica

ICAFE (1989b) Manual de Recomendaciones para el Cultivo del Café, 6th Ed. Programa Cooperativo Instituto del Café de Costa Rica (ICAFE) – Ministerio de Agricultura y Ganaderia (MAG). San José, Costa Rica

Imbach AC, Fassbender HW, Beer J, Borel R and Bonneman A (1989a) Sistemas agroforestales de café (*Coffea arabica*) con laurel (*Cordia alliodora*) y café con poro (*Erythrina poeppigiana*) en Turrialba, Costa Rica. IV. Balances hídricos y lixiviación de elementos nutritivos. Turrialba 39: 400–414

Imbach AC, Fassbender HW, Borel R, Beer J and Bonnemann A (1989b) Modeling agroforestry systems of cacao (*Theobroma cacao*) with laurel (*Cordia alliodora*) and *Erythrina poeppigiana* in Costa Rica. IV. Water balances, nutrient inputs, and leaching. Agrofor Syst 8: 267–287

INC (1942) Algunas Observaciones Prácticas acerca del Cultivo y Beneficio del Café. Instituto Nacional del Café. Coop. de Artes Gráficas Caracas, Venezuela

Jaramillo A and Valencia G (1980) Los elementos climáticos y el desarrollo de *Coffea arabica* L. en Chinchina. Cenicafe 31: 127–143

Jiménez AE (1979) Estudios ecológicos de agroecosistemas cafetaleros: I. Estructura de los cafetales de una fina cafetalera en Coatepec, Veracruz, México. Biotica 4(1): 1–2

Kajomulo-Tibaijuka A (1985) Factors influencing the cultivation of firewood trees on peasant farms: a survey on smallholder banana-coffee farms, Kagera Region, Tanzania. Swedish University of Agricultural Sciences, International Rural Development Centre #18. Uppsala, Sweden (ICRAF acc.N°5799)

Kirkpatrick TW (1935) The Climate and Eco-Climates of Coffee Plantations. East African Agricultural Research Station, Amani

Kozlowski TT (1982) Water supply and tree growth: Part I. Water deficits. For Abstr 43: 57–95

Kumar D and Tieszen LL (1980) Photosynthesis in *Coffea arabica* L. I. Effects of light and temperature. Exp Agric 16: 13–19

Kursten E and Burschel P (1993) CO_2 mitigation by agroforestry. Water, Air and Soil Pollution 70: 533–544

Lagemann J and Heuveldop J (1983) Characterization and evaluation of agroforestry systems, the case of Acosta – Puriscal, Costa Rica. Agrofor Syst 1: 101–115

Lee AK and Kasbi H (1980) Intercropping cocoa and oil palm. In: International Conference on Cocoa and Coconuts, 1978, pp 158–171. The Incorporated Society of Planters, Kuala Lumpur, Malaysia

Leite RM de O, Alvim R and Alvim P de T (1981) Effects of wind and solar radiation on the mechanical rupture of the cacao pulvinus (Abstract) In: Proceedings of the VII International

Cocoa Research Conference, Cameroun, 1979, p 129. Lagos, Nigeria. Cocoa Producers Alliance

Leon RA (1990) Determinación de la hidroerosión laminar, utilizando varillas metalicas asociadas a parcelas de escorrentia, La Suiza, Cuenca del Rio Tuis, Costa Rica. MS Thesis. CATIE, Turrialba, Costa Rica

Lindblad P and Russo R (1986) C_2H_2 reduction by *Erythrina poeppigiana* in a Costa Rican coffee plantation. Agrofor Syst 4: 33–37

Ling AH (1984) Litter production and nutrient cycling in a mature cocoa plantation on inland soils of Peninsular Malaysia. In: Pushparajah E and Soon CP (eds) International Conference on Cocoa and Coconuts: Progress and Outlook, pp 451–467. The Incorporated Society of Planters, Kuala Lumpar, Malaysia

Lock Charles GW (1888) Coffee: Its Culture and Commerce in All Countries. E & FN Spon, London, England

Luján R (1992) Dinámica de doseles de tres especies de leguminosas de sombra y efectos sobre la fenología de seis cruces interclonales de cacao. MSc Thesis CATIE, Turrialba, Costa Rica

Machado A (1959) Estado actual de las investigaciones sobre el uso de la sombra en los cafetales. Cenicafé 10(1): 5–15

Maestri M and Barros RS (1977) Coffee. In: Alvim P de T and Kozlowski TT (eds) Ecophysiology of Tropical Crops, pp 249–278. Academic Press, New York, USA

Meléndez L (1993) Microambiente, cantidad de esporas en el aire e incidencia del hongo *Moniliophtora roreri* (Cif & Par) Evans et al. bajo tres sistemas de manejo de sombra leguminosa en cacao (*Theobroma cacao*). M.Sc.thesis, CATIE, Turrialba, Costa Rica

Mitchell HW (1963) Results of coffee and banana interplanting trial in Bukoba. Ann Res Rep 1963, pp 1–6. Coffee Research Station, Lyamungu, Tanganyika

Montoya LA, Sylvain PG and Umaña R (1961) Effect of light intensity and nitrogen fertilization upon growth differentiation balance in *Coffea arabica* L. Coffee 3: 97–104

Muñoz F (1993) Dinámica de raíces finas en los sistemas agroforestales de cacao con sombras de poró o laurel en Turrialba, Costa Rica. M.S. Thesis. CATIE, Turrialba, Costa Rica

Murray DB (1954) A shade and fertilizer experiment with cacao III. In: A Report on Cacao Research, 1953, pp 30–37. Imperial College of Tropical Agriculture, St Augustine, Trinidad

Muschler RG (1995) Efectos de diferentes niveles de sombra de *Erythrina poeppigiana* sobre *Coffea arabica* vars. Caturra y Catimor. In: II. Semana Científica del Centro Agronómico Tropical de Investigación y Enseñanza (CATIE), pp 158–160. CATIE, Turrialba, Costa Rica

Muschler RG (1997) Shade or sun for ecologically sustainable coffee production: a summary of environmental key factors. In: III. Semana Científica del Centro Agronómico Tropical de Investigación y Enseñanza (CATIE), pp 109–112. CATIE, Turrialba, Costa Rica

Mussak MF and Laarman JG (1989) Farmer's production of timber trees in the cacao-coffee region of coastal Ecuador. Agrofor Syst 9: 155–170

Nair PKR (1979). Intensive multiple cropping with coconuts in India: Principles, programmes and prospects. Parey Berlin, Germany

Nair PKR (1993) Introduction to Agroforestry. Kluwer, Amsterdam, The Netherlands

Nair PKR, Kang BT and Kass DCL (1995) Nutrient cycling and soil erosion control in agroforestry systems. In: Juo ASR (ed) Agriculture and the Environment: Bridging Food Production and Environmental Protection in Developing Countries, pp 117–138. Amer Soc Agron Spec Publ 60. Madison, USA

Nair SK and NSS Rao (1977). Microbiology of the root region of coconut and cacao under mixed cropping. Plant and Soil 46: 511–519

Nataraj T and Subramanian S (1975) Effect of shade and exposure on the incidence of brown-eye-spot of coffee. Indian Coffee 39(6): 179–180

Nutman FJ (1937a) Studies on the physiology of *Coffea arabica*. I. Photosynthesis of coffee leaves under natural conditions. Ann Bot 1: 353–367

Nutman FJ (1937b) Studies on the physiology of *Coffea arabica*. II. Stomatal movements in relation to photosynthesis under natural conditions. Ann Bot 1: 681–693

Nygren P and Ramírez C (1995) Production and turnover of N_2 fixing nodules in relation to foliage development in periodically pruned *Erythrina poeppigiana* (Leguminosae) trees. For Ecol Man 73: 59–73

O'kting'ati A and Mongi HO (1986) Agroforestry and the small farmer: a case study of Kilema and Kima Vunjo in Kilimanjaro. Intern Tree Crops J 3: 257–265

OFICAFE (1978) Manual de Recomendaciones para Cultivar Café. 3rd ed. Oficina de café-Ministerio de Agricultura y Ganaderia, San Jose, Costa Rica

Ortega P (1951) El Cultivo del Cafeto en Honduras, 4th ed. Tegucigalpa: Talleres Tipográficos Nacionales

Palm CA (1995) Contribution of agroforestry trees to nutrient requirements of intercropped plants. Agrofor Syst 30: 105–124

Peoples MB and Herridge DF (1990) Nitrogen fixation by legumes in tropical and subtropical agriculture. Adv Agron 44: 155–223

Perfecto I, Rice R, Greenberg R and Van der Voort M (1996) Shade coffee: a disappearing refuge for biodiversity. BioScience 46: 598–608

Picado W (1985) *Mimosa scabrella*: especie con potencial para sombra y producción de leña en cafetales de Costa Rica In: Salazar R (ed) Técnicas de Producción de Leña en Fincas Pequeñas, pp 227–239. CATIE, Turrialba, Costa Rica

Platen, HH (1993) Evaluación económica de sistemas agroforestales de cacao con laurel y poró en Costa Rica. In: Phillips-Mora W (ed) Sombras y Cultivos Asociados con Cacao, pp 163–172. Serie Técnica Informe Técnico No. 206. CATIE, Turrialba, Costa Rica

Porras VH and Sánchez JA (1991). Enfermedades del Cacao. Serie: Tecnología, Comunicación y Desarrollo No. 5. IICA-PROCACAO/FHIA, La Lima, Honduras

PROMECAFE (1995) Antecedentes, conclusiones y recomendaciones. El seminario regional de consulta sobre caficultura sostenible. Boletín PROMECAFE 66–67: 7–9

Purseglove JW (1968) Tropical Crops: Dicotyledons. Longman, Harlow, UK

Purseglove JW (1972) Tropical Crops: Monocotyledons. Longman, Harlow, UK

Pushparajah E and Soon CP (1986) Cocoa and Coconuts: Progress and Outlook. United Selangor Press, The Incorporated Society of Planters, Kuala Lumpur, Malaysia

Ramírez LG (1993) Producción de café (*Coffea arabica*) bajo diferentes niveles de fertilización con y sin sombra de *Erythrina poeppigiana* (Walpers) O.F. Cook. In: Westley S B and Powell MH (eds) Erythrina in the New and Old Worlds, pp 121–124. Nitrogen Fixing Tree Association, Paia, Hawaii

Revista Forestal Centro-Americana (1996) 16: 45–52

Reynolds JS (1991) Soil nitrogen dynamics in relation to groundwater contamination in the Valle Central, Costa Rica. Ph.D. Thesis. Univ of Michigan. Ann Arbor, MI, USA

Robinson JBD (1961) The influence of interplanted bananas on *Arabica coffee* yields. Ann Res Rep 1961, pp 31–38. Coffee Research Station, Lyamungu, Tanganyika

Rodríguez LE (1982) Estudio económico del nogal cafetero (*Cordia alliodora*) en el suroeste antioqueño. BS Thesis, Universidad Nacional de Colombia, Medellín, Colombia

Rodríguez RM, Carvajal JF, Machicado M and Jiménez M (1963) Nutritional requirements of the cocoa plant during a yearly cycle. Cacao 8(4): 1–7

Rojas Cubero R (1996) Modelo de costos de producción de café. Instituto del Café de Costa Rica, San José, Costa Rica

Rosero P and Gewald N (1979) Growth of laurel (*Cordia alliodora*) in coffee and cacao plantations and pastures in the Atlantic region of Costa Rica, In: De las Salas G (ed) Workshop Agro-forestry Systems in Latin America, pp 205–208. CATIE, Turrialba, Costa Rica

Roskowski JP and Van Kessel C (1985) Annual seasonal and daily variation in nitrogen fixing activity by *Inga jinicuil*, a tropical leguminous tree. Oikos 44: 306–312

Sabogal C (1983) Observaciones sobre la combinación de *Cedrela odorata* con café en Tabarcia-Palmichal (Cantón Puriscal). In: Heuveldop J and Espinoza L (eds) El componente árboreo en Acosta y Puriscal, Costa Rica, pp 90–101. CATIE, Turrialba, Costa Rica

Sáenz N (1895) Cultivo del Cafeto. Casa Editorial de JJ Pérez, Bogotá, Colombia

Salas, A (1994) Crecimiento de *Eucalyptus deglupta* y *E. grandis* bajo tres sistemas de plantacion a nivel de finca, en la zona de Turrialba, Costa Rica. M.Sc. thesis. CATIE, Turrialba, Costa Rica

Salazar R (1984) Producción de leña en árboles de *Gliricidia sepium* usados como sombra en cafetales en Costa Rica. Silvoenergía # 2, CATIE, Turrialba, Costa Rica

Salazar R (1985) Producción de leña y biomasa de *Inga densiflora* Benth en San Ramon, Costa Rica. Silvoenergía # 3, CATIE, Turrialba, Costa Rica

Sanchez PA (1995) Science in agroforestry. Agrofor Syst 30: 5–55

Schlönvoigt M (1993) Aufwuchsentwicklung von *Cordia alliodora* (Ruiz & Pavon) Oken in Abhängigkeit verschiedener Pflanzmethoden in agroforestlichen Systemen in der Atlantikzone von Costa Rica. Ph.D. thesis, Georg-August-Universitat, Gottingen, Germany

Schroeder R (1951) Resultados obtenidos de una investigación del micro-clima en un cafetal. Cenicafe 2(18): 33–43

Schroth G (1995) Tree root characteristics as criteria for species selection and systems design in agroforestry. Agrofor Syst 30: 125–143

Silva NTMH de, Tisdell CA and Byrne PN (1990) Evaluating techniques for weed control in coffee in Papua New Guinea. Internat Tree Crops J 6(1): 31–49

Smith ESC (1981) The interrelationships between shade types and cocoa pest and disease problems in Papua New Guinea. In: Proceedings of the VII International Cocoa Research Conference, Cameroun 1979, pp 37–43. Cocoa Producers Alliance, Lagos, Nigeria

Smith NJH, Williams JT, Plucknett DL and Talbot JP (1992) Tropical Forests and their Crops. Cornell University Press, Ithaca, New York

Somarriba E (1990) Sustainable timber production from uneven-aged shade stands of *Cordia alliodora* in small coffee farms. Agrofor Syst 10: 253–263

Somarriba E (1992) Timber harvest, damage to crop plants and yield reduction in two Costa Rican coffee plantations with *Cordia alliodora* shade trees. Agrofor Syst 18: 69–82

Somarriba E and Beer J (1987) Dimensions, volumes and growth of *Cordia alliodora* in agroforestry systems. For Ecol Man 18: 113–126

Somarriba E, Beer J and Bonnemann A (1996) Arboles leguminosos y maderables como sombra para cacao: el concepto. Serie Técnica Informe Técnico No 274. CATIE, Turrialba, Costa Rica

Swinkels RA and Scherr SJ (Compilers) (1991) Economic analysis of agroforestry technologies: an annotated bibliography. ICRAF, Nairobi, Kenya

Tapley RG (1961). Crinkly-leaf of coffee in Tanganyika. Kenya Coffee 26: 156–157

Trejos S and Platen H von (1995) Sombras maderables para cacaotales: aspectos económicos. Serie Técnica Informe Técnico No. 266. CATIE, Turrialba, Costa Rica

Vernon AJ (1967a) New developments in cocoa shade studies in Ghana. J Sci Food and Agr 18: 44–48

Vernon AJ (1967b) Yield and light relationship in cocoa. Trop Agric 44: 223–228

Wellman (1961) Coffee: Botany, Cultivation, and Utilization. Leonard Hill, London, England

Wiersum KF (1984) Surface erosion under various tropical agroforestry systems. In: O'Loughlin CL and Pearce AJ (eds) Symposium on Effects of Forest Land Use on Erosion Control and Slope Stability, pp 231–239. East-West Center, Honolulu, USA

Wille C (1994) The birds and the beans. The coffee fields of Mexico and Central America may be the last best habitat for migrating birds. Audubon Nov/Dec 94: 58–64

Willey RW (1975) The use of shade in coffee, cocoa and tea. Horticult Abstr 45(12): 791–798

Wood GAR and Lass RA (1985) (eds) Cacao. 4th ed. Longman, New York, USA

Wrigley G (1988) Coffee. Longman, New York, USA

Young AM (1994) The Chocolate Tree. Smithsonian Institution Press, Washington, DC, USA

Zamora G and Soto B (1976) Arboles usados como sombra para café y cacao. Revista Cafetalera (Guatemala) Oct-Nov (1976): 27–32

Agroforestry Systems **38**: 165–176, 1998.

The domestication and commercialization of indigenous trees in agroforestry for the alleviation of poverty

R. R. B. LEAKEY[1,*] and A. J. SIMONS[2]
[1] *Institute of Terrestrial Ecology, Bush Estate, PENICUIK, Midlothian, Scotland, UK, E-mail: r.leakey@ite.ac.uk;* [2] *International Centre for Research in Agroforestry (ICRAF), P.O. Box 30677, Nairobi, Kenya. E-mail: t.simons@cgnet.com (* Corresponding author)*

Key words: fruit trees, non-timber forest products, tree improvement

Abstract. New initiatives in agroforestry are seeking to integrate into tropical farming systems indigenous trees whose products have traditionally been gathered from natural forests. This is being done in order to provide marketable products from farms that will generate cash for resource-poor rural and peri-urban households. This poverty-alleviating agroforestry strategy is at the same time linked to one in which perennial, biologically diverse and complex mature-stage agroecosystems are developed as sustainable alternatives to slash-and-burn agriculture.

One important component of this approach is the domestication of the local tree species that have commercial potential in local, regional or even international markets. Because of the number of potential candidate species for domestication, one crucial first step is the identification of priority species and the formulation of a domestication strategy that is appropriate to the use, marketability and genetic potential of each species.

For most of these hitherto wild species little or no formal research has been carried out to assess their food value, potential for genetic improvement or reproductive biology. To date their marketability can only be assessed by their position in the local rural and urban market-places, since few have attracted international commercial interest. To meet the objective of poverty alleviation, however, it is crucial that market expansion and creation are possible, hence for example it is important to determine which marketable traits are amenable to genetic improvement. While some traits that are relatively easy to identify do benefit the farmer, there are undoubtedly others that are important to the food, pharmaceutical or other industries that require more sophisticated evaluation.

This paper presents the current thinking and strategies of ICRAF in this new area of work and draws on examples from our program.

Introduction

Agroforestry practices come in many forms but have traditionally been categorized into two groups – those that are sequential, such as fallows, and those that are simultaneous, such as alley cropping (Cooper et al., 1996). In all, some 18 different agroforestry practices have been recognized by Nair (1993), although each has an infinite number of variations.

Leakey (1996), however, has suggested that agroforestry practices should be seen as stages in the development of an agroecosystem such that the increasing integration of trees into land-use systems can be seen as the passage towards a mature agroforest of increasing ecological integrity. In this way, with increasing scale, the integration of various agroforestry practices into the landscape is like the formation of a complex mosaic of patches in an ecosystem, each of which is composed of many niches. These niches are occupied by different organisms, making the system ecologically stable and biologically diverse. Filling some of these niches with indigenous species that provide important environmental services or economically valuable products traditionally obtained from natural forest, or both, should result in land use that is both sustainable and productive.

Increasing the quality, number and diversity of domesticated trees that provide a wide array of non-timber forest products (NTFPs) to fill these niches should enhance agroforestry's capacity to fulfil its ultimate potential as a way to alleviate poverty and to mitigate deforestation and land depletion. In the humid tropics this could lead to the development of viable alternatives to slash-and-burn agriculture. The large-scale adoption of such an approach should be especially beneficial, since the ecological and social benefits of diversity on a landscape-scale are considerably greater than the sum of the individual farm-scale benefits. There is great urgency to achieve these benefits if the severity of the current man-made episode of species extinction – the so-called 'Sixth Extinction' – which was so dramatically described by Leakey and Lewin (1996), is to be defused.

In this respect, Sanchez and Leakey (1997) see domestication as one of the three determinants for balancing food security with natural resource utilization in sub-Saharan Africa. The other two determinants are the need for an enabling policy environment that favors smallholder rural development, and the means to reduce soil fertility depletion.

Discussion of the role of domestication can not, however, be divorced from that of commercialization, since without an expanded or a new market the incentives to domesticate intensively for self-use are insufficient. Conversely, if the market explodes, the incentive for large-scale producers to establish monocultural plantations may sweep away the benefits that agroforestry could deliver to small-scale, resource-poor farmers around the tropics (Leakey and Izac, 1996). Thus policies that promote the linkages between the domestication and commercialization of NTFPs are one of the important areas in which policy scientists need 'to stretch their conceptual framework . . . and to consider more carefully the links between markets, the environment, household production and household welfare' (Dewees and Scherr, 1996).

Domestication strategies

The term 'Cinderella trees' (Leakey and Newton, 1994) is now widely accepted as a phrase applicable to traditionally important indigenous species that have been overlooked by science for agroforestry and forestry, as evidenced by the term's use in numerous articles and conference proceedings (see Leakey et al., 1996). Similarly, the need to rapidly domesticate the Cinderella trees has been accepted, and is now one of the three pillars of ICRAF's research program. In genetic terms, domestication is accelerated and human-induced evolution. Domestication, however, is not only about selection, as domestication integrates the four key processes of the identification, production, management and adoption of agroforestry tree genetic resources (ICRAF, 1997).

The Working Group on 'Product Domestication and Adoption by Farmers' in the recent ICRAF Conference on the 'Domestication and Commercialization of Non-timber Forest Products in Agroforestry' (Leakey et al., 1996) defined the domestication of trees producing NTFPs as 'a progression from collection and utilization of products, through protection, management and cultivation, which culminates with genetic manipulation.' Two extreme strategies (Figure 1) were envisaged:

1) making incremental improvement through management on farm; and
2) making major leaps in improvement by genetic selection and breeding.

The first strategy is farmer-oriented, while the second is market-oriented. The question is whether the market-oriented strategy will automatically lead to monocultures? History would suggest that this is likely, but history is not always the precedent people expect. It seems likely that between these extremes there are many scenarios that could benefit farmers and the environment. For example, the two strategies are not mutually exclusive: farmer-domesticated trees can be inputs into more formal genetic improvement programs, just as selected cultivars from genetic improvement programs can be grown on smallholder farms (Figure 1). Furthermore, there is the precedent of 'jungle rubber' production in Indonesia, where rubber produced in complex agroforestry systems is overtaking that from monocultural, clonal plantations (Michon and de Foresta, 1996). To encourage this development

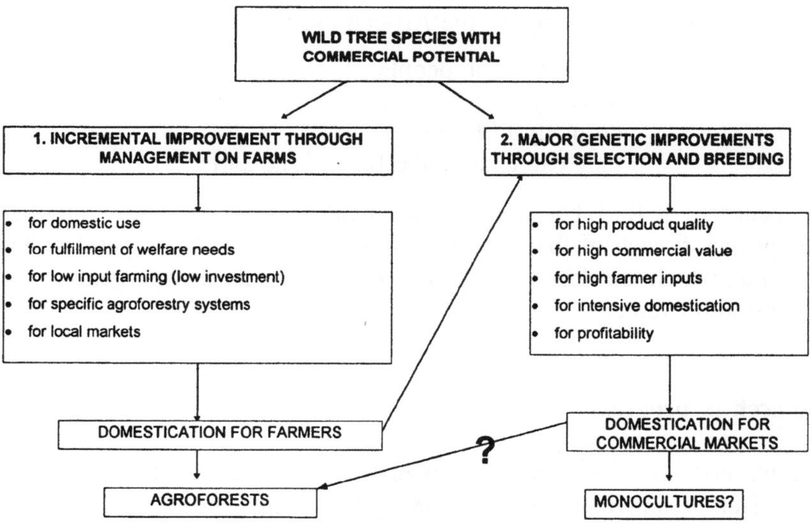

Fig. 1. Two extreme pathways in the domestication and commercialization of non-timber forest products (modified from Leakey et al., 1996).

for other tree species producing non-timber forest products, incipient markets require good information flows so that domestication may attract a price premium for high quality products and better marketing practices. In this way, small-scale farmers will be able to respond to, and benefit from, domestication opportunities, reducing the risk of displacement by monoculture plantations. With regard to market information, Ndoye (1995) has called for price surveillance systems for minor non-timber forest products, similar to those for international commodity crops, so that farmers will be more aware of marketing opportunities and the benefits of domestication will be maximized.

The Working Group on Policy and Institutional Aspects of Domestication at the conference referred to earlier (Leakey et al., 1996) considered the economic and social implications of domestication and commercialization on indigenous peoples, and the trade-off, if any, between the objectives of raising farmers' incomes and biodiversity. It was concluded that the domestication of trees to produce NTFPs for agroforestry must benefit both the farmer and the environment, and that there must be no policy barriers that discriminate against smallholder farmers. The recommendations charged ICRAF with developing guidelines for the domestication of NTFPs that would ensure benefits to the farmer. It was recognized that local traders and processors, and hence the community, would also benefit. It was felt that a farmer-oriented approach to domestication would also benefit the environment and biodiversity. It should enhance biodiversity: (1) within the domesticated species by deliberate genetic conservation, so ensuring the maximization of the genetic base of new cultivars, and (2) within the land-use system, by rehabilitating abandoned farmland through agroforestry with cultivars of locally important trees (Leakey, in press).

ICRAF's philosophy of domestication and commercialization of trees for agroforestry is closely allied to the African proverb:

> 'If many little people,
> in many little places,
> do many little things,
> they can change the face of the earth.'

Consequently, ICRAF has started domestication programs with a number of tree species in each of the six ecoregions in which ICRAF operates. These programs, which are described below, involve farmer-driven, genetic selection and improvement.

Research in progress

To date, more than 2500 tree species have been documented for their use on farms (Burley and von Carlowitz, 1984). It is essential for ICRAF to develop priorities to select the species most suitable for domestication. There is a move away from domesticating only fast-growing, nitrogen-fixing trees towards wild indigenous trees with potential to generate cash for farmers. The germplasm of the new priority species is, almost without exception, unimproved. Given the restrictions on funds and research manpower, the emphasis of ICRAF's approach to domestication is on developing decision-making frameworks which first ascertain whether domestication should proceed, and then determine its direction and intensity (Simons, 1996b).

In the domestication of 'Cinderella trees' for agroforestry, ICRAF is dealing with a very imperfect knowledge base, since these species have often been virtually overlooked by science, and are little known commercially, except in their local area. The situation is further complicated by the fact that the most appropriate domestication methods will vary depending on the uses of the trees, the value of their products or environmental services, the length of time they will be retained in agroforestry systems, and their biological traits, such as genetic variability.

The majority of domestication methods under consideration have been adapted from those used for agricultural/horticultural crops or industrial forest species. It is now recognized that the Northern-temperate bias to tropical tree improvement needs to be redressed by the development of novel approaches that take into consideration the requirements of small-scale, resource-poor farmers and their farming systems. Given that improvement is as much a social and political challenge as a biological one, it will only be through experimental implementation of a range of approaches that methods and strategies will progress. Consequently, ICRAF uses a few priority species as models, hoping that others will then utilize these models to domesticate a much wider array of species.

The first two steps in the domestication process are common to most tree species, thereafter species will vary in the methods applied (Simons, 1996a). The first step is the determination of which species should receive priority for domestication. This step of priority setting has already been widely reported (Leakey and Jaenicke, 1995; Jaenicke et al., 1995; Franzel et al., 1996; Simons, 1996b). It involves household interviews to determine farmer preferences, the assessment of market potential and, finally, the inputs of researchers on technical points relating to genetic variability. A shortlist of priority species for a region can then be assembled. The second step is the likely need to make rangewide germplasm collections of the chosen species. These collections must conform to the requirements of the Convention on Biological Diversity and the rights of sovereign states and of the rural people. The implementation of this priority setting process by ICRAF and her partners has already led to the preliminary selection of 18 priority species from the humid and semiarid lowlands of West Africa, as well as in the Peruvian Amazon (Table 1).

Less rigorous means than those described above have been used to select other species for domestication in the southern Africa plateau or Miombo woodland and for the East African Highlands (Table 2). The application of systematic priority setting has been beneficial in some cases and less than satisfactory in others. For example, although it has clearly identified a number of species of great interest to farmers, it failed to identify any fodder trees in the Sahelian region. In an earlier Diagnosis and Design exercise in this region, farmers clearly identified the need for dry season fodder as one of their major constraints to livestock farming (Djimde, 1991). Why did farmers not identify this need the second time? Perhaps at the time of the interview, fodder was not in short supply. Alternatively, the interviewers might have shown more interest in other species and forgotten to ask specifically about fodder. A different problem emerged in Amazonia, where many farmers of migrant origin did not know the pool of local species from which they could choose.

Currently, germplasm collection by ICRAF is focused strongly, but not exclusively, on the species identified through farmer input and the priority setting exercise. However, in the Sahel, work is also in progress with *Pterocarpus erinaceus*, *Bauhinia rufescens* and *Combretum aculeatum*, all of which have fodder potential but were added to the list of priority species by researchers aware of the need for dry season fodder. Similarly, in Cameroon, two commercially important medicinal species, *Prunus africana* and *Pausinystalia johimbe*, were added to the priority list for the humid lowlands of West Africa because of opportunities for farmer income, fears about the future of the resource for industry, and the perceived needs for conservation of the species and their habitats.

Table 1. Priority tree species selected for domestication by ICRAF and partners through implementation of farmer preference surveys and priority setting guidelines.

Priority order	West Africa		Latin America
	Humid lowlands	Semiarid lowlands*	Peruvian Amazon*
1	*Irvingia gabonensis/I.wombolu*	*Adansonia digitata*	*Bactris gasipaes*
2	*Dacryodes edulis/D. klaineana*	*Vitellaria paradoxa*	*Cedrelinga catenaeformis*
3	*Ricinodendron heudelottii*	*Parkia biglobosa*	*Inga edulis*
4	*Chrysophyllum albidum*	*Tamarindus indica*	*Calycophyllum spruceanum*
5	*Garcinia kola/G. afzelii*	*Zizyphus mauritiana*	*Guazuma crinita*

* Preliminary, awaiting full economic evaluation.
Source: Franzel et al. (1996).

Table 2. Priority fruit tree species selected for domestication by ICRAF and partners through market and ethnobotanical surveys.

Priority order	Southern Africa plateau
1	*Uapaca kirkiana*
2	*Sclerocarya birrea*
3	*Zizyphus mauritiana*
4	*Vangueria infausta*
5	*Azanza garckeana*

Rangewide germplasm collections, the second step in the domestication process, are both expensive and time consuming. Furthermore, once germplasm has been collected, especially for recalcitrant species such as *Irvingia gabonensis* and *Uapaca kirkiana*, there is a large commitment to nursery work and the establishment and maintenance of living genebanks. Consequently, the germplasm of only a few priority species has so far been extensively collected by ICRAF and its partners, the National Agricultural Research Systems (NARS) of countries in the different regions.

Prior to the present phase of ICRAF's domestication program, seeds were collected mainly from species used primarily for wood. These included *Markhamia lutea* in Kenya and Uganda, *Sesbania sesban* (and a few related species) in southern Africa, and *Faidherbia albida*, *Combretum aculeatum*, *Balanites aegyptiaca* and *Prosopis africana* in the Sahelian zone. The collections of *Sesbania sesban* were the first to be done by ICRAF in a fully participatory manner with the NARS (Ndungu and Boland, 1994). This involved detailed planning and extensive training of field teams in each country and then, through Material Transfer Agreements, the exchange of germplasm between participating countries for storage in national and regional seed banks.

More recently, seeds of three indigenous fruit-tree species were collected using similar collaborative procedures. These collections include *Irvingia gabonensis* (and some related species, notably *I. wombolu*) in West Africa, and *Uapaca kirkiana* and *Sclerocarya birrea* in southern Africa. For the *Irvingia* species, genebanks have been established at Onne and Ibadan in Nigeria, and at Mbalmayo in Cameroon. Each genebank contains approximately 60 accessions (single-tree progenies) of each species, from a wide range of sites in Ghana, Nigeria, Cameroon and Gabon. The collection strategy used for *Irvingia* involved farmers in the targeting of superior trees. In each village visited, farmers selected 20 to 30 trees with desirable traits (fruit shape, fruit size, kernel size, tree form, etc.). Further selection by the collection team (NARS, NGOs and ICRAF) narrowed these down to two trees per village. During this period of exploration and collection, studies were made of the phenotypic variation in tree form, phenology, and fruit characteristics such as shape, color and sweetness (Ladipo et al., 1996). Variation in fruit characteristics is very extensive in wild populations, as has been reported for *Dacryodes edulis*, the species of second priority in humid West Africa (Leakey and Ladipo, 1996).

In addition to studies on the morphological variation of these collections, work is in progress to characterize genetic variation using Polymerase Chain Reaction (PCR) techniques. Aspects of genetic structure, amount of variation, ecogeographic partitioning and hybridization are being examined. Molecular characterization is also in progress for *Prunus africana* populations.

Building upon the experience of collecting West African fruit species, ICRAF's latest collection efforts with national partners focused on two indigenous fruit species (*S. birrea* and *U. kirkiana*) of the Miombo Woodlands of southern Africa. Although conservation and utilization of the genetic resources are intimately linked, collection missions usually assign one of these objectives higher priority than the other by carrying out either random or targeted collections, respectively. In the absence of knowledge on the efficacy of phenotypic selection, it was decided to carry out collections from the same populations, using both random and targeted selective procedures.

For *U. kirkiana*, seeds of 25 trees have been collected from each of 30 provenances by ICRAF and its partners in five countries, through close collaboration with the Southern Africa Development Community (SADC) Tree Seed Centre Network. These have been exchanged between Malawi, Mozambique, Tanzania, Zambia and Zimbabwe, and nurseries are overflowing in advance of setting up field genebanks. A similar exercise has been conducted for *S. birrea* with collections in Botswana, Malawi, Mozambique, Namibia, Swaziland, Tanzania, Zambia, and Zimbabwe.

In the next phase of the work on indigenous fruit trees, ICRAF plans to use the standard horticultural techniques of grafting, air-layering, and rooted cuttings, to capture the additive and non-additive genetic variation between individual plants (Leakey and Jaenicke, 1995), a strategy that is highly appropriate for the genetic improvement of long-lived, high-value tree species (Leakey and Simons, in press). Interestingly, *U. kirkiana* and *S. birrea* are both dioecious. Therefore, female trees (fruit-bearing) can be preferentially multiplied, while recognizing the need to retain a certain number of male trees.

Commercialization

Throughout the tropics the products of indigenous trees are marketed locally on a small scale, as a means of generating cash to supplement a subsistence lifestyle (Falconer, 1990; Arnold, 1995, 1996; Lamien et al., 1996; Melnyk, 1996). Frequently these products are collected in natural forest or from wild trees retained on farmland. Some of these products enter regional trade (Ndoye, 1995) and, increasingly, a few are marketed in the USA and Europe. It has been recognized that expanded markets for these products would increase the value of natural forests and benefit forest dwellers (Peters et al., 1989). Similarly, markets for non-timber forest products produced on areas already deforested could improve the income of subsistence farmers, and provide an alternative to slash-and-burn agriculture, one of the major causes of deforestation.

There is, however, a problem. The market for most of these NTFPs is fairly small, and the supply is often seasonal. Furthermore, the wild, unimproved product may not have enough market appeal to encourage greater commercial interest. As would be expected, there is great genetic diversity in the products of each species (see Ladipo et al., 1996) affecting, for example, fruit size, shape, color, flavor, and pulp:seed ratio. It was interesting to note in *Dacryodes edulis* that large size and high pulp:seed ratios were not the only factors determining the prices of different fruitlots in Yaoundé market. Presumably, therefore, flavor and other non-visible attributes are important contributors to price determination (Figure 2). These genetic traits provide the opportunity for selection to enhance the appeal to both the market and the growers. In the case of fruits, traditional horticulture provides the methodology for domestication. For other products (e.g., food additives and spices), the selection of appropriate traits is more difficult to determine. For instance, what characteristics need to be improved to make a better food thickening agent? Here there is a need for dialogue between the food industry and the field scientists – a dialogue that does not exist at present. Similar problems relate to the pharmaceutical and cosmetic industries, where again appropriate domestication could lead to higher quality natural products if the desired traits were known to field scientists. Furthermore, there are regulations affecting the food, pharmaceutical and cosmetic industries that determine the toxicological acceptability of products. Scientists involved in the domestication of species producing potential new products need to be aware of the limits and opportunities imposed by these regulations.

Another commercial problem is the seasonality of production, especially for perishable products such as fresh fruits. Domestication can almost certainly have considerable impact on the seasonality of production. For example, in the fruit trees *I. gabonensis* and *U. kirkiana*, there is considerable within-species variation in phenology, with some trees flowering and fruiting outside the main season or fruiting several times per year instead of the normal once per year. Since kernel extraction is a very labor-demanding process, which

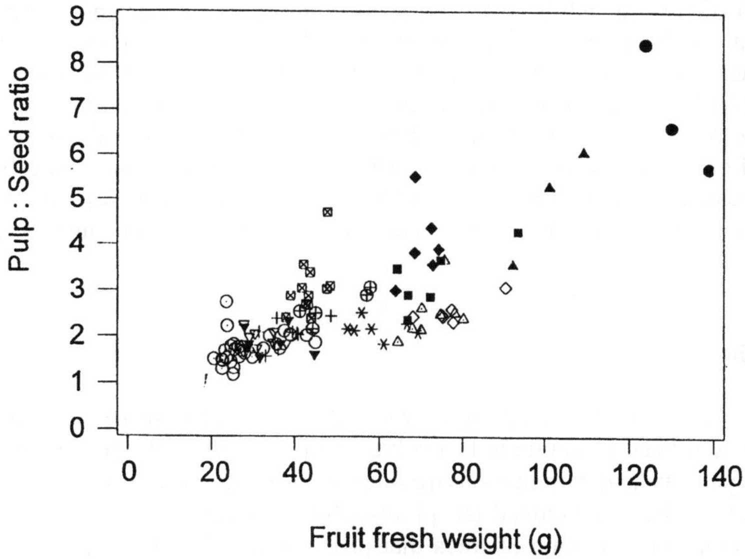

Fig. 2. Genetic variation in fruit characteristics of *Dacryodes edulis* in a Yaoundé market, Cameroon (after Leakey and Ladipo, 1996). Different symbols represent different fruitlots. Solid symbols represent fruits selling for more than 500 CFA (US $1.00) per kg of pulp.

potentially reduces market appeal, a remarkable discovery for *I. wombolu* in Gabon was a self-cracking seed. Assuming that this is genetically determined variation, these trees are strong candidates for selection to become cultivars through cloning (Leakey and Jaenicke, 1995).

Returning to the problem of market size, one is confronted with the question of which comes first, the demand or the supply? This is a 'chicken or egg' type of problem, which is further complicated by the price elasticity of supply and demand.

In one sense, the relatively small market for these products is perhaps an asset, at least at this stage in the development of domestication activities. If demand for the products of improved cultivars were too high, large-scale company interests might swamp the production before small-scale farmers practicing agroforestry had developed marketing infrastructures. Ideally, for the purpose of poverty alleviation, and incidentally for environmental benefits through agroforestry, small-scale operations are preferable. However, a small, seasonal supply of goods is not attractive to traders. Thus, it would be more desirable commercially to have a year-round supply of products from a number of species coming in and out of production at different times of the year. It is clear that such developments require two conditions: an appropriate policy environment (Leakey and Izac, 1996) and commercial interests sympathetic to small-scale production (e.g., cooperatives of farmers), such as those associated with organic farming. Fortunately, there are already examples of highly successful international marketing of produce grown by small-scale farmers, e.g., French beans grown in Kenya (see also Sanchez and Leakey, 1997). Furthermore, whilst integrating a range of tree crops into farming systems at the landscape scale, there is scope for farmers to consider segregating these from staple food crops (van Noordwijk et al., 1995). The contiguous blocks of damar agroforests in Sumatra are a good example of this approach (Michon and de Foresta, 1996). Numerous other configurations of trees and crops are, however, possible and at a later stage of the domestication process there will be a need to match tree cultivars with crop varieties, in a way already well advanced for linear tree/crop plantings (Ong and Huxley, 1996).

Conclusion

The concepts, strategies and policies associated with agroforestry are rapidly evolving, and need to continue to evolve, if the promises of agroforestry to alleviate poverty and mitigate environmental degradation are to be met. In this paper, we have developed the philosophy of integrating the domestication and commercialization of trees that produce valuable non-timber forest products into agroecosystems involving a number of agroforestry practices. This strategy gives hope of improving the well-being of the rural poor, while also promoting better land-use practices.

References

Arnold JEM (1995) Socio-economic benefits and issues in non-wood forest product use. Report of the International Expert Consultation on Non-wood Forest Products. Non-wood Forest Products No. 3, pp 89–123

Arnold JEM (1996) Economic factors in farmer adoption of forest product activities. In: Leakey RRB, Temu AB, Melnyk M and Vantomme P (eds), Domestication and Commercialization of Non-timber Forest Products in Agroforestry Systems, pp 131–146. Non-wood Forest Products No. 9. FAO, Rome, Italy

Burley J and von Carlowitz P (1984) Multi-purpose tree germplasm. Proceedings of a planning workshop at National Academy of Sciences, Washington, DC, June 1983. ICRAF, Nairobi, Kenya

Cooper PJ, Leakey RRB, Rao MR and Reynolds L (1996) Agroforestry and the mitigation of land degradation in the humid and sub-humid tropics of Africa. Experimental Agriculture 32: 235–290

Dewees PA and Scherr SJ (1996) Policies and markets for non-timber tree products. EPTD Discussion Paper No. 16. IFPRI, Washington DC, USA

Djimde M (1991) Synthese zonal des systemes d'utilisation des terres et des potentialites agroforestieres dans le cadre de la Planification du Reseau Collaboratif de Recherche en Agroforesterie de l'ICRAF pour les Zones Semi-Arides du Burkina Faso, Mali, Niger et Senegal. AFRENA Report No. 48. ICRAF, Nairobi, Kenya

Falconer J (1990) The Major Significance of 'Minor' Forest Products. The local use and value of forests in the West African humid forest zone. Forests, Trees and People, Community Forestry Note No. 6, FAO, Rome, Italy

Franzel S, Jaenicke H and Janssen W (1996) Choosing the right trees: setting priorities for multipurpose tree improvement. ISNAR Research Report 8. ISNAR, The Hague, The Netherlands

Jaenicke H, Franzel S and Boland DJ (1995) Towards a method to set priorities among species for tree improvement research: a case study from West Africa. Journal of Tropical Forest Science 7: 490–506

Ladipo DO, Foundoun JM and Ganga N (1996) Domestication of the Bush Mango (*Irvingia* spp.): some exploitable intraspecific variations in West and Central Africa. In: Leakey RRB, Temu AB, Melnyk M and Vantomme P (eds), Domestication and Commercialization of Non-timber Forest Products for Agroforestry, pp 193–205. Non-Wood Products No. 9. FAO, Rome, Italy

Lamien N, Sidibe A and Bayala J (1996) Use and commercialization of non-timber forest products in Western Burkina Faso. In: Leakey RRB, Temu AB, Melnyk M and Vantomme P (eds), Domestication and Commercialization of Non-timber Forest Products in Agroforestry Systems, pp 51–64. Non-Wood Forest Products No. 9. FAO, Rome, Italy

Leakey RE and Lewin R (1996) The Sixth Extinction: Biodiversity and it's Survival. Weidenfeld and Nicholson, London, 271 pp

Leakey RRB (1996) Definition of agroforestry revisited. Agroforestry Today 8(1): 5–7

Leakey RRB (In press). The use of biodiversity and implications for agroforestry. In: Proceedings of FORUM BELEM. POEMA, Belém, Brazil

Leakey RRB and Izac A-MN (1996) Linkages between domestication and commercialization of non-timber forest products: implications for agroforestry. In: Leakey RRB, Temu AB, Melnyk M and Vantomme P (eds), Domestication and Commercialization of Non-timber Forest Products for Agroforestry, pp 1–7. Non-Wood Forest Products No. 9. FAO, Rome, Italy

Leakey RRB and Jaenicke H (1995) The domestication of indigenous fruit trees: opportunities and challenges for agroforestry. In: Suzuki K, Sakurai S, Ishii K and Norisada M (eds). Procedings of 4th International BIO-REFOR Workshop, pp 15–26. BIO-REFOR, Tokyo, Japan

Leakey RRB and Ladipo DO (1996) Trading on genetic variation – fruits of *Dacryodes edulis*. Agroforestry Today 8(2): 16–17

Leakey RRB and Newton AC (1994) Domestication of 'Cinderella' species as a start of a woody plant revolution. In: Leakey RRB and Newton AC (eds), Tropical Trees: Potential for Domestication and the Rebuilding of Forest Resources, pp 3–6. HMSO, London

Leakey RRB and Simons AJ (In press) When does vegetative propagation provide a viable alternative to propagation by seed: opportunities and strategies for forestry/agroforestry in tropical and subtropical countries. In: Wolf H (ed) The Problems of Forestry in Tropical and Subtropical Countries: The Procurement of Forestry Seeds with Special Reference to Kenya. Ecomed Publishers, Landsberg/Lech, Germany

Leakey RRB, Temu AB, Melnyk M and Vantomme P (eds) (1996) Domestication and Commercialization of Non-timber Forest Products in Agroforestry Systems. Non-Wood Forest Products No. 9. FAO, Rome, Italy

Melnyk M (1996) Indigenous enterprise for the domestication of trees and the commercialization of their fruits, In: Leakey RRB, Temu AB, Melnyk M and Vantomme P (eds), Domestication and Commercialization of Non-timber Forest Products in Agroforestry Systems, pp. 97–103. Non-Wood Forest Products No. 9. FAO, Rome, Italy

Michon G and de Foresta H (1996) The agroforest model as an alternative to the pure plantation model for domestication and commercialization of NTFP's. In: Leakey RRB, Temu AB, Melnyk M and Vantomme P (eds), Domestication and Commercialization of Non-timber Forest Products in Agroforestry Systems, pp 160–175. Non-Wood Forest Products No. 9. FAO, Rome, Italy

Nair PKR (1993) An Introduction to Agroforestry. Kluwer Academic Publishers, Dordrecht, The Netherlands

Ndungu JN and Boland DJ (1994) *Sesbania sesban* collections in Southern Africa: developing a model for co-operation between a CGIAR Centre and NARS. Agroforestry Systems 27: 129–143

Ndoye O (1995) The markets for non-timber forest products in the humid forest zone of Cameroon and its borders: Structure, conduct, performance and policy implications. Report to CIFOR, Bogor, Indonesia

Ong CK and Huxley P (1996) Tree-Crop Interactions: A Physiological Approach. CAB International, Wallingford, UK

Peters CM, Gentry AH and Mendelsohn RO (1989) Valuation of an Amazonian rainforest. Nature 339: 655–656

Sanchez PA and Leakey RRB (1997) Land use transformation in Africa: three determinants for balancing food security with natural resource utilization. European Journal of Agronomy (in press)

Simons AJ (1996a) Delivery of improvement for agroforestry trees. In: Dieters MJ, Matheson AC, Nikles DG, Harwood CE and Walker SM (eds), Tree Improvement for Sustainable Tropical Forestry, Vol 2, pp 391–400. Queensland Forestry Research Institute, Gympie, Australia

Simons AJ (1996b) ICRAF's strategy for domestication of non-wood tree products, In: Leakey RRB, Temu AB, Melnyk M and Vantomme P (eds), Domestication and Commercialization of Non-timber Forest Products in Agroforestry Systems, pp 8–22. Non-Wood Forest Products No. 9. FAO, Rome, Italy

van Noordwijk M, van Schaik CP, de Foresta H and Tomich TP (1995) Segregate or integrate nature and agriculture for biodiversity conservation. In: Proceedings of Biodiversity Forum, 4-5 November 1995, Jakarta, Indonesia

Agroforestry Systems **38**: 177–193, 1998.

Socioeconomic research in agroforestry: progress, prospects, priorities

D. E. MERCER[1,*] and R. P. MILLER[2]
[1] *Southern Research Station, US Forest Service, P.O. Box 12254, Research Triangle Park, North Carolina, USA, E-mail: emercer@rtpmail.emapfhm.gov;* [2] *School of Forest Resources & Conservation, University of Florida, Gainesville, Florida, USA (* Corresponding author)*

Key words: anthropology, economics, geography, policy, sociology

Abstract. Fourteen years after the birth of the journal *Agroforestry Systems*, biophysical studies continue to dominate agroforestry research while other important areas have not received the attention they deserve. This paper reviews the progress in one of these under-researched areas, socioeconomics. A quantitative and qualitative analysis of published socioeconomic research papers and a survey of agroforestry socioeconomic researchers are used to evaluate the achievements in agroforestry socioeconomic research over the past 14 years: what are the major advances, gaps in knowledge, and constraints for closing those knowledge gaps? Although agroforestry socioeconomics remains a field in its infancy, both the scope and the quality of socioeconomic research are slowly improving. The recent trend away from literature reviews, qualitative, and purely descriptive quantitative research based on small sample sizes, and toward more rigorous statistical analyses of better and larger data sets is encouraging. Priority areas for future research include theoretical and empirical analyses of agroforestry adoption decisions, improved economic analyses, and policy studies at local, national, and regional levels.

Introduction

Although existing for centuries (maybe millennia) as an array of traditional land-use practices, agroforestry emerged in the late 1970s as a modern, improved land-use system suitable for scientific study, replete with its own international research center and journal. Like the traditional land-use disciplines of agriculture and forestry, agroforestry science spans the disciplinary spectrum from the biological and physical sciences to the social sciences. Understandably, the biophysical sciences have dominated the first two decades of agroforestry research and development because the interest in agroforestry as a land use emerged from observations of the impacts of nonsustainable farming systems on tropical soils and forests (Nair, 1996). Concerns over the inadequacy of socioeconomic research in agroforestry began to grow, however, as improved agroforestry systems were transferred from research institutions to rural development projects.

During the 1980s, agroforestry became an established focus of international rural development efforts. For example, in 1988 and 1989 ICRAF identified 166 agroforestry projects supported by development organizations and government agencies (Müller and Scherr, 1990), and by the early 1990s the US Agency for International Development alone supported 28 agroforestry projects in 16 countries (Mercer, 1993). Despite some impressive scientific and technological advances, agroforestry rural development efforts were frequently unsuccessful (Nair, 1996). Although agroforestry projects failed for a number of different reasons, one common factor was the inadequate attention given to socioeconomics in the development of systems and projects. As Current et al. (1995) point out, many projects failed because producing benefits for farmers was rarely an important objective of agroforestry. To remedy this, many agroforestry institutions are now calling for increased socioeconomic research. In this paper, we evaluate the current status of socioeconomic research in agroforestry to answer three basic questions: what has been done, what are the gaps in our knowledge, and what are the priorities for future research? To answer these questions, we quantitatively and qualitatively analyze socioeconomic papers published in *Agroforestry Systems* during the period from 1982 to 1996 and present the results of a survey of agroforestry socioeconomic researchers.

What has been done? A review of socioeconomic articles in *Agroforestry Systems*

In order to assess the current status of agroforestry socioeconomic research, we reviewed all papers published in *Agroforestry Systems* since its inception in 1982. Since socioeconomic research results are published in a variety of other journals, books, conference proceedings, and research reports, restricting our analysis to *Agroforestry Systems* may overlook many important contributions to the field. Nevertheless, we felt that as the major outlet for agroforestry research (of all disciplines), *Agroforestry Systems* should provide a reasonable overview of the relative emphasis and progress made in the many disciplines that comprise socioeconomics (including all disciplines in the social sciences, but specifically, sociology, economics, anthropology, human ecology, geography, and public policy studies). Therefore, using *Agroforestry Systems* as a sample of current and past agroforestry research efforts, we examine what has been accomplished, where the socioeconomic side of agroforestry is headed, what methodologies are being employed, and what major questions are being addressed.

Methods

An initial screening of all issues of *Agroforestry Systems* was carried out to identify articles that had a substantial focus in one of the socioeconomic disciplines (i.e., sociology, economics, anthropology, human ecology, geography, or public policy). Abstracts of all articles published between 1982 and 1996 were examined, and those that mentioned any type of social, cultural, or economic content were selected for further analysis. These articles were then read in detail to ascertain the extent of socioeconomic content. Many of these only briefly mentioned socioeconomic information, such as family size, population density, and village size, as background information for biophysical studies, so were deleted from the set of socioeconomic papers. This process resulted in the identification of 113 papers with a primary focus on socioeconomics. To facilitate analysis, the articles were divided into three groups, based on five-year time periods: 1982–86, 1987–91 and 1992–96. The selected articles were then re-read and classified into categories, according to the following criteria:

- discipline;
- type of research
- data collection method(s);
- type of analytical method(s);
- type of agroforestry system(s); and
- geographic and eco-region of study site(s).

Results

A count of articles published in *Agroforestry Systems* between 1982 and 1996 (and for each five-year period) is presented in Table 1.[1] Out of a total of 517 articles published, 113 (22%) were classified as predominately socioeconomic. The percentage of socioeconomic papers has remained fairly constant: 24% for 1982–86, 20% for 1987–91, and 23% for 1992–96. This suggests that despite recent efforts to increase socioeconomic research, the dominance of the biophysical sciences in *Agroforestry Systems* has remained virtually unchanged over the past 14 years.

Table 2 presents the relative numbers and percentages of socioeconomic articles by primary disciplinary focus. Economics, whether theoretical or applied, has consistently made up the largest single category of socioeconomic papers, representing 50% (11 articles) of the total in 1982–86, 59% (25 articles) in 1987–91, and 49% (24 articles) in 1992–96. The general category of Project Development follows in second place, as the focus of about one-fifth of all socioeconomic papers (9% in 1982–86; 29% in 1987–91; 16% in 1992–96). The Project Development category includes all papers that examine agroforestry projects in general terms without any specific academic focus. The majority of these are qualitative evaluations and descriptions of the design and implementation of projects and extension methodologies.

Table 1. Number and percentage of socioeconomic articles in *Agroforestry Systems* from 1982 to 1996.

Agroforestry Systems articles	1982–86	1987–91	1992–96	Total 1982–96
Number of socioeconomic articles	22	42	49	113
Total number of articles	92	215	210	517
Percentage of socioeconomic articles	24	20	23	22

Table 2. Number and percentage of socioeconomic articles in *Agroforestry Systems*, by discipline.

Discipline	1982–86		1987–91		1992–96		Total 1982–1996	
	n	%	*n*	%	*n*	%	*n*	%
Project development	4	18	12	29	8	16	24	21
Sociology	3	14	4	9	5	10	12	11
Human ecology/ Anthropology/Geography	4	18	1	2	4	8	9	8
Economics	11	50	25	59	24	49	60	53
Policy	0	0	0	0	8	16	8	7

Only 12 sociology papers (11% of all socioeconomic papers) were published during the 14 years. The nine papers examining cultural aspects of agroforestry systems (counted in the combined discipline of human ecology, human geography and anthropology) represent only 8% of the total. There are several possible explanations for this paucity of sociology and anthropology papers in *Agroforestry Systems*. First, as one reviewer noted, sociologists and anthropologists may be publishing much of their agroforestry research in more disciplinary-focused, peer-reviewed journals due to academic pressures. However, these academic pressures are no different for economists or biophysical scientists. Second, perhaps sociological and anthropological studies are being disproportionately published in the non-peer-reviewed, grey literature produced by donor agencies and non-governmental organizations (NGOs). For example, one anonymous reviewer reported that the Food and Agriculture Organization, the Overseas Development Institute, and Intermediate Technology have been publishing sociology and anthropology case studies for at least five years. However, much biological and economic research is also initially published in this type of literature prior to publication in peer-reviewed journals. There is no *a priori* reason to believe that sociology and anthropology studies are disproportionately represented in the grey literature. Nevertheless, *Agroforestry Systems* may need to make a greater effort to convince sociologists and anthropologists to publish in peer-reviewed fora, such as *Agroforestry Systems*, that reach the broader agroforestry community. Third, there may indeed be significantly fewer sociological and anthropological studies of agroforestry, which would suggest a considerable need for increased sociological and cultural research.

Finally, it is noteworthy that no papers with policy as the primary focus were published until 1992. Prior to 1992 there were only four papers published in which policy analysis was identified as even a secondary disciplinary focus. It is encouraging, however, that the number of policy papers has increased dramatically in the past few years.

Type of research and analysis

The percentages of papers by type of research and analysis are presented in Tables 3 and 4, respectively. It appears that agroforestry may be entering a third generation of socioeconomic research. In the first generation (1982–86), there was an emphasis on general theoretical and review papers of agroforestry and the problems that socioeconomics might address. The majority of these were based on literature reviews and qualitative analyses. In the second generation of papers, we see an increase in empirical, quantitative papers. The importance of such empirical studies increased over time, representing 45% of the papers in 1982–86, 58% in 1986–92, and 70% in 1992–96. Most of these empirical studies, however, rely primarily on qualitative and simplistic quantitative methods, and cost-benefit analysis. During 1987–91 and 1992–96, there was a disconcerting increase in papers relying on qualitative analyses and a decrease in simulation and modeling papers. In recent years, however, a third generation of research appears to be emerging, research that is primarily empirical but that utilizes increasingly sophisticated techniques based on multiple regression analyses. These studies increased from 0% of all studies

Table 3. Type of research of socioeconomic papers in *Agroforestry Systems*.

Type of research	Percent of all socioeconomic articles			
	1982–86	1987–91	1992–96	Total 1982–96
Theoretical/modeling	9	7	6	7
Empirical/descriptive	27	37	39	35
Empirical/analytical	18	21	31	25
Literature review	32	21	16	21
Simulation	5	12	6	8
Case study	9	9	2	4

Table 4. Type of analysis used in socioeconomic articles published in *Agroforestry Systems*.

Type of analysis	Percent of All socioeconomic articles			
	1982–86	1987–91	1992–96	Total 1982–96
Qualitative	54	33	44	41
Quantitative/descriptive	14	19	14	16
Quantitative/means comparison	5	5	6	5
Quantitative/cost-benefit	22	26	24	25
Quantitative/regression	–	–	6	3
Quantitative/Simulation	–	10	2	4
Quantitative/optimization	–	2	4	3
Quantitative/modeling	5	5	–	3

in 1991 to 6% from 1992 through 1996. This trend must continue for agroforestry socioeconomic research to have significant impact.

Types of surveys and sample sizes
As indicated in Table 5, the number of papers based strictly on literature declined sharply between 1982 and 1996, from 54% to 18% of all socioeconomic papers. The number of papers using formal survey techniques (based on scientific sampling methodologies and structured survey instruments to collect primary data suitable for statistical analysis) increased significantly over the past 15 years, from 14% in 1982–86 to 38% in 1992–96 (see Table 5). The percentage of studies using informal survey techniques (based on non-scientific sampling and unstructured/conversational approaches to collect data suitable for qualitative analysis), however, remained much the same and even increased recently (1982–86: 14%; 1987–91: 12%; 1992–96: 16%).

As shown in Table 6, sample sizes used in surveys ranged from an average of 39 (median = 36) in 1982-86 to 101 (median = 95) in 1987–91 and 124 (median = 66) in 1992–96. Overall, 42% of all surveys were based on 50 or fewer observations, and 63% were based on 100 or fewer observations. These small sample sizes suggest that conclusions drawn from past empirical work in socioeconomics are rarely generalizable, substantially reducing their reliability and impact. As one reviewer noted, there are certain types of research for which small sample sizes are appropriate, for example, participatory, qualitative research which is not necessarily seeking global or widely generalized outcomes. However, the agroforestry literature is dominated by studies that have little significance beyond the individual study, and from which little can be concluded for other sites or situations. If agroforestry socioeconomic research is to progress, it must move toward research intended to produce more generalizable results, and these usually require larger sample sizes. Thus, the recent trend toward larger sample sizes is encouraging.

Table 5. Type of data collection methods used in socioeconomic articles in *Agroforestry Systems*.

Data collection method	Percent of all socioeconomic articles			
	1982–86	1987–91	1992–96	Total 1982–96
Literature	54	28	18	29
Formal survey[a]	14	29	38	29
Informal survey[b]	14	12	16	14
Secondary sources	9	21	14	16
Field Plots[c]	9	10	14	12

[a] Formal surveys are defined as surveys using scientific sampling methodologies and structured survey instruments to collect data suitable for quantitative statistical analysis.
[b] Informal surveys are defined as surveys using non-scientific sampling methods and unstructured, conversational approaches to collect data suitable for qualitative analysis.
[c] Field plots refers to data, collected from research stations' or farmers' fields, usually used for inputs and outputs for cost-benefit analysis.

Table 6. Number of socioeconomic articles in *Agroforestry Systems* based on different survey sample sizes.

Survey sample size	1982–86	1987–91	1992–96	Total 1982–96
≤ 25	2	2	4	8
26–50	2	2	6	10
51–100	1	3	5	9
101–200	–	2	3	5
201–300	–	3	1	4
301–400	–	–	5	5
≥ 401	–	2	–	2
N =	5	14	24	43
Mean	39	101	124	139
Median	36	95	66	72

Geographic and eco-regions

Table 7 presents the percentage of papers focusing on various geographic and eco-regions. With the exception of the first five-year period (1982–86), in which papers dealing with unspecified regions made up 26% of the total, socioeconomic papers have concentrated on Africa, Southeast Asia, and South Asia. Latin American papers are under-represented, declining from 14% in 1982–86 to 10% in the last two five-year periods (1987–91 and 1992–96). The paucity of Latin American papers may be due to language barriers or preferences for publishing in regional journals, or alternatively to a real decline in agroforestry socioeconomic research in Latin America relative to Africa and Asia.

Table 7. Eco- and geographic regions of socioeconomic studies in *Agroforestry Systems.*

	Percent of all socioeconomic papers			
	1982–86	1987–91	1992–96	Total 1982–96
Ecoregion:				
Temperate	5	7	13	9
Humid tropics	45	33	24	31
Subhumid tropics	5	17	25	18
Semiarid	9	12	15	12
Tropical highlands	–	5	10	6
Unspecified	36	26	13	24
Geographic region:				
Africa	22	14	38	25
Southeast Asia	23	21	10	16
South Asia	5	26	20	20
Latin America	14	10	10	11
Temperate	5	7	12	9
Various/unspecified	31	22	10	19

As expected, the humid tropics were the most important ecoregion represented in the period 1982–86, with 45% of the total, while papers dealing with unspecified ecoregions made up 36%. However, both of these regional categories gradually decreased in importance as temperate, tropical highland, semiarid and subhumid ecoregions were increasingly represented. During the last time period (1992–96), the subhumid tropics were the most studied (25%), followed by the humid tropics (24%), semiarid regions (15%), temperate areas (13%) and tropical highlands (10%). Although the percentage of papers from temperate regions has increased over the past 15 years (from 5% in 1982–86 to 13% in 1992–96), agroforestry socioeconomics continues to focus on less-developed, tropical regions.

Type of agroforestry system
The period 1982–86 is characterized by a substantial number of theoretical articles that do not refer to any specific agroforestry system, as can be seen in Table 8. However, the 'No specified system' category decreases gradually over time from 24% of the articles in 1982–86 to only 14% in 1992–96. The remaining papers deal with a wide range of agroforestry systems, a demonstration of the broad universe of interest to researchers. Particular emphasis occurs, however, in the categories 'trees in fields or field boundaries (bunds)' which constitutes 19% of the 1987–91 papers, and 'alley cropping,' which accounts for 14% of the papers in 1992–96. In both of these cases, the relative importance can be correlated with a regional emphasis for each of these periods. In 1987–91, the most important region represented in the socioeconomic papers was South Asia (26%), where (particularly in India) research

Table 8. Type of agroforestry system examined in *Agroforestry Systems* socioeconomics papers.

Type of agroforestry system	Percent of all socioeconomic articles			
	1982–86	1987–91	1992–96	Total 1982–96
No specified system	24	17	14	17
Taungya	8	7	2	5
Alley cropping	8	4	14	9
Silvopastoral	4	11	12	10
Multipurpose trees	8	4	12	9
Woodlots	4	7	7	6
Swidden agriculture	12	7	3	6
Mixed perennials	4	4	7	5
Trees in fields/bunds	–	19	10	12
Non-timber forest products	8	–	3	3
Small farm (unspecified)	12	–	2	3
Homegarden	4	9	7	7
Erosion control	4	2	–	2
Other	–	9	2	4
Various agroforestry systems	–	–	5	2

emphasized the planting of eucalyptus in farmers' fields. Likewise, in articles from 1992–96, the most important region was Africa (38%), where alley cropping received considerable attention on research stations and also in on-farm trials.

What are the major gaps, advances, and priorities? A survey of socioeconomic researchers

A survey of agroforestry socioeconomics researchers was performed in August 1996. The primary objective of the survey was to obtain a cross-section of opinion on the status of and priorities for socioeconomic research. Agroforestry researchers around the world were asked their opinions on the advances and gaps in agroforestry socioeconomic knowledge, constraints to closing those gaps, and the most important priorities for future socioeconomic research. Methods and results are presented in this section.

Methods

Developing a sample frame for a survey of agroforestry socioeconomics researchers proved difficult. The authors contacted several agroforestry institutions (e.g., ICRAF, *Agroforestry Systems*, USDA National Agroforestry Center, Association for Temperate Agroforestry) in an attempt to develop a mailing list of agroforestry socioeconomists. Unfortunately, not a single agroforestry institution was able to provide a listing of agroforestry scientists by discipline. Finally, a sample frame was developed using lead authors of recent (since 1990) agroforestry socioeconomic publications and lists of presenters at recent agroforestry conferences and symposia. The final sample frame consisted of 93 scientists (from 23 countries) who had been involved in agroforestry socioeconomics research during the past five years. Forty-seven percent were current residents of the United States or Canada, while fifty-three percent were current residents of other countries.

Diskettes containing a computer-assisted questionnaire, developed with Sawtooth Software's Ci3 software package[2], were mailed to each scientist. Respondents were asked to provide background information on their education, current employment, agroforestry experience, and to list the following:

- The three most important advances in agroforestry socioeconomics during the past 20 years;
- The three most crucial gaps in agroforestry socioeconomic knowledge;
- The three most important constraints to closing the gaps in agroforestry socioeconomic knowledge; and
- The three highest priorities for future agroforestry socioeconomic research.

Results

Of the 93 surveys mailed, 33 usable surveys (35%) were returned. The relatively low response rate can be attributed to the international nature of the survey and the fact that limited funds did not allow subsequent re-mailings to non-respondents. The respondents' current residences reflected the initial sample frame, with 45% of responses from scientists in the US and Canada, 15% from Europe, and the remaining 40% from Africa, Asia, Latin America, Australia, and New Zealand.

Characteristics of the respondents are presented in Table 9. Seventy-eight percent are PhDs, and seventy percent have more than five years of experience in agroforestry research (12% have more than 15 years of agroforestry research experience). Similar to the disciplinary focus of *Agroforestry Systems* articles, the majority of respondents (54%) were economists, followed by

Table 9. Characteristics of survey respondents.

Highest education:			
PhD 78%, Masters 19%, Bachelors 3%			
Agroforestry research experience:			
More than 15 years:	12%	5–10 years:	40%
10–15 years:	18%	Less than 5 years:	30%
Disciplines:			
Economics:	54%	Sociology:	3%
Forestry:	18%	Agriculture:	3%
Geography:	9%	Other:	12%
Institutional affiliation:			
Universities:	67%		
Int. research org.:	18%		
Other:	15%		
Number of agroforestry socioeconomic publications:			
All publications:	Total = 218, Mean = 6.36, Median = 5, Range 1–40		
Peer-Reviewed:	Total = 84, Mean = 2.5, Median = 2, Range 0–20		
Number of agroforestry projects:			
Total (all respondents):	210		
Mean:	6.36		
Median:	4		
Range:	1–25		
Ecoregions of agroforestry projects:		Past	Current
Temperate-Mediterranean:		18%	27%
Tropical:		62%	57%
Arid/Semiarid:		19%	10%
Disciplinary focus of past projects:			
Economics:	30%	Anthropology:	10%
Policy:	18%	Biophysical:	9%
Sociology:	15%	Geography:	3%
Extension:	14%		

foresters (18%), other (12%) and geographers (9%). Sixty-seven percent were employed by universities, while 18% worked for international research organizations such as ICRAF. Respondents had published a total of over 200 agroforestry publications (mean = 6.36, median = 5) of which 84 were in peer-reviewed journals (mean = 2.5, median = 2). Respondents identified 210 previous agroforestry projects with which they had been involved, 62% of which were in the tropics, 18% in temperate or Mediterranean regions, and 19% in arid or semiarid areas. Although the majority of respondents were economists, the disciplinary foci of their past projects were more evenly distributed among the major socioeconomic disciplines.

Major advances

The most important advances in agroforestry socioeconomics identified by the survey respondents are presented in Table 10. Economic analysis methods (cost-benefit, optimization, mathematical programming) were the most frequently cited advance. They were cited 11 times and accounted for 21% of all responses. Of these, cost-benefit analysis (cited seven times) predominated while optimization tools such as mathematical programming were cited four times. Although no anthropologist and only one sociologist chose to respond to the survey, understanding traditional knowledge and gender issues, and incorporating them in agroforestry research and development were the second and third most common advances identified. Understanding factors influencing adoption behavior, identifying impacts of land- and tree-tenure, developing methods for integrating socioeconomics and biophysical approaches, and participatory research techniques were the other frequently mentioned advances in agroforestry socioeconomic knowledge.

Table 10. Most important advances in agroforestry socioeconomics research.

Advances	Number of listings*	Percentage
Economic analysis (cost-benefit, optimization, mathematical programming)	11	21
Traditional knowledge	7	13
Gender issues	6	11
Adoption behavior	6	11
Systems approaches to research (integrating socioeconomics with biophysical)	5	9
Participatory research	5	9
Land- and tree-tenure and tree tenure issues	4	8
Understanding bioeconomic relationships	2	4
Intra-household time allocation	2	4
Market impacts	2	4
Credit issues	1	2
Multipurpose trees	1	2
Role in rural development	1	2

* Based on survey responses.

Knowledge gaps

As illustrated in Table 11, many of the most crucial gaps in agroforestry socioeconomic knowledge are identical to the major advances cited by survey respondents. For example, understanding factors influencing adoption behavior was the most frequently listed gap in socioeconomic knowledge (19% of all listings), while improving economic models and tools (principally cost-benefit analysis) placed a close second with 14%. Participatory research methods, land- and tree-tenure, and market analyses also appear on both the 'advances' and 'gaps' lists. Furthermore, several of the other gaps listed are crucial areas for improving cost-benefit analyses. These include valuation of non-market goods and services produced by agroforestry systems, risk and uncertainty, and market analyses.

The constraints to closing the gaps in socioeconomic knowledge which were listed by survey respondents are shown in Table 12. The four most important constraints are closely related to one another. The lack of strong agroforestry institutions and the short-sighted focus and priorities of governments and donor organizations result in limited funding opportunities for agroforestry socioeconomic research. Inadequate funding, the narrow disciplinary focus of researchers and institutions, and the lack of institutional support for interdisciplinary studies produce difficulties in attracting socioeconomics professionals and students to agroforestry research, resulting in the human resources constraint. This suggests that increasing and improving agroforestry socioeconomic research will require a concerted and multifaceted effort that will be neither simple nor easy.

Table 11. Most crucial gaps in agroforestry socioeconomic knowledge.

Knowledge gaps	Number of listings*	Percentage
Adoption (household decision-making)	13	19
Better economic models and analytical methods	10	14
Impacts on households (welfare and equity)	7	10
Macro-economic and policy studies	7	10
Non-market valuation	6	9
Sustainability (long-term studies and impacts)	6	9
Land- and tree-tenure	4	6
Improved extension strategies and participatory research	3	4
Risk and uncertainty	3	4
Comparative, cross-site studies	3	4
Integrating anthropology and cultural studies with economics	2	3
Better data sets	2	3
Market analyses	2	3
Incorporating GIS with socioeconomic research	1	1

* Based on survey responses.

Table 12. Constraints to closing gaps in agroforestry socioeconomic knowledge.

Constraints	Number of listings*	Percentage
Funding	13	27
Human resources	8	17
Narrow disciplinary focus of researchers and institutions	8	17
Priorities and short-term focus of government institutions and aid agencies	8	10
Lack of agroforestry institutions	7	8
Complexity of agroforestry	2	4
Lack of appropriate methodologies	2	4
Lack of institutional support for interdisciplinary studies	2	4
Systems for delivering research results	2	4
Inadequate biophysical data for socioeconomic studies	1	2
Fragmentation and duplication of research efforts	1	2

* Based on survey responses.

Priorities for future socioeconomic research

The most important future research priorities reported by the survey respondents are listed in Table 13. These follow closely, but are not identical to, the crucial gaps identified in Table 11. Understanding adoption behavior again leads the list (19% of all listings), followed by improving economic models and methods (11%), analyzing impacts on farm households (11%), and integrating farm-level research with policy and regional analyses (10%). The most important priorities tend to be interrelated, for example, understanding and predicting adoption behavior and analyzing impacts on farm households. Furthermore, the most commonly used tool for impact assessment is cost-benefit analysis. Some of the major areas in need of research for improving

Table 13. Priorities for future socioeconomic research.

Research priority	Number of listings*	Percentage
Adoption behavior	10	19
Impacts on farm households	6	11
Improved cost-benefit and profitability analysis	6	11
Integrate farm, policy, and regional level research	5	10
Non-market valuation	4	8
Participatory research methods	4	8
Land- and tree-tenure issues	3	6
Risk and uncertainty	3	6
Criteria for sustainability	3	6
Institutionalize agroforestry	2	4
Improve data (bioeconomic, culturally informed)	2	4
Market research	2	4
Gender research	2	4

* Based on survey responses.

agroforestry cost-benefit analysis include methods for measuring and incorporating non-market values, evaluation of risk and uncertainty, improvement of bioeconomic data sets, and increased understanding of the interactions between food, wood, labor, and capital markets in semi-subsistence rural economies.

Discussion

As emphasized in the survey of agroforestry socioeconomics scientists (Tables 10–13), the areas in which the most important advances have been made remain the priority areas for future research. This suggests that agroforestry socioeconomics remains a field in its infancy. One might be tempted to speculate that the greatest advance to date is the growing recognition (among non-socioeconomists) of the need for increased socioeconomic research. This research should focus on three principal areas: (1) understanding the agroforestry adoption decision-making process, (2) improving economic analyses of agroforestry systems, and (3) analyzing the impacts of alternative policies (at local, regional and national levels) on the potential of agroforestry-based, rural development initiatives. Each will be briefly discussed below.

As Sanchez (1995) aptly points out, developing and empirically testing models that accurately predict farm household land-use decisions is as essential to achieving success in agroforestry as developing models that accurately predict how trees and crops will interact. Unfortunately, most of the socioeconomic modeling efforts to date have not adequately incorporated many basic and distinguishing features of agroforestry systems. These features include the multiple output nature of agroforestry, the intertemporal and spatial variability in production of the multiple outputs, the multiple economic roles of trees on farms, the impacts of off-farm and extra-household opportunities on agroforestry decisions, and the social and cultural context of farm household decision-making. Both theoretical and empirical models of decision-making should focus on the multi-attribute nature of land-use systems, the ways in which farmers value different attributes of land use, and how they make trade-offs in choosing between land-use systems with varying attribute combinations. Under-studied attributes include production of non-market goods and services; aesthetics; long-term benefits such as soil conservation; risk and uncertainty; the role of markets; gender issues and impacts on intra-household allocation of labor; and religious and other social and cultural attributes.

Recent reviews of economic evaluations of agroforestry projects (e.g., Swinkels and Scherr, 1991; Scherr, 1995) confirm our analysis that improving agroforestry cost-benefit analysis remains an area that requires significant additional research effort. The majority of previous cost-benefit studies consist of ex-ante financial analyses lacking sufficient detail and geographic scope to provide reliable and robust conclusions on the relative profitability of agroforestry. Rarely have they reflected farmers' perspectives or used empirical field data. Even from a planner's or government's perspective, most previous studies are deficient, since the vast majority ignore environmental and other non-market benefits that are often the primary motivations for most agroforestry projects. Furthermore, a number of cost-benefit analyses report unbelievably high rates of return to agroforestry investments. If agroforestry systems truly produced internal rates of return of 30% to 50% (as have been reported in several studies), farmers (and other investors) would be flocking to agroforestry. Extension and incentive programs would be unnecessary. More likely, analysts are either relying on inaccurate data that exaggerate the benefits or overlooking some of the costs associated with agroforestry.

Recent efforts are beginning to improve agroforestry cost-benefit analysis, as evidenced by the collection of studies presented in Sullivan et al. (1992) and Current et al. (1995). Nevertheless, the majority of these improved studies still rely on inadequate sample sizes and ignore non-market benefits and services of agroforestry systems. Critical research needs for improving agroforestry cost-benefit and impact analyses include incorporating risk and uncertainty, estimating and incorporating non-market goods and services produced by agroforestry systems, multi-input and output production function analysis, and linking economic analyses to primary, field and household survey data. Now that many agroforestry projects have 10- to 15-year histories, it is time to move from over-reliance on ex-ante analyses to thorough and rigorous ex-post cost-benefit analyses based on extensive farmer surveys. These analyses should also include evaluations of the inter-household and inter-generational equity impacts of agroforestry projects.

A wide variety of policies directly and indirectly influence the ability of agroforestry systems and projects to deliver benefits to individual farmers and the larger society. These include, but are not limited to, policies affecting labor, capital and goods markets, land- and tree-tenure policies, and energy policies. Understanding how this morass of interacting public policies influences the agroforestry decisions of small, semi-subsistent farmers in rural agricultural communities with varying social and cultural environs remains a daunting challenge. Yet, without this understanding we will mostly be shooting in the dark in designing agroforestry systems and implementing agroforestry projects.

Conclusions

Our analysis of the first 14 years of articles in *Agroforestry Systems* suggests that both the scope and the quality of socioeconomic research is slowly improving. Although the absolute numbers of socioeconomics research papers have increased significantly, the percentage of socioeconomic papers relative to the total number of papers in *Agroforestry Systems* has remained virtually unchanged since 1982. This suggests either that recent efforts to encourage more socioeconomic research have yet to bear fruit or that socioeconomics researchers are publishing much of their agroforestry research elsewhere. The recent trend away from literature reviews, qualitative, and purely descriptive quantitative research methods based on small sample sizes, and toward more rigorous, statistical analyses of better and larger data sets is encouraging.

Economics has consistently dominated agroforestry socioeconomics research. This may result from the influences of donor and funding agencies that emphasize cost-benefit and profitability analyses for project decision-making. Unfortunately, most of the economic analyses have applied rather simplistic analytical techniques to questionable data. We need to encourage the move toward more rigorous analyses that we have witnessed during the past couple of years.

Despite the considerable efforts that have been expended toward promoting and implementing participatory research methods, the dearth of sociology and anthropology based agroforestry research (as reflected in *Agroforestry Systems*) is disturbing. Even if sociologists and anthropologists are publishing the bulk of their work in their own disciplinary journals or in the grey literature produced by donor agencies and NGOs, this is still problematic since it is doubtful that agroforestry planners (particularly in less-developed countries) have access to or regularly review those publications. It is difficult to envision effective participatory research or agroforestry project and policy initiatives without substantial social and cultural understanding of how individuals and communities relate to, make, and react to changes associated with replacing current agricultural systems with modern agroforestry systems. Combined with inadequate economic analyses, the paucity of social, cultural, and policy research is likely responsible for much of the ineffectiveness associated with past agroforestry development projects.

Notes

[1] A list of all the identified socioeconomics articles may be obtained by sending an e-mail request to *emercer@rtpmail.emapfhm.gov*.

[2] The Ci3 System is a software package with capabilities for composing computer-assisted interviews, administering interviews in the field, and doing simple data analysis for facilitating computer-assisted interviewing.

References

Current D, Lutz E and Scherr S (eds) (1995) Costs, Benefits, and Farmer Adoption of Agroforestry: Project experience in Central America and the Caribbean. World Bank Environment Paper 14. World Bank, Washington, DC

Mercer DE (1993) A framework for analyzing the socioeconomic impacts of agroforestry projects. Southeastern Center for Forest Economics Research, Research Triangle Park, NC. FPEI Working Paper No. 52, 34 pp

Müller EU and Scherr SJ (1990) Planning agroforestry interventions in extension projects. Agroforestry Systems 11: 23–44

Nair PKR (1996) Agroforestry directions and literature trends. In: McDonald P and Lassoie J (eds) The Literature of Forestry and Agroforestry, pp 74–95. Cornell University Press, Ithaca, NY, USA

Sanchez PA (1995) Science in Agroforestry. Agroforestry Systems 30: 5–55

Scherr SJ (1995) Economic analysis of agroforestry systems: the farmers' perspective. In: Current D, Lutz E and Scherr S (eds) Costs, Benefits, and Farmer Adoption of Agroforestry: Project experience in Central America and the Caribbean, pp 28–44. World Bank Environment Paper 14. World Bank, Washington, DC

Sullivan GM, Huke SM and Fox JM (eds) (1992) Financial and Economic Analyses of Agroforestry Systems. Nitrogen-Fixing Tree Association, Paia, Hawaii, USA

Swinkels R and Scherr SJ (1991) Economic Analysis of Agroforestry Technologies: An Annotated Bibliography. ICRAF, Nairobi, Kenya

Agroforestry Systems **38**: 195–222, 1998.

Policy issues in agroforestry: technology adoption and regional integration in the western Brazilian Amazon

S. A. VOSTI[1,*], J. WITCOVER[1], S. OLIVEIRA[2] and M. FAMINOW[3]
[1] *International Food Policy Research Institute, Washington, D.C., USA*; [2] *EMBRAPA-Rondônia, Brazil*; [3] *University of Manitoba, Canada (* Author for correspondence: e-mail s.vosti@cgnet.com)*

Keywords: agricultural development, livestock, socioeconomic characteristics

Abstract. Agroforestry systems (AFS) have received much attention from policymakers and others for their perceived ability to contribute significantly to economic growth, poverty alleviation, and environmental sustainability. Yet in many areas of the developing world where AFS have proven their agronomic potential, they have yet to be widely adopted. This paper approaches the issue with a focus on one such area – the western Brazilian Amazon, an area where economic integration within the region itself, across national boundaries, and between the region and the rest of Brazil ('regional integration') may soon vastly shift incentives to natural resource users. The paper addresses AFS adoption via a conceptual framework that argues for scrutinizing agronomic and socioeconomic characteristics of AFS and their potential adopters. This framework facilitates *a priori* assessment of whether specific AFS will be adopted, and if so, by whom and with what likely impact on the development objectives at hand. It highlights characteristics that lend socioeconomic and agronomic 'agility' – connoting flexibility to easily shift production systems and accompanying marketing strategies in response to unforeseen or rapid change (due to qualities of the AFS itself, or the producer). Financial performance indicators (net present value of AFS during establishment and for the life of a project, and years to positive cash flow) are presented for selected AFS, and juxtaposed to a livestock production system that is already gaining wide adoption in the area. This juxtaposition illustrates the practical effects of socioeconomic and agronomic characteristics posited as influential. Against this backdrop, trends in 'regional integration' are discussed, along with implications this integration process has for AFS adoption. The paper ends with a discussion of the roles and limitations of policy in promoting AFS adoption.

Introduction

This paper discusses technology adoption issues in the context of agroforestry systems (AFS) in the western Brazilian Amazon, and the potential impact of regional integration as a *source* (perhaps *the* major source) of changes that may greatly influence directly and indirectly the adoption of AFS in that region. It does so against the backdrop of the following questions, which are critical to policymakers in that region and elsewhere:

(1) *The Big Picture* – What role can AFS play in increasing growth, reducing poverty, and protecting the environment: the 'critical triangle' of development objectives? What factors condition the potential role of AFS in achieving these development objectives?

(2) *The Farm-level View* – If AFS do have a role to play in achieving these multiple development objectives, why are AFS not being widely adopted in some key areas of the developing world, especially where they have a demonstrated agronomic comparative advantage over alternative land uses? What farm-level issues, economic and non-economic, might slow AFS adoption rates?
(3) *Search for Policy Guidance* – What, if anything, can policymakers at different levels do to increase the potential role of AFS in achieving rural development objectives? What, if any, policy action might be taken (again, at different levels) to speed adoption rates in areas where AFS have potentially important roles to play?

The specific geographic focus for this paper, the western Brazilian Amazon, may limit its generalizability across the rest of the Amazon basin, and to other agroecological settings. However, the Brazilian Amazon is the largest remaining tropical moist forest on earth, and efforts to establish AFS both as buffers to contain deforestation and as a means to recover degraded areas have not been generally successful. Moreover, the region appears to be undergoing some profound changes in infrastructure and market links that could have dramatic effects on land use throughout the Amazon Basin. We therefore need to reexamine not only definitions of AFS and the potential for AFS to play a dominant role in conserving the western Amazon and improving welfare there, but also what researchers and policymakers might do (if anything) to expand and promote that role, in the context of the regional integration – here defined broadly to embrace heightened trade among partners in the same state or nation as well as among partners in different nations – currently taking place. That said, many of the issues addressed here will be relevant in other areas where AFS play, or may someday play, important roles.

As implied above, research on technology adoption needs to take, and generally does take, a farm-level 'view': changes in product mix and production technology occur (or fail to occur) at farm level, with research focusing on factors that influence farmers' decisions, such as profitability, labor constraints, and access to credit. However, if the research focus remains exclusively at farm level, and uses a pre-set analytical framework for technology adoption issues as often is the case, two types of problems arise. First, since methodologies for technology adoption may need to be modified to reflect the characteristics of the products to be adopted and/or their potential adopters (and gaps remain in our knowledge about the extent to which, or under what conditions, this is the case), applying a pre-set analytical framework may lead to key issues being overlooked in the search for ways to promote agricultural change. Second, and perhaps more important, an exclusively farm-level focus tends to not distinguish among *sources* of factors (such as input and output prices) known to influence farmers' decisions; these are assumed to be linked somehow to 'policy' at one level or another. As a result, policymakers are called upon to 'fix' everything, even things far beyond their control, and the consequences of likely trends in the absence of action go unconsidered. This paper addresses both sets of issues.

We proceed as follows. The next section refines the concept of AFS as relevant to the technology adoption and regional integration issues under discussion. The section on *AFS technology adoption with growth, poverty, and environment in mind* briefly reviews some known technology adoption issues and places some new issues specific to AFS into the debate. This section also presents the results of field data on the profitability of a particular set of AFS being adopted in the western Amazon, and compares these to the financial returns to small-scale livestock production – a land-use system that *is* gaining wide acceptance among small farmers in the western Amazon. The section on *Regional integration* further defines that term, identifies the chief driving forces of regional integration in the western Brazilian Amazon, and discusses its potential impact on AFS adoption. Finally, the last section concludes the paper and discusses policy implications.

Selecting relevant subsets of AFS for adoption research

Most studies on AFS focus on specific subsets of the systems or important components of these subsets, usually emphasizing the agronomic characteristics of the forestry component – what might grow in specific geographic areas, and what contributions particular AFS might make to poor soils there. But should other characteristics and components of AFS also be taken into account when deciding where to focus? Rarely is any systematic thought given to the implications for technology adoption assessments of focusing on one component or group of components or another of an AFS. Different tools for studying technology adoption may apply under different circumstances, and policy action may be specific to subsets of AFS.

In narrowing the focus of AFS studies, a second type of criterion, perhaps more important in some socioeconomic settings than in others, should, but rarely does, come into play: the characteristics of potential adopters of AFS systems. In a given region, there may be a variety of economic actors, with different potentials for adopting AFS. Many of the characteristics of potential adopters are known and can be used *a priori* to select relevant subsets of AFS for adoption research (as well as extension strategy) appropriate to the actors. Dominant, area-specific concerns about poverty, growth, or environmental sustainability could be used to help decide which group(s) AFS technology adoption should target, and which AFS systems or components are most relevant for each.

An additional criterion important for, but rarely included in, the process of refining the set of AFS for technology adoption study is current land-use patterns. The extent to which some 'woody perennials' *already* exist on farmers' lands, including private holdings of primary or secondary forest, for example, may affect future AFS adoption (Vosti et al., 1998). This aspect will be particularly important when assessing the marginal benefits of AFS adoption to the environment – one 'point' on the critical triangle of development objectives.

In addition, some assessment of current or likely future trends in relevant characteristics should be part of the selection/refinement process. In the case discussed in this paper, regional integration may imply a raft of socioeconomic and land-use changes that shifts the focus from one component or type of AFS or AFS user, to another. Moreover, if changes can be expected to come rapidly, the ability (agronomically, or socioeconomically) to shift production patterns or behavior in accordance with new incentive signals (often price) – what we call 'agility' in this paper – also matters. As is discussed in more detail later in this paper, it is useful to distinguish between two types of agility that may influence a producers' ability to shift production systems rapidly: (1) 'agronomic' agility, which describes production systems' characteristics, such as long establishment periods, that limit producers' ability to switch out of them, and (2) 'socioeconomic' agility, which describes producer characteristics, among them marketing savvy, market information, access to credit, and risk aversion. Note that the ability to change *location* in response to changed signals – towards either an urban center or a new forest margin – must also be taken into consideration.

In sum, the comprehensive set of AFS must be pruned to make technology adoption research, and assessments of likely changes (in the case of this paper, the impact of regional integration), practical. The pruning should be done through consideration of the characteristics of AFS (or their components), potential adopters, and the lands they occupy – not just today, but in the future. The specific characteristics that will be used in the pruning can be expected to shift from one place to another, and over time. Figure 1 depicts that some subsets of AFS might meet both the agronomic and socioeconomic characteristics necessary to achieve desired goals (overlap areas),[1] and should thus receive priority consideration in technology adoption studies. It also suggests that not just current areas of overlap should be considered, but future ones as well. Applied beyond farm level, Figure 1 could also embrace broader environmental effects of technologies (an *agronomic* characteristic), as valued by producers and consumers alike (a *socioeconomic* characteristic).

Agronomic and socioeconomic characteristics, so interpreted, are useful components to a refinement of AFS focus in our context – the western Brazilian Amazon – in a way likely generalizable to other areas. The general definition above, for example, would include both a complex AFS involving mahogany (*Swietenia* sp.), peach palm (*Bactris gasipaes*), coffee (*Coffea canephora*) and intercropped annual crops, and an AFS comprised of a degraded pasture where woody perennials are purposely allowed to grow. The potential for the former AFS to contribute to growth and poverty alleviation (at least) is expected to be far greater than the latter. What's more, the degree of purposiveness associated with the perennials' growth may be relevant in

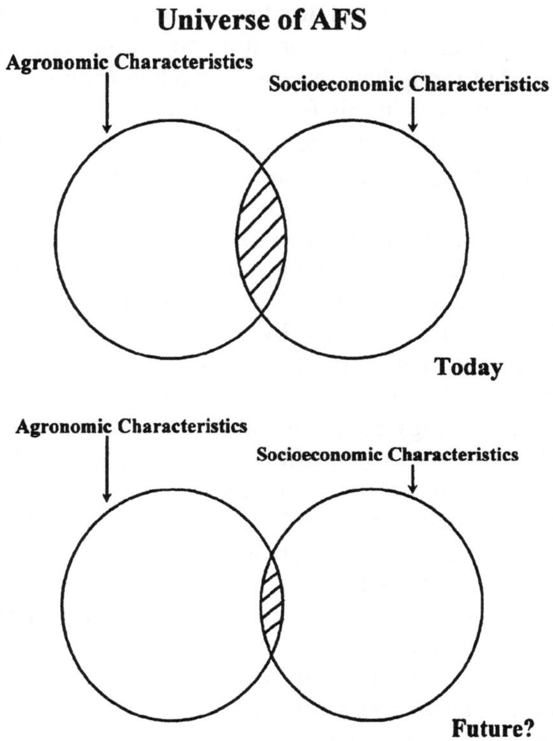

Fig. 1. Representation of all agroforestry systems (AFS) and subsets having desirable agronomic and socioeconomic characteristics – today and in the future.

gauging their durability in the face of alternative land uses that may arise with sweeping change (for example, the regional integration expected here). Therefore, the former system would be given greater weight when deciding where along the AFS continuum – from woody perennials appearing spontaneously on abandoned land to complex AFS comprised of planted and carefully managed perennial species – to concentrate technology adoption research and possible policy action.

The distinction suggested between different potential user groups is also relevant in our context. There are at least three prominent types of land users in the western Brazilian Amazon – extractivists, small farmers (usually linked to colonization projects), and large farm enterprises. The relevant set of AFS for technology adoption research may vary by user type, as do the socioeconomic and agronomic characteristics describing them and their production systems. Extractivists, for example, are quite literally surrounded by woody perennials, seasonally labor constrained, and usually removed from markets. While fewer in number than other user types, they use large tracts of land. Large farm enterprises, on the other hand, are technologically sophisticated and mechanized, have high capital/labor and land/labor ratios, and (currently) focus on livestock production. Smaller farmers occupy intermediate positions in terms of land/labor and capital/labor ratios, sometimes enjoy relatively easy access to market, but rarely have access to mechanization. A common characteristic across land users in this forest margin area is engagement in some degree of slash-and-burn agriculture. It is essential to target subsets of AFS for technology adoption by type of land user, in part because adopter characteristics vary so strongly, and in part because current land uses (and the role of AFS in production activities) vary as well. The weighting of poverty alleviation, growth, and environmental concerns would influence the balance of target users of AFS, and, in turn, which AFS (or components) would come under study for technology adoption by that target group.

Among the types of land users, this paper will focus on small farmers in the western Brazilian Amazon and will use the set of AFS relevant to this group to discuss technology adoption and regional integration issues, in that order. The small farmer focus stems from a simultaneous emphasis on *poverty alleviation* and *environmental concerns*; this group includes the largest number of poor land-users, accounts for a substantial portion of land area, and may include subpopulations likely to migrate into new forest margin areas. The other two land user groups (extractivists and large farm enterprises) are not addressed further, but this exercise could be repeated in order to select a relevant set of AFS for them.

We begin with a look at farm and farm household characteristics. Tables 1 and 2 provide a brief overview of farm household characteristics for one sample of households in the western Brazilian Amazon (Witcover et al., 1996). Data are drawn from a 1994 field survey of approximately 150 rural households, approximately one half of which were located in the colonization project of Pedro Peixoto, Acre, and the remainder in the colonization project of Theobroma, Rondônia. Household characteristics can influence agricultural change. Note that household size (number of members) is not large, especially since average farm size is about 80 hectares, and there are few female-headed households. Perhaps the most relevant demographic trend vis-a-vis agricultural change (and specific AFS, given their labor demands), though, is that, although dependency ratios are currently relatively low, reductions in fertility rates and age-specific out-migration will likely change this, leaving an aging rural population. Finally, more than half of sample households have already diversified their income through off-farm sources – often outside the agricultural sector – signaling a search for alternative income sources and a possible window of opportunity for AFS.

But if current land-use patterns matter to future adoptability of AFS, what are the current land-use patterns of these small farmers, and where do AFS already appear in them? Figure 2 depicts the percentage of land that an average sample household dedicated to different types of uses, including primary forest (Witcover et al., 1996). Note that just a small fraction (about 4 percent) of an average household's land was dedicated to perennial tree crops – and most of that, to pure stands of coffee and cacao (*Theobroma cacao*).

Table 1. Household composition and literacy in two colonization projects in the western Brazilian Amazon: Pedro Peixoto, Acre and Theobroma, Rondônia.

Variables	Mean	*PP* *Theo*	S.D.	Min	Max	N	Variable labels
HHSIZE	5.38		2.36	1	12	155	Household size (members)
ADULT EQUIV	4.22		1.80	1	10.46	152	Adult equivalent (@ 2723 kcals/day)
ADULT MALES*	1.77	*1.60* *1.95*	1.19	0	5	155	HH male adults (number)
ADULT SEX RATIO	0.41		0.22	0	1	151	HH adult males/HH non-dependents
DEPENDENCY RATIO**	0.35	*0.39* *0.30*	0.24	0	1	155	HH dependents/HH size
ELDERLY/WORKER RATIO	0.13		0.26	0	1	151	HH elderly/HH non-dependents
CHILD/WORKER RATIO*	0.55	*0.64* *0.45*	0.62	0	3	151	HH members < 12 yrs/ HH non-dependents
LITRATIO**	0.66	*0.61* *0.71*	0.31	0	1	155	Literate HH members/ HH members > 5 yrs
HHH AGE	47.57		12.60	19	92	142	Age of HH head
FEMALE HHH	0.06		–	–	–	155	Female-headed HH (0 = No; 1 = Yes)

Source: Witcover et al. (1996).
Note: *, **, and *** denote significantly different means for a given variable across projects at the 10%, 5%, and 1% confidence levels, respectively; when significantly different for a given characteristic, project means are reported in italics alongside sample means.
PP = Pedro Peixoto; Theo = Theobroma.

Several points in Figure 2 merit mention. In every land-use category depicted, AFS (using the broadest definition) may have already existed, but were not identified in the survey. Why is this true, and is it important? This survey (like most land-use surveys) allocated land to primary activities; systems in which woody perennials were not predominant fell outside the AFS category and were subsumed in other land-use categories. For example, Brazil nut (*Bertholletia excelsa*) trees are almost always retained in all land-use categories – it is illegal to cut them down. Perhaps more relevant, in the pastures of small farmers it is common to find patches of shrubs or occasional shade trees, almost all of which occur naturally. Bushes, shrubs and trees (other than Brazil nut trees) can also be found in areas dedicated primarily to annual crop production. Once one allows for inclusion of woody perennials over time in a land-use system, then the likelihood of an AFS (of some sort or another) appearing on a particular piece of land is quite high. Indeed, the

Table 2. Wealth and off-farm income of households in two colonization projects in the western Brazilian Amazon: Pedro Peixoto, Acre and Theobroma, Rondônia.

Variables	Means	*PP* *Theo*	S.D.	Min	Max	N	Variable Labels
GOODS INDEX	2.64		2.05	0	9	155	Index of goods on farm
OFF-FARM INC (*ANY* TYPE)	0.52		–	–	–	155	Off-farm income of any above types (0 = No; 1 = Yes)
OFF-FARM AG	0.19		–	–	–	155	Off-farm income (ag or livestock sector; 0 = No; 1 = Yes)
OFF-FARM NON-AG	0.15		–	–	–	155	Off-farm income (non-ag sector; 0 = No; 1 = Yes)
PENSIONS	0.16		–	–	–	155	Off-farm income (pension, remittance; 0 = No; 1 = Yes)
COMMERCE	0.05		–	–	–	155	Off-farm income (commercial; 0 = No; 1 = Yes)
OFF-FARM OTH	0.01		–	–	–	155	Off-farm income (unclassified; 0 = No; 1 = Yes)

Source: Witcover et al. (1996).
Note: *, **, and *** denote significantly different means for a given variable across projects at the 10%, 5%, and 1% confidence levels, respectively; when significantly different for a given characteristic, project means are reported in italics alongside sample means.
PP = Pedro Peixoto; Theo = Theobroma.

fact that many land uses *can* be characterized as AFS, if a general definition is used, reinforces the point made earlier that, for purposes of technology adoption study that takes dynamic forces into account, some narrowing of focus is required.

If naturally occurring (indeed, invading) woody species appear sooner or later in almost all land-use systems in the western Brazilian Amazon, then how should the set of AFS be narrowed? The focus here is not all AFS, but primarily those that contribute to growth and poverty alleviation goals (that is, those that might be adopted by farmers and would be contained in the overlap sections of Figure 1), and those making important contributions to environmental sustainability (which may lie in the overlap area of Figure 1, viewed at a level beyond that of the farm). How such concerns change the list and interpretation of relevant characteristics of AFS (agronomic and socioeconomic) for technology adoption study are discussed in the next section.

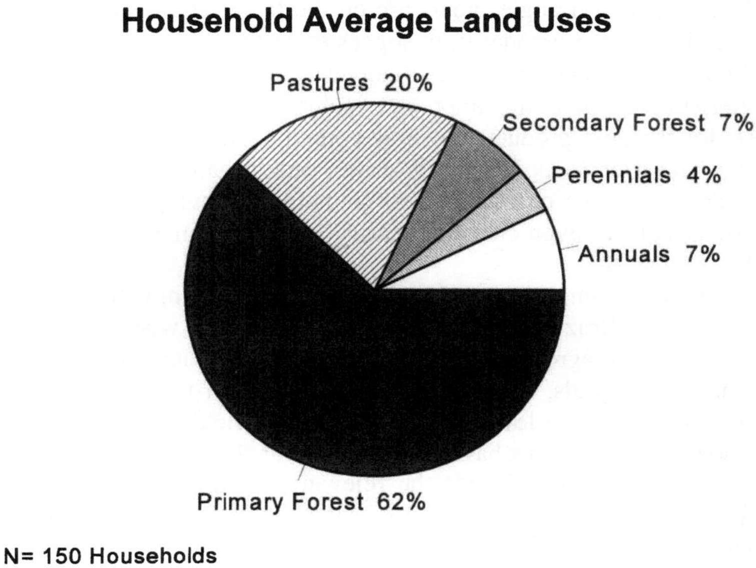

Source: Witcover et al. (1996).

Fig. 2. Land-use patterns of small farmers in the western Brazilian Amazon, based on survey data for the 1993/94 agricultural year.

AFS technology adoption with growth, poverty, and environment in mind

All modern technology adoption research takes as a point of departure farm household needs, constraints, and opportunities, including off-farm opportunities that do not involve agriculture (Reardon and Vosti, 1992; 1995). It is worth restating that farmers make technology adoption decisions based on the private costs and benefits of alternative production systems, and subject to the financial, labor, and other constraints they face. By definition, then, farmers do *not* include in their calculations the social benefits or costs associated with these alternative systems. So, while farmer's decisions may be based, to some extent, on the incentives of poverty alleviation and environment sustainability (at least with respect to maintaining the productivity of the natural resource base upon which the farmers depend for their livelihood), such decisions on many farms need not add up to societal goals such as broader poverty alleviation, economic growth, or preservation of public goods provided by the environment (Vosti et al., 1998). Furthermore, farmers' time horizons for planning – not unrelated to their wealth level – can be expected to influence how environmental sustainability fits into their on-farm decisions (Vosti and Witcover, 1996).

To tackle the issue of how technology adoption research tools may change once these 'critical triangle' goals are on the table requires a look at farm-level decisions in the broader light suggested in the last section (that is, with a spotlight simultaneously on the agronomic characteristics *and* socioeconomic characteristics of potential adopters, including consideration of relevant markets for farmers' products and how potential adopters are linked to these markets). This exercise points out 'positives' and 'negatives' among AFS characteristics (with some characteristics, Janus-faced, having both positive and negative aspects, or even changing 'faces' over the course of the production cycle).

This section examines how the set of basic technology adoption issues, for the western Brazilian Amazon at least, expands when one considers simultaneously the agronomic and socioeconomic characteristics of AFS and small farm households, keeping in mind the 'critical triangle' development objectives beyond farm level. Data derived from field research on financial returns to selected AFS that have been adopted, or show potential for adoption, are then presented to highlight the relevance of the 'new issues' put on the table through the consideration of such agronomic and socioeconomic characteristics.

Technology adoption – basic issues

The extensive literature on technology adoption in agriculture (for example, Feder et al., 1985; CIMMYT Economics Program, 1993) is based on the common 'departure point' of household needs mentioned above. Within this framework, the literature has generally focused on profitability (that is, returns to land, labor, and capital), risk (both production risk and price risk), the short- and long-term investments required for engaging in particular agricultural activities, and more recently property rights and the impact of changes in product mix on on-farm asset accumulation (Griliches, 1957; Gittinger, 1982; von Braun and Webb, 1989; Lipton with Longhurst, 1989; Hazell and Ramasamy, 1991; Reardon and Vosti, 1995; Nerlove et al., 1996; Vosti and Witcover, 1996). Some authors have also sought to identify the combined factors – biophysical, economic, and others – that condition both the expected and the realized returns to technology adoption (Barbier, 1996). Frequently, the literature moves beyond the farm-level view to discuss or speculate on broader implications of technology adoption for development objectives such as poverty alleviation, growth, and environmental sustainability (but rarely all three together).

Rather than review the literature here, this section highlights a few important points that illustrate how the AFS agronomic and socioeconomic characteristics that are considered important for technology adoption (following work by Current et al. (1995) among others) can, when considered jointly, yield new insights regarding the adoption of production systems that contain woody perennials, without losing sight of the effects these systems might have on poverty alleviation, growth, and environmental sustainability.

AFS technology adoption – additional issues?

While other agricultural and extractive activities can be multi-component and multi-product in nature, AFS stand out as being necessarily so (see, for example, Amacher et al., 1993; Traxler and Byerlee, 1993; and Renkow and Traxler, 1994). Indeed, it is in the multi-component, multi-product nature of AFS that many find the most virtue for AFS; but from these same characteristics sometimes stem limits to AFS adoption by small farmers, or caveats about adoption. Joint consideration of AFS agronomic and socioeconomic characteristics brings out these implications more clearly.

The multi-product/multi-component nature of AFS, first and foremost, requires that the term 'technological change' be precisely defined since the term can cover a number of situations: changes in *crop mix*, changes in *products extracted from a given crop mix*, changes in *input mix* (for a given crop/product mix), or some *combination of the three*. This in itself can be considered a plus for AFS; they offer a variety of avenues via which each type of technological change can occur, through 'tinkering' with subcomponents of the system (by scientists, or farmers themselves). For our purposes, though, it is necessary to take a closer look at the effect tinkering may have on the socioeconomic and agronomic characteristics of interest. Changes associated with tinkering may affect the overall adoptability of the AFS in question; entail a shift in characteristics *away* from those desired to meet poverty alleviation, growth, and/or environmental sustainability goals; or both.

As noted earlier, the agronomic acceptability of technological change is more likely than its socioeconomic implications to receive attention from technology adoption studies *a priori*. Some examples illustrate how bringing in socioeconomic aspects – markets for products, the producer's access to markets, the producer's production capacity – complicates yet clarifies the technology adoption picture.

Some AFS generate intermediate and/or final products for off-farm sale, while others produce inputs for on-farm production systems. The type of AFS 'tinkered with' and the type of AFS it becomes will influence the extent to which producing households are directly or immediately immersed into new sets of market arrangements – a factor that relates to both poverty alleviation and growth impacts.

Or, by including in an AFS a component that is itself multi-product in nature (e.g., a tree or bush that generates several products over time), a farmer may ease his/her risk by multiplying diversification effects via addition of a single component. Still, the socioeconomic characteristics of the user may show that the farmer will be able to harvest only one among the multiple products (perhaps because a more critical household activity requires labor at harvest time). Or, the products produced may be in markets so inter-related that diversification entails little reduction in risk. Or, the marketing channels may be so different or so differently regulated that, depending on level of producer marketing savvy, the additional product may be marketed only with great difficulty. An agronomic characteristic (provision of multiple products) combined with socioeconomic characteristics (household labor patterns, product markets, producers' knowledge of marketing) can influence the farmer's decision to adopt, the effects of adoption, or both (these situations are derived from Guimarães, 1998).

The ability to tinker with AFS certainly leads to farmers adapting crop and inputs mixes in technology packages to their own needs. While this might encourage more widespread adoption of AFS, AFS in farmers' fields are likely to differ substantially from recommendations regarding plant density, component distribution, and/or rotation scheme determined on an experimental station, or some average of practices observed on farmers' fields. If farmers do not know (at least in the short run) what mix of components or inputs agronomic gains depend on, they may lose those gains as they modify the original AFS packages to meet their needs. Thus, a better understanding of the relative contribution of different components (or combinations of components) to desired goals – agronomic gains being only one possibility – will help in the assessment of the likely vulnerability of particular AFS to evolve during adaptation into a system less likely to meet those goals.

In summary, the multi-component/multi-product nature of AFS perhaps enhances opportunities to develop and diffuse effective technology packages. By the same token, it may complicate the task of evaluating whether an AFS package is likely to be adopted, with what changes, by whom, and to what effect. The scenarios mentioned above focus on how characteristics relating to adoptability or impact may shift as AFS components themselves change. Specific AFS examples taken from the study area bear this out.

AFS technology adoption – examples from the western Brazilian Amazon

For evaluation of AFS (and other production systems), it is critical to know the physical and financial returns to all factors of production used in the establishment, production, processing, and marketing/distribution phases of products. Several of the factors are described here in the context of the study area, along with implications for the profitability of AFS. Before starting, it is necessary to mention some specifics regarding the western Brazilian Amazon study area.

First, in the labor- and credit-constrained environments in which most small farmers exist in this area and throughout the developing world, cash flow and the ability for household labor to meet the needs of particular production systems are vital to technology adoption. Among the household needs that smallholders (in the forest margin and elsewhere) consider when making technology adoption decisions, food security can be the most critical when compromised. In the forest margin area under study here, this is especially true during the early phases of farm establishment.

Second, it is important to reiterate that technology adoption studies must incorporate the farm-level view taken by the primary decision makers whom we seek to understand. Because alternative land-use choices 'compete' with AFS, when we examine specific AFS characteristics in this paper, the AFS potentially relevant for the western Brazilian Amazon will be juxtaposed against the land-use system that farmers in the area increasingly rely on to meet food and livelihood security needs – livestock production systems (LPS) and pastures (described in detail for Brazil in Faminow et al., 1998, and more generally in Faminow, 1998). Successful AFS will have to be at least as profitable as LPS, and fit as well as LPS into the overall activities/investment strategies of small farm households if they are to be adopted. They also need to be as 'agile' as LPS in order to respond to changes likely in the wake of regional integration. Furthermore, where primary forest is on private land, as is the case here, forested land should be viewed as another possible land use. Thus, even if AFS were more profitable than other uses for already cleared land, they need not act as a buffer against additional clearing of primary forests.

That said, a series of factors affecting AFS technology adoption is described below. The factors center around profitability for intermediate and final products sold off farm. How production is organized; how products are prepared, priced, promoted, and distributed; and what particular segment(s) of the market is targeted by producers will determine (with segment-specific demand) the profitability of each activity (see Figure 3). The factors below were chosen not only for their importance for adoption *per se*, but because they may shed light on how AFS adoption might be affected by other shifts in the economic milieu.

- *Scale of Production* – How the returns to factors of production *change* as the *scale* of production increases is vital information for those deciding to adopt or change technology (particularly changes in crop mix), and for assessment of who is likely to adopt a given technology, and when. Labor costs of establishing and extracting output from one of the AFS systems

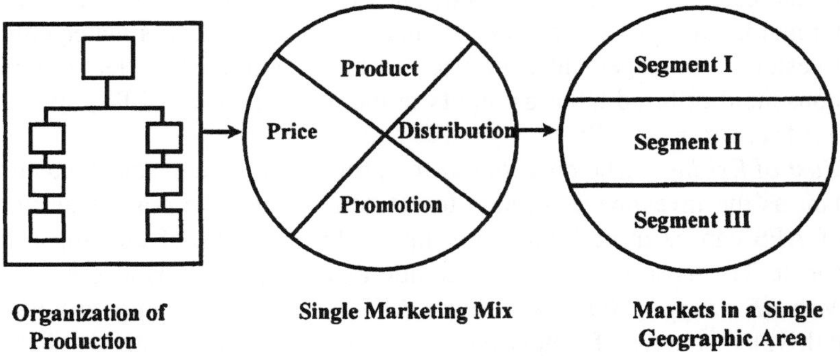

Source: Derived from Pride and Ferrell 1989.

Fig. 3. Factors affecting agroforestry system adoption and product profitability: production, marketing, and market segments for single geographic areas.

chosen for study here – coffee/bandarra (*Schizolobium amazonicum*) (described later in this paper) – vary proportionately with the amount of land dedicated to this system. While labor costs of establishing livestock production systems can also be high (especially when pasture establishment is required), they do not vary proportionately with the amount of land dedicated to these systems or the amount of product harvested. That is to say, labor requirements per unit of land dedicated to livestock production and per unit of product (milk and/or meat) both decrease as size of operational holding increases, at least up to a scale of operational holding that is relevant for small farmers. Diseconomies of scale in livestock production systems may begin to appear in large operations, and are probably due to increases in overall farm management costs rather than to costs of labor directly involved in milk or beef production (J. Valentim, personal communication). Issues of scale are not limited to production. Even in the absence of scale economies, for instance, farmers may need to form associations to meet product processing, marketing, and distribution needs (Vosti et al., 1998).

- *Timing/Size of Investment* – The costs to production systems of *under*investing or *delaying* investment may be quite high for some AFS. For example, initial investments of labor and purchased inputs in establishing a cupuaçu (*Theobroma grandiflorum*)/freijó (*Cordia* spp.)/black pepper (*Piper nigrum*) AFS are high and critical to long-term system productivity – failure to effectively establish this system can only be addressed by replanting. The same productivity drop-off is not expected for poorly timed investments in pasture formation, and subsequent recovery is possible without complete reinvestment (J. Valentim, personal communication).
- *Maintenance Costs* – For AFS, maintenance costs (especially labor costs) can be quite high, and may *not* vary with the amount of product extracted from these systems, as is the case for control of weeds in coffee/bandarra systems (labor cost is high and necessary, regardless of whether any coffee is harvested). On the other hand, once established, pastures require little human labor input and can be maintained by the judicious management of livestock – management expertise combined with animal 'labor' replaces household or hired labor as inputs in the maintenance of LPS; the same is not true for most AFS.
- *Cost of Exiting* – The costs of exiting production systems can be as important as the investments needed to establish them. The cost of getting *out* of AFS can be quite high, even higher than the cost of clearing primary forest. This cost warrants consideration by the farmer prior to adoption.
- *Non-AFS Supply Sources* – Some AFS may have to compete with non-cultivated sources of supply (such as natural forests) for market share. For example, heart of palm in Brazil can (in some areas) either be cultivated or extracted from the forest. The presence of multiple sources can reduce incentives to cultivate heart of palm in AFS, especially if extraction costs are low relative to cultivation costs, including opportunity costs of family

land and labor (Guimarães, 1998). By the same token, cultivation may actually lower production costs vis-a-vis the non-AFS sources.

- *Sources of Risk* – Compared to annual crops, AFS are generally better able to withstand production risks associated with weather. In the western Brazilian Amazon, however, production risk associated with pests (weeds and insects) can also be important, and, with the exception of a few important diseases such as 'witches broom', the implications for AFS are not well known. Furthermore, many AFS in the western Brazilian Amazon may face high price risk due to the underdeveloped nature of the markets for many of the component products (Guimarães, 1998). The double reliance on markets that AFS bring, particularly if they replace basic food grain production – a reliance on markets for the AFS products, and a reliance on the market for food, both of which may function imperfectly, and perhaps can even fail seasonally (Witcover and Vosti, 1995) – may also raise an AFS adopter's risk profile. Among the additional sources of risk that may be specific to AFS in this slash-and-burn area, fire is of chief concern. Both annual crops and woody perennials can be destroyed by fire, but the cost associated with AFS reestablishment is much higher. On the other hand, fire in pastures is often perceived to be beneficial, since it controls weed and insect infestations. While research suggests this perception is probably false in most cases, the marginal damage done to pastures by unintentional burning is far less than that done to systems such as coffee/bandarra. Finally, little is known about the production losses or increased pest risk associated with adjustments to the intensity or the component populations of AFS. For example, we do not know what happens to insect populations as cupuaçu density is increased, nor do we know the implications for soil fertility.
- *Responses to Risk* – The potential for production systems to *respond* to risk also varies substantially – the greater the scope for responding, the better from the point of view of small farmers. LPS can respond to weather and price shocks by varying product mix without changing crop mix. For example, if milk prices fall, milk can be processed for cheese or the herd can be culled. As noted earlier, AFS can sometimes be harvested in different ways to extract different products, but not always with the same degree of temporal flexibility that is seen in LPS.

When interpreting the profitability figures presented in the next subsection, it is important to keep in mind that all these factors affect profitability.

Profitability of selected AFS in the western Brazilian Amazon

Having highlighted profitability and other factors that affect technology adoption generally and AFS in particular, we now examine the financial returns to selected AFS, again in the context of the western Brazilian Amazon. Note that the profitability analyses that follow are based on typical productivity in an area known for having some of the best soils in the Amazon (Ouro Preto de Oeste, Rondônia). Sensitivity analysis to highlight changes in profitability depending on soil type (holding management practices constant) still needs to be done.

The five AFS systems examined are (to one degree or another) currently in farmers' fields, and include: cacao/bandarra, cacao/rubber (*Hevea brasiliensis*), coffee/bandarra, coffee/rubber and cupuaçu/freijó/black pepper. They would generally be classified as simple AFS, with the possible exception of the cupuaçu/freijó/black pepper system, which when mature would be considered a complex AFS. These AFS were chosen for analysis because their production and other characteristics are known, they produced products for which markets currently exist, production has been profitable, small farm production is possible, family labor is a key input, and future demand conditions are promising, especially for coffee.

The profitability of each AFS is examined over a 20-year production cycle, which may include several replanting sub-cycles, and reflects the returns to one-hectare investments in each AFS. Family labor is valued at its opportunity cost, taken to be the daily wage for short-term agricultural labor. Family land is valued at the market rental price, including the asset value of the AFS on the land – market values for land were derived from expert interviews. Outputs and inputs are valued at market prices.[2, 3, 4] Input and output prices remain constant over the 20-year horizon of this analysis.

Table 3 reports profitability indicators for each of the five AFS during their establishment periods and for the entire 20-year period. Several points merit mention. First, with the exception of the cacao/bandarra system (which uses no purchased inputs during the establishment phase), the net present value (NPV) for all systems of returns to out-of-pocket expenditures during the establishment period is negative, indicating the need for cash outlays by rural households, sometimes for several years. Second, financial returns during establishment periods become even more negative when family land and labor are included, though the impact of including family labor and land on profitability varied substantially across AFS. Third, the time to positive cash flow varies from one to four years, and three of the five systems experience negative cash flow during at least one year thereafter.

The AFS that stands out in terms of financial performance over the entire project cycle (cupuaçu/freijó/pepper) owes most of its returns to one component, black pepper. Black pepper alone would financially dominate the AFS at current prices (although little is known about the market potential for black pepper). Price sensitivity analyses undertaken (but not presented here) suggest that a 10-percent drop in pepper prices (plausible) dramatically reduces the financial performance of this apparently most promising AFS. What's more, black pepper's vulnerability to disease may make long-term productivity a real concern. Profitability improves if cupuaçu is sold outside the harvest period, as the market for this emerging product is highly seasonal (but unstable, rapidly changing from one year to the next).

Table 3. Financial performance indicators for selected agroforestry systems, Ouro Preto de Oesto, Rondônia, Brazil.

Agrofestry Systems[a]	Financial Performance Indicators					
	Returns to Out-of-Pocket Expenses Only			Returns to All Expenses – Out-of-Pocket Expenses, Family Land, and Family Labor		
	NPV[b] during establishment period (R$)	NPV for entire project (R$)	Years to Positive Cash Flow (Years)	NPV during establishment period (R$)	NPV for entire project (R$)	Years to Positive Cash Flow (Years)
Cacao and bandarra	0	4,865	1	–375	930	2[c]
Cacao and rubber	–658	4,371	2	–1,641	315	7
Coffee and bandarra	–167	3,227	2[d]	–595	1,311	3[d]
Coffee and rubber	–538	3,204	3	–1,234	195	4[e]
Cupuaçu, freijó, and black pepper	–1,378	7,522	3	–2,103	5,299	3

Source: Oliveira and Vosti (1997).

[a] The scientific names for the components of these agroforestry systems are: *Theobroma cacao* (cacao), *Schizolobium amazonicum* (bandarra), *Hevea brasiliensis* (rubber), *Coffea canephora* (coffee), *Theobroma grandiflorum* (cupuaçu), *Cordia goeldiana*, and *C. alliodora* (freijó) and *Piper nigrum* (black pepper). All systems include annual intercropping with maize (*Zea mays*) and beans (*Phaseolus vulgaris*) during the first and possibly the seecond years. Systems that include cupuaçu or cacao also include bananas (*Mus sp*), which are planted one year prior to planting these crops to provide shade.

[b] Net present value.

[c] Negative cash flows occur in years 4 and 5.

[d] Negative cash flows occur in years 11 and 12.

[e] Negative cash flow occurs in year 11.

Financial performance indicators for a small-scale dairy production system for the same 20-year production period provide an interesting counter-example to these AFS figures.[5] The analysis confirms a small but dependable annual return for a traditional dairy system, and, at reasonable discount rates, shows an improved technology package to be a good economic investment (at a 10% discount rate, NPV goes to R$5,289). This technology package includes investments in improved pasture, new corral, additional fencing, and better breeding stock, plus improved health management.

Depending on initial herd composition and timing of investments envisioned in the livestock improvement package, there are about three years until positive cash flow (comparable to the cupuaçu/freijó/pepper AFS), with small negative cash flows in several subsequent years, depending upon the need to replace/improve existing farm infrastructure, such as fences, dams, and corrals. The LPS has other features attractive to the smallholder. Cattle production is an established technology in the region (with yields fairly well known from both experience and experiment station testing), in contrast to AFS, where the risk from disease, pests, and price fluctuations, and optimal management practices are less well understood. In addition, the labor demands for the LPS system are low (again in contrast to AFS). The traditional dairy system requires little major cash outlay, but this LPS investment package (at normal discount rates) can involve substantial losses in the initial three years of investment. It will therefore require large cash reserves or access to credit, neither of which the small farmers are likely to have. If this barrier can be overcome, the analysis suggests a substantial pay-off.

Any comparison of the adoptability of AFS and LPS also needs to consider the profitability effects of their substantial differences in scale of operation (and changes in scale of operation), flexibility (in terms of timing) of key investments, and patterns and flexibility of product 'off-take'. Moreover, the indicators presented in Table 3 for AFS and the profitability calculations for small-scale dairy production do *not* include changes in the asset values of rural households – land improvements and production of perennial tree crops in the case of AFS, and livestock or land improvements in the case of dairy production. While both may be substantial, the general perception is that asset values increase more quickly and are more liquid for LPS than for AFS, but more research will be required to confirm this. If true, it would mean that the above figures, in a sense, underestimate differences between LPS and AFS profitability.

In short, this financial analysis needs to be supplemented by further research to refine the manner in which it incorporates other factors (listed above) considered critical to profitability and adoption (for example, sources of and response to various types of risk, cost of exiting, competing sources of supply). Some of these are difficult to assess quantitatively given current information. Nevertheless, the observations on and comparisons between LPS and AFS presented above provide a realistic look at likely changes in production patterns and land-use systems.

Up to this point, we have focused on system-specific financial performance indicators and taken as a given the underlying supply and demand factors that influence the prices of inputs and outputs, and perhaps more importantly, the role of macroeconomic forces in determining these prices. In the next section, we take up some of these macroeconomic issues – specifically, regional integration – as a way of assessing the future characteristics of successful AFS, especially those generating products that will be traded.

Regional integration

Some factors influencing technology adoption are 'macro' in nature (for example, exchange rates, interest rates, relative prices of broadly traded commodities, and long-distance travel costs), and beyond the control of local, state and sometimes even national policymakers. One of these factors, regional integration, is taken up in this section, with particular emphasis on the forces shaping regional integration in the western Brazilian Amazon, and the potential impact on AFS adoption (and the effects of adoption on growth, poverty alleviation, and environmental sustainability). Note that we are *not* attempting a general analysis of the links between trade and growth or environment, but continue to focus on technology adoption and the effects of adoption on development objectives. For more on these links and the methods used to analyze them, see Bredahl et al. (1996).

Regional integration is generally referred to in the context of international trade, and is increased when countries agree to reduce or eliminate barriers to trade (for example, tariffs, quotas, and domestic policies that distort product prices). (See Anderson and Blackhurst (1993) and Anderson and Norheim (1993) for an historical and global view of economic integration, and Lahera (1992) and Ffrench-Davis (1995) for Latin American perspectives.) Here, the notion of regional integration is expanded to include investments in infrastructure (important policy decisions in their own right) that have the potential to dramatically alter trade and hence production patterns. In addition, the notion of regional integration is refocused to allow for increased trade among partners within the same state, or in different states but within the same nation – a reasonable refocusing given the massive geographic scope of the western Brazilian Amazon.

In our view, *the* most important macroeconomic process in the western Amazon over the next quarter century will be regional integration. This regional integration will be (indeed, is being) driven by investments in physical infrastructure and reductions in bureaucratic and legal obstacles to trade. Both of these actions decrease transactions costs, thereby increasing or promoting the establishment of trade flows among trading partners – both existing trading partners and new trading partners, and for currently traded products and new products.

Several examples of this phenomenon currently exist in the western Amazon. First, *within*-state integration is already underway. For example, the state of Rondônia, under a regional development/integration program (PLANAFLORO), is investing heavily in rural infrastructure, communication links and market development, as a way of linking forest margin areas with the major market center of Porto Velho, and elsewhere. Second, *inter*-state integration is underway, for example, the paving of the Acre/Rondônia segment of the road BR364, which dramatically facilitates flows of inputs and outputs between these two states (and between Acre and the rest of Brazil). Finally, and perhaps most importantly, *intra-regional* integration efforts are progressing. For example, several 'avenues to the Pacific' are on the drawing board or in the process of construction, two of which would link Acre/Rondônia with Bolivia and Peru. Another example is the massive grain (and other products) transfer/transport facility recently inaugurated in Rondônia, which will more efficiently link some of Brazil's savannah areas (in Mato Grosso and parts of Rondônia, for example) to regional and global markets.

One driving force behind this regional integration is the search for new trading partners, often looking beyond traditional north-south links within Brazil towards east-west links with other Latin American trading partners. The pace and direction of this driving force is guided by comparative advantage in production and differences in levels and patterns of consumption. A second driving force is the need to reduce transportation costs among established international trading partners, especially those linked to soybeans and soybean byproducts.

The implications of these changes are several. First, new demand patterns for finished and intermediate products/services will emerge. Second, producers will be confronted with new levels and types of production and transportation costs. These two forces will combine to increase trade in finished products and intermediate inputs, and bring about a new spatial distribution of agricultural production – a distribution that will be far less sensitive than current regional production patterns to (increasingly porous) local, state, or even national boundaries. Third, who produces what and where will depend on comparative advantage in getting products to consumers. This comparative advantage will depend not only on how and where production is undertaken, but also on the ways in which products are processed, packaged, priced and distributed, and marketing strategies are deployed.[6] Perhaps most important, comparative advantage will require rapid identification of consumer needs and the organizational, production, processing, and marketing agility (in both agronomic and socioeconomic senses) to respond to these new needs.

Figure 4 extends the notion of market segmentation introduced earlier to incorporate additional markets, and segments within them. Quickly and efficiently identifying and responding to the needs of larger sets of new consumers (and 'target market' segments) will be the major challenge to producers and processors in the western Brazilian Amazon. But what will these changes mean for AFS in general, and for the small farmers who are currently engaged in (or who may in the future be engaged in) their production? Some clues are provided when the characteristics of AFS and their producers are set alongside the expected changes in demand and supply patterns.

Some of the changes will affect AFS via market shifts that concern factors other than the products themselves. One input shortage that plagues many AFS in the western Brazilian Amazon – labor – will probably not be resolved by regional integration, at least in the short term, since labor (unlike other factors of production, such as fertilizers and capital) is less easily transportable across national boundaries. An economic boom in the area, however, could be

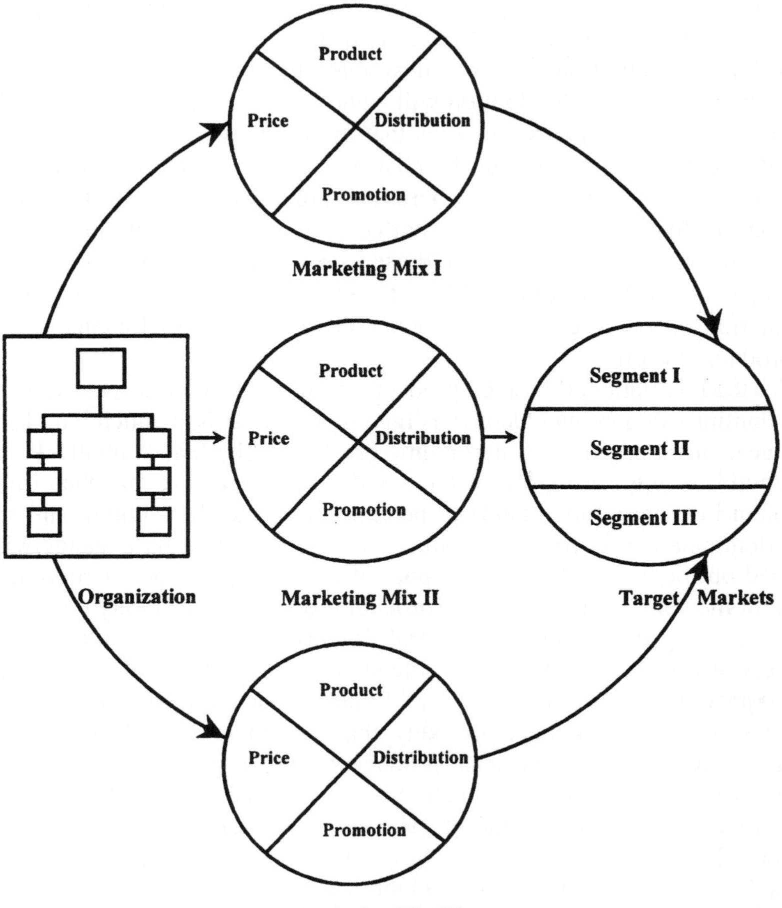

Source: Derived from Pride and Ferrell 1989.

Fig. 4. Factors affecting agroforestry system adoption and product profitability: production, marketing, and market segments for multiple geographic areas.

expected to attract additional migrants from areas less well off. Or, new input markets might supply a reasonably priced, more accessible, and more flexible substitute for existing inputs – for example, chemical fertilizers might provide the nitrogen that is presently contributed to soils by nitrogen-fixing trees in AFS. Hence, the agronomic benefits of some AFS may be provided more cheaply through trade. Likewise, the food security benefits of some AFS may lose importance with a more highly developed market in food grains. By the same token, regional integration should hasten development of private credit markets, which could help finance the establishment of AFS in the future.

In short, regional integration and the associated development of markets will tend to decrease or remove some of the non-product factors that are currently responsible for making AFS more difficult or more attractive to adopt under current conditions, or even desirable in terms of development objectives. Increasingly, AFS adoption will stand or fall on the profitability of the system's own components, vis-a-vis that of alternatives.

Other scenarios grow out of the changes expected in the markets for AFS products. Traded AFS products generally fall into two categories: those traded internationally (such as cacao and coffee), for which local producers cannot expect to influence prices, and those consumed/traded locally (such as cupuaçu), for which prices are likely to respond to changes in local/regional production. Examining the supply/demand issues separately for each category of product is useful.

For the internationally traded products, production and transportation costs will continue to dominate decisions regarding where, how much, and how to produce. In the absence of agronomic breakthroughs that dramatically alter productivity (e.g., an agroforestry-based response to the 'witches broom' problem in cacao), pure stands of perennial crops such as coffee and cacao will dominate export-focused production systems. These pure stands will be located on the best soils and near ports that facilitate export.[7] Economies of scale in the value added phases of product processing will likely dominate as well, reducing the number and overall importance of smallholders in these aspects of production. At least in the short term, however, labor shortages are expected to persist and household and farmer associations will remain sources of labor, making it unlikely that smallholder production will be entirely displaced by large-scale producers, even if profitability rises. This view does not preclude the formation of some niches of AFS-based production for the international market, perhaps supported by strategies such as 'green labeling'; this mechanism, however, is not without problems, and is unlikely to encourage widespread adoption of AFS in relatively remote areas (Witcover and Vosti, 1995).

For locally produced/traded components of AFS, surges in demand brought about by regional integration are possible, but the response to that increased demand by *existing* local production is uncertain. Since transportation costs will loom large in deciding where to locate production to meet (potential) new demand, existing local producers may enjoy short-term increases in product prices, but in the longer term lose market share (even local market share) if production 'migrates' closer to new demand centers, especially if production conditions (say, better soils) reduce production costs in new areas. For example, an all-weather road to Peru may increase demand for cupuaçu produced in Acre, assuming Peruvians chose to consume it over some locally produced substitute. This would increase the price to producers in Acre in the short term, but would probably motivate Peruvian farmers to produce cupuaçu in the medium term. If Peru turns out to enjoy a comparative advantage in cupuaçu production (due to climate conditions), then demand for cupuaçu from Acre (and eventually production) could fall. To the extent that demand increases locally (engendered by rising incomes, for example), local suppliers stand a better chance of thriving.

There will likely be new AFS products introduced (for example, from Peru into western Brazil) by regional integration, and the introduction of these new products may have both positive and negative effects on current production patterns and human welfare. New products may displace traditional ones in local diets, and there may be long lags (due to issues addressed in the technology adoption section) before local production responds to new local demand (if it does). Or, supply response may be swift and local producers may enjoy a comparative advantage (vis-a-vis traditional, non-local producers) in producing new products, and thereby gain market share and improved incomes.

The bottom line in the cases of both the internationally and locally consumed/traded products of AFS is that successful responses to regional integration require agility – agility in moving products within (now expanded) systems of regional trade, and in shifting production among various AFS or components. But, as noted earlier, because of the nature of some factors affecting AFS profitability, this sort of agronomic and/or socioeconomic agility may be elusive, especially among small farmers. But is this necessarily the case?

Conclusions

The evidence on the spread of AFS that generate income and create wealth in the tree-abundant western Brazilian Amazon is less than encouraging. This paper sets out some reasons to explain why AFS appear so seldom on the landscape there, and what might happen given regional integration trends. Some characteristics of AFS (investment requirements, negative cash flow for the first few years, uncertain local or international demand) reduce the possibility that AFS will be adopted, particularly by smallholder users, even in the absence of major changes, such as those fostered by regional integration.

Regional integration may create new opportunities for AFS, but some may not be seized due to the lack of agility caused by AFS characteristics (e.g., long and costly establishment periods) and/or producer characteristics (e.g., labor and capital constraints). Some local AFS products may even lose ground to those produced by competitors in other states or nations. This is especially true for AFS that make intensive use of labor and provide products that are produced in neighboring countries with relatively low rural wage rates.

And, likely economies of scale in processing and distribution for internationally traded AFS products may well mean that AFS production will give way to pure perennial stands (with a consequent decline in the environmental benefits associated with AFS in these areas). In any case, production (in an AFS or other system) might be increasingly concentrated on better soils and in areas close to final centers of demand. Finally, given the high labor and capital requirements associated with many AFS production systems, the rural poor would not be expected to move widely into these production systems (although they might benefit by selling their labor to local, larger-scale producers of AFS products).

By laying out specific reasons for this rather gloomy set of projections for AFS in the western Brazilian Amazon, this paper hopes to point to the scope for, and limits of, technology development and policy action to improve the contribution of AFS to the alleviation of poverty, economic growth, and environmental sustainability.

To increase adoption rates of AFS in the forest margins, and the likelihood that the rural poor will be involved, technology development must first increase and sustain the productivity of AFS, with particular emphasis on increasing the productivity of *labor* in these systems. This may mean mechanizing or otherwise modernizing the most labor-intensive aspects of production and processing. Other changes to the AFS technology package would help increase the agronomic and socioeconomic agility of AFS, to make them more responsive to changes in economic/market conditions *without* resorting to wholesale changes in production systems. They include investigating the diversity of input mixes that might go into given AFS crop mixes without sacrificing agronomic benefits, as well as raising the number and types of products that can be extracted with few trade-offs from a given AFS crop mix.

The comparison with LPS suggests some other pathways to improve AFS' competitiveness that may involve technologies, policies, or both: (1) find ways to reduce the processing and transport costs associated with AFS products, particularly in post-harvest processing and storage (and do so in ways that are scale neutral); and (2) make AFS closer economic substitutes for livestock production systems, for example, by assessing the potential for establishing futures markets for AFS products to allow 'off-take' from these systems between seasons or in the event of crop failures, without resorting to their complete destruction or abandonment.

Other policy actions can increase the likelihood of AFS adoption and the involvement of the poor, and influence the geographic location of AFS production (in accordance with development objectives). While some such measures mentioned below may appear to be part and parcel of regional integration trends, policy still has a role in the timing, location, and scope of the change. For the profitability of products produced in AFS to stabilize and rise, investments in infrastructure (particularly roads and electricity) will be critical for increasing the returns to and reducing costs of AFS production systems.[8] Second, policy action aimed at diversifying AFS product forms will make them more attractive to consumers, and focus attention on promoting the use of AFS products as intermediate inputs into (non-AFS) final products. Credit for AFS establishment will always be needed, and government will be a key provider directly and indirectly. Finally, governments can underwrite some of the costs of factors needed to increase agility, especially those factors for which scale economies in provision are great.

The role and limitations of policy

This paper began by noting the importance of clear policy objectives as a prelude to policy action. Policy action, in turn, occurs via specific policy instruments. Policymakers and researchers alike must maintain these distinctions (often 'policy' is used to signify any one or combination of these three meanings): decisions need to be taken on all three – confusing them can lead to inaction, or worse.

Because technology development and promotion is expensive, policymakers must have *a priori* estimates of the aggregate impact of technological change on environmental sustainability, rural growth, and poverty alleviation and set these against the costs of effecting change. The sum total of social and private benefits associated with a particular technology may be quite small, even if it is widely adopted. The costs of effecting change can be realistically assessed only if an effort is made to identify and distinguish among the various macroeconomic and other forces affecting AFS adoption patterns, and to determine the extent to which these forces *can* be addressed by policy (at any level, using any available policy instrument).

If regional integration is the driving force behind changes in land-use patterns (including AFS adoption patterns) and if regional integration is driven by economic forces truly beyond the control of policymakers at local, state, or regional levels, then policy action undertaken at these levels to counteract such forces may be futile or even damaging in the long run. Research and policy wisdom can help identify where and how policy action can be directed to achieve development objectives *given* major economic forces. Ignoring such forces is counterproductive. Taking action with them in mind can enhance chances for successful outcomes in terms of poverty alleviation, growth, and the environment.

Acknowledgments

This paper benefitted from the helpful comments of Dale Bandy, Sara Scherr, Frank Merry, David Kaimowitz, Judson Valentim and an anonymous reviewer. Thanks also to Lourdes Hinayon for formatting of text and graphics. The lead author also benefitted greatly from the real estate and encouragement provided by Erennio Giacomazzi, and the patience and support of his family.

Notes

1. Relevant AFS *agronomic* characteristics here may include not only what will grow in a given area, but with what inputs, with what effect on soils (possibly improving them), and with what products as outputs. It may also include a measure of flexibility to change components of the system without disrupting it ('agronomic agility'). Relevant AFS *socioeconomic* characteristics will probably include, among other things, profitability, risk (including produc-

tion and price risk), ability to shift production systems quickly ('socioeconomic agility'), and a special focus on the amount and types of human input required to produce and process traded and non-traded AFS products. It is understood that sets of desired characteristics will likely change, depending upon targeted 'user' groups.

2. Product prices (reported in Brazilian Reais, BZR1.10 = $1.00 in February 1997) are as follows:

   Banana (bunch) = R$0.80 (price, May–July 1996)
   Coffee (50 kilograms of processed beans) = R$67.00 (price in October 1996)
   Natural rubber (kilogram of latex) = R$0.98 (average price over past 3 years)
   Maize (kilogram) = R$0.09 (average price for the period February–October 1996)
   Beans (kilogram) = R$0.38 (average price for the period June–September 1996)
   Cacao (kilogram) = R$1.20 (average price over the past 3 years)
   Cupuaçu (fruit) = R$0.25 (harvest price 1997)

   For further detail on prices used, please refer to Oliveira and Vosti, 1997.
3. Market prices (collected in the fall of 1996) are used to value all inputs, including labor.
4. A discount rate of 10% was used in all NPV calculations.
5. Care should be taken in making direct NPV comparisons across LPS and AFS from data presented here. The AFS examined begin with a one-hectare investment starting with newly deforested land, while the dairy production system begins with an existing, low-productivity livestock production system (which may produce some milk), on 75 hectares of pasture and advances due to investments in herd quality, fencing, etc. Still, these differences in points of departure do represent the investment decisions confronted by thousands of small farmers in the western Brazilian Amazon. Both examples use a discount rate of 10%; NPV calculations are made using slightly different methodologies.
6. For an overview of how spatial economics illuminates local demand and supply changes for cattle products in the Amazon, see Faminow, 1997.
7. Note, however, that other high value products may appear alongside perennials and vie with them for 'best soils', mitigated by the product's agronomic needs and the availability of external inputs. Plus, work on the productivity 'cost' of moving from best to worse soils, for pure perennial stands vis-a-vis AFS, is still in its infancy.
8. The reader should note that such investments will also reduce the costs of, and therefore increase the returns to, alternative agricultural activities in these areas.

References

Amacher GS, Hyde WF and Joshee BR (1993) Joint production and consumption in traditional households: fuelwood and crop residues in two districts in Nepal. J of Dev Studies 30(1): 206–225

Anderson K and Blackhurst R (eds) (1993) Regional Integration and the Global Trading System. St. Martin's Press, New York, USA

Anderson K and Norheim H (1993) History, geography and regional economic integration. In: Anderson K and Blackhurst R (eds) Regional Integration and the Global Trading System. St. Martin's Press, New York, USA

Barbier B (1996) A method of bioeconomic modeling at the microwatershed level for the Central American hillsides. Report prepared by the Environment and Production Technology Division (EPTD), International Food Policy Research Institute (IFPRI) for the Inter-American Development Bank. IFPRI, Washington, DC, USA

Braun, J von and Webb PJR (1989) The impact of new crop technology on the agricultural division of labor in a West African setting. Econ Dev and Cult Change 37(3): 513–534

Bredahl ME, Ballenger N, Dunmore JC and Roe TL (eds) (1996) Agriculture, Trade, and the

Environment: Discovering and Measuring the Critical Linkages. Westview Press, Boulder, CO, USA

Centro Internacional de Mejoramiento de Maiz y Trigo (CIMMYT) Economics Program (1993) The Adoption of Agricultural Technology: A Guide for Survey Design. CIMMYT, Mexico, DF, Mexico

Current D, Lutz E and Scherr SJ (eds) (1995) Costs, Benefits, and Farmer Adoption of Agroforestry: Project Experience in Central America and the Caribbean. World Bank Environment Paper Number 14. Washington DC, USA

Faminow M (1997) Spatial economics of local demand for cattle products in Amazon development. Agri Ecosyst and the Environm 62 (March, No. 1): 1–11

Faminow M (1998) Cattle Deforestation and Development in the Amazon: An Economic, Agronomic and Environmental Perspective. CAB International, Wallingford, UK (In press)

Faminow M, Sá C and Oliveira SJ (1998) Development of an investment model for the smallholder cattle sector in the Western Amazon – I. Preliminaries. Documento Series, Empresa Braziliera de Pesquisa Agropecuaria (EMBRAPA), Rondônia, Brazil (in press)

Feder G, Just RE and Zilberman D (1985) Adoption of agricultural innovations in developing countries: a survey. Econ Dev and Cult Change 33(2): 255–298

Ffrench-Davis R (1995) Trends in regional cooperation in Latin America: the crucial role of intra-regional trade. In: Teunissen JJ (ed) Regionalism and the Global Economy. Forum on Debt and Development (FONDAD), The Hague, The Netherlands

Gittinger JP (1982) Economic Analysis of Agricultural Projects, 2nd edition. Johns Hopkins University Press, Baltimore, MD, USA

Griliches Z (1957) Hybrid corn: an exploration in the economics of technological change. Econometrica 25(4): 501–522

Guimarães V Di Addario (1998) Pesquisa de mercado de produtos vegetais não madeireiros. Pesquisa de mercado – estudos de caso: Pimenta longa (*Piper hispidinervium*) e pupunha (*Bactris gasipaes H.B.K.*). Documento Series, Empresa Brasiliera de Pesquisa Agropecuaria (EMBRAPA), Acre, Brazil (in press)

Hazell PBR and Ramasamy C (1991) The Green Revolution Reconsidered: The Impact of High-yielding Rice Varieties in South India. Johns Hopkins University Press (for the International Food Policy Research Institute), Baltimore, MD, USA

Lahera PE (1992) Integration today: bases and options. CEPAL Review 47 (August): 63–76

Lipton M with Longhurst R (1989) New Seeds and Poor People. Unwin Hyman, London, UK

Nerlove M, Vosti SA and Basel W (1996) Role of Farm-level Diversification in the Adoption of Modern Technology in Brazil. Research Report No. 104. International Food Policy Research Institute, Washington, DC, USA

Oliveira S and Vosti SA (1997) Aspectos financeiros de sistemas agroflorestais em Ouro Preto do Este, Rondônia. In: Administração rural: Instrumento de sucesso numa economia estável. Anais do 2^{0} congresso Brasileiro de administração rural: Tema central, pp 115–130. Uberaba, MG, Brazil

Pride WM and Ferrell OC (1989) Marketing: Concepts and Strategies, 6th edition. Houghton Mifflin, Boston, MA, USA

Reardon T and Vosti SA (1992) Issues in the analysis of the effects of policy on conservation and productivity at the household level in developing countries. Quart J of Inter Agricult 31(4): 380–396

Reardon T and Vosti SA (1995) Links between rural poverty and the environment in developing countries: asset categories and investment poverty. World Dev 23(9): 1495–1506

Renkow M and Traxler G (1994) Grain fodder tradeoffs and varietal adoption in mixed farming systems. Paper presented at the 1994 Agricultural Economics summer meeting, San Diego, California, USA

Traxler G and Byerlee D (1993) Joint production analysis of the adoption of modern cereal varieties in developing countries. American Journal of Agricultural Economics 75(4): 981–989

Vosti SA and Witcover J (1996) Slash-and-burn agriculture: household perspectives. Agriculture, Ecosystems and Environment 58: 23–38

Vosti SA, Witcover J, Gockowski J, Tomich T, Carpentier CL, Faminow MD and Oliveira S (1998) Socioeconomic issues linked to *best bets* – Modelers' workshop. Documento Series, Empresa Brasiliera de Pesquisa Agropecuaria (EMBRAPA), Acre, Brazil (in press)

Witcover J, Vosti SA, de Almeida Barbosa FR, Batista J, Beatriz V, Böklin G, de Franca SB, Castilla C, Fujisaka S, Figueiredo Galo SL, Garcia JH, Gomes Cabral W, Hurtado L, Leite A, Nogueira Leite FM, Souza DA, de Souza Marinho JT, Rocha K and Neto CR (1996) Alternatives to slash-and-burn agriculture (ASB): a characterization of Brazilian benchmark sites of Pedro Peixoto and Theobroma, August/September 1994. MP-8 Working Paper No. US96-003. Environment and Production Technology Division, International Food Policy Research Institute (IFPRI), Washington, DC, USA

Witcover J and Vosti SA (1995) Workshop on non-timber tree product (NTTP) market research. Environment and Production Technology Division Workshop Summary Paper No. 3, International Food Policy Research Institute (IFPRI), Washington, DC, USA

Agroforestry Systems **38**: 223–245, 1998.

Directions in tropical agroforestry research: past, present, and future*

P. K. R. NAIR
School of Forest Resources and Conservation, University of Florida, Gainesville, FL 32611, USA. E-mail: pkn@gnv.ifas.ufl.edu

Key words: deductive research, inductive research, science, technology

Abstract. Reflections on the past two decades of organized research in tropical agroforestry raise several issues. Research efforts started with an inductive and experiential approach but have subsequently followed a deductive and experimental approach that includes hypothesis testing and the development of predictive capability; agroforestry research is thus being transformed into a rigorous scientific activity. The research agenda, so far, has given high priority to soil fertility and other biophysical interactions, less priority to anthropological and sociological aspects, and little priority to evaluating costs and returns, pests and diseases, and the so-called non-timber forest (tree) products. Moreover, larger-spatial-scale issues, such as carbon sequestration, water quality, and biodiversity conservation, have been neglected because of the emphasis on field- and farm-scale studies.

Overall, the high expectations that were raised about the role and potential of agroforestry as a development vehicle have not been fulfilled. In order to overcome this, it is imperative that research be focused on the generation of appropriate, science-based technologies of wide applicability, especially under resource-poor conditions and in smallholder farming systems. Future research agendas should entail a judicious blending of science and technology. Applied research should build upon the findings of basic research to generate technologies for application at the farm, regional and global levels. Such research should place increased focus on previously neglected subjects, for example, the exploitation of indigenous fruit-producing trees, the agronomic components of agroforestry systems, and the global issues mentioned above. Furthermore, an appropriate methodology that embodies economic, social, and environmental costs and benefits needs to be developed to realistically assess the impacts of agroforestry, and an enabling policy environment that will facilitate agroforestry adoption needs to be made available.

Agroforestry research of the 21st century should strive to build bridges from the inductive phase of the past, through the deductive phase of the present, to the future phase of harnessing science and generating technologies for the benefit of the land and its present and future users.

Introduction

Organized research in agroforestry started two decades ago with the establishment of ICRAF (International Council – Centre since 1991 – for Research in Agroforestry). The image of agroforestry and the directions in agroforestry research have changed considerably since then. Agroforestry was conceived and promoted as a sustainable approach to land management, capable of

* The paper reflects only the author's personal views and thoughts; it does not represent the official policy of his institution, nor of *Agroforestry Systems*.

providing solutions to problems and issues that were not being satisfactorily addressed by modern agriculture and forestry. Early initiatives were based on data and assumptions derived from past research in related areas such as agronomy, forestry, soil science, ecology, economics, and anthropology, and were supported by information, albeit of a largely descriptive nature, collected from successful, existing agroforestry systems (Nair, 1991).

In the wake of frustrations arising from the Green Revolution's failure to benefit poor farmers, and from escalating land-management problems, such as tropical deforestation, fuelwood shortage, and soil degradation, the development community embraced this 'new' concept with unprecedented enthusiasm as a magical development vehicle, with a perceived relevance to 'difficult' or 'fragile' environments and resource-poor conditions (Nair, 1993). Consequently, some relatively major investments were made in agroforestry research during the past two decades – often in support of hastily prepared development efforts – with the expectation of producing results with immediate practical application. Inevitably, these efforts focused on 'applied' research (Nair, 1993, 1994), and progress in the basic science of agroforestry was slow (Sanchez, 1995). In any applied science – be it medicine, engineering, or agriculture – scientific progress cannot be attained unless there is a strong and rigorously developed theoretical foundation. Because of its development-driven agenda, agroforestry research, unfortunately, had little emphasis on building such a foundation.

Agroforestry, which involves the application of a vast array of biophysical and socioeconomic disciplines, is both an artful science and a scientific art. The implication is that researchers have a dual obligation: (1) to develop better solutions to land-management problems (applied research), and (2) to advance the science of agroforestry (basic research). Agroforestry researchers and research-funding agencies have faced, and still face, a serious dilemma as to how to allocate their limited resources to fulfill these obligations and expectations. The purpose of this paper is to analyze this dilemma and think aloud as to what directions research ought to follow. The scope of the paper is limited to agroforestry in tropical and subtropical regions.

Agroforestry research: building bridges

If agroforestry, the age-old practice, is to be transformed into a modern-day scientific practice, we need to formulate plausible hypotheses about it through inductive reasoning, test and validate the hypotheses through deductive experimentation, and strive to attain predictable understanding that will be of wide applicability (Sanchez, 1995). Given the existence of a multitude of complex and traditional agroforestry systems (Nair, 1989), it is inevitable that traditional knowledge about the systems be integrated with scientific (biophysical) research for the development of improved practices. Agroforestry research must be firmly grounded on existing knowledge obtained not only by the relatively few researchers, but more so by the millions of farmers who have traditionally practiced agroforestry. In other words, agroforestry research should 'build bridges' *from* the past, *through* the present, *to* the future. A discussion on the directions in agroforestry research should therefore involve an examination of the 'past,' an evaluation of the 'present,' and a prognosis for the 'future'.

The past

For the purpose of this paper, the 'past' of agroforestry research refers to the period until 1991 when ICRAF, the major research entity in agroforestry, was formally transformed from a Council to a Centre, implying a major shift in its functional strategy and focus. Nair's (1993) analysis showed that the main categories of agroforestry research during that period were three: methodological, component and system descriptions, and empirical. Among these, methodological investigations and system/component descriptions formed the major categories in the early 1980s, but their relative importance had declined by late 1980s, when empirical (quantitative and experimental) research became more dominant. On the empirical side, a major share of the studies were related to soil improvement by trees and soil fertility management. Screening and evaluation of multipurpose trees and shrubs, and evaluation of prototypes of new technologies such as alley cropping were the main areas of applied research during that period. Many such studies evaluated improved agroforestry practices in comparison with the local land-use systems, usually sole-crop-agricultural stands, and were of relatively short durations of two to four years. The issue of research methods and sites – i.e, on research stations versus on farmers' fields – also became important; 'on-station' and 'on-farm' research were recognized as mutually complementary aspects of agroforestry field research.

The overall output of research during this early (or 'past') phase can be summarized as the development of: (1) the concepts and principles of agroforestry, (2) databases, (3) methodologies, and (4) prototype technologies and their performance evaluations (Box 1). These efforts provided the sound conceptual framework that was critically important to agroforestry during those early days. A number of concepts were identified and developed, for example, the formation of systems by integrating crops and/or animals around the central component of trees; the existence of increased diversity and complexity within these systems; the emphasis on exploitation of ecological and economic interactions among components in order to achieve optimum, rather than maximum, benefits; and the recognition of the importance of the interplay of sociocultural values in the manifestation of these benefits. All of these concepts are perhaps taken for granted today, but the efforts that went into their development, and their importance in providing a solid foundation upon which to build future research strategies, should not be underestimated. Equally important are the efforts of the 1980s in building up databases – for

TROPICAL AGROFORESTRY RESEARCH

"THE PAST"

Conditions before the mid-1970s leading to the development of agroforestry

- Existence of agroforestry as a practice, but not as a concept or even a word
- Commercial, monocultural, emphasis of agricultural and forestry developments (Green Revolution and Plantation Forestry)
- Serious problems of hunger and malnutrition
- Scientific developments in a number of disciplines and social compulsions in support of integrated land-use
- Serious rethinking on development alternatives.

Developments during 1975 - 1990

- Institutionalization of agroforestry
- Initiation of organized research
  - Concepts and potentials
  - Databases
  - Methodologies
  - Empirical research
    - Soil-related
    - Biophysical interactions
    - Systems
    - Multipurpose trees

Box 1.

example, on agroforestry systems (Nair, 1989), multipurpose trees and shrubs (von Carlowitz et al., 1991), and nitrogen-fixing trees (MacDicken, 1994) – and developing research methodologies of a general nature, such as the Diagnosis and Design (Raintree, 1987), or on specific subject areas (Avery et al., 1990).

Among the efforts to develop prototype technologies, the major focus has been on alley cropping (Kang et al., 1990). The expectations that this technology aroused during the 1980s are somewhat comparable to those aroused by the 'miracle' varieties of cereal crops in the 1970s during the height of the Green Revolution. But the comparison between the two stops there because the success of alley cropping and the extent of research in alley cropping research pale when contrasted with those of high-yielding varieties of food crops. This has resulted in much recent criticism of alley cropping. To be fair, however, it must be noted that, from the very beginning, the potential role of alley cropping as a land-management technology was exaggerated. The technology has been tested in numerous locations representing a wide range of physical (soil, rainfall, relief, and altitude) and socioeconomic conditions, and using a large number of tree and crop species, and system-management techniques (Rao et al., 1997). In some of these situations, there was no logical chance for the technology to succeed, e.g., in areas with serious limitations of climate, relief (slope), and soil conditions (such as high acidity). Although cautions about the inapplicability of the technology in such situations had been expressed (for example, Nair, 1990; Tejwani and Lai, 1992), expectations always remained high, based not on reason but on rote. Little wonder that some results have been disappointing. Criticisms against alley cropping have continued unabated and have been exaggerated to the extent that the benefits of the technology – e.g., in soil conservation (Lal, 1989; Young, 1989; Kiepe and Rao, 1994) and in food-crop production (Kang, 1993; Kass, 1996) – have been vastly overshadowed by the extreme criticisms of a 'pie-in-the-sky' (Ong, 1994) type of unfairness. In retrospect, it is clear that at least some of the criticisms could have been avoided had rational thinking preceded the enthusiastic promotion of this technology. Several lessons – procedural as well as scientific – can be learned from the experience with alley cropping during the past 10 to 15 years: (1) there is insatiable craving in the technology-starved field of agroforestry for any technology that appears new and promising; (2) realistic assessment of the potential, role, and limitations of any new land-use innovation should be carefully undertaken before it is promoted; and (3) technology generation and testing should be pursued based on rigorous scientific procedures, not on wishful speculations.

An unfortunate consequence of this obsessive preoccupation of the agroforestry research community (as well as its critics) with alley cropping has been that other promising technologies – with the exception perhaps of shaded perennial systems (Tejwani, 1994; Beer et al., 1997) and agroforestry for soil conservation (Young, 1989, 1997) – have received relatively little attention during this period. The potential of agroforestry for other applications, such as tree fodder and silvopastoral systems, sequential systems for fallow improvement (ICRAF, 1997), and fuelwood-production systems, has remained largely underexploited.

The present

In the context of this paper, the 'present' represents the period since ICRAF became a research center in 1991. During this period, empirical research and the *science* of agroforestry grew in importance, and the mostly descriptive information of earlier years gave way to a more robust and predictive understanding of the theoretical foundations of agroforestry.

Sanchez (1995) wrote an elaborate paper describing the steps involved in the process of this transformation, and reviewed developments in agroforestry research to explain how the new research direction was being implemented. He argued that the development of agroforestry as a science should be based on two principles: competition and complexity, and that they in turn determined two desirable properties: profitability and sustainability. It could be argued that, semantically, competition and complexity are not principles, but reflect the system's properties, which when managed could lead to the other listed properties (profitability and sustainability). According to Sanchez's (1995) argument, complex agroforestry systems could become more profitable economically and sustainable ecologically when interactions between (or among) components for growth resources are managed based on scientific principles; the author draws heavily from the literature to substantiate his arguments. Although some aspects of Sanchez's claims have been questioned (Kass, 1996), the paper provides an excellent account of the pangs of agroforestry research in the past and at present.

What significant changes have happened in agroforestry research since the early 1990s? An analysis of the topics of research publications in agroforestry during the period should be informative. Table 1 presents information on the nature of agroforestry research papers that were published during 1991 to 1996 in four major international journals. In spite of some shortcomings (for example, agroforestry research results are published in several other publications, and the grouping of articles into various categories is arbitrary), the Table shows that:

- a vast number of papers (55%) deal with biophysical aspects (plants and soils and their interactions);
- papers on animal-related and socioeconomic aspects have been relatively few (4% and 10%, respectively); and,
- very little has been reported (researched) on issues such as pests and diseases, indigenous trees, and agronomic crops.

These journal articles, along with a vast number of other scientific and technical publications that have become available during this period, indicate that there has been an upsurge in empirical research in agroforestry (Nair, 1996). As a part of that, several hypotheses have been formulated, notably in soil-related aspects (initiated by Young, 1989) and later in the socioeconomic arena (Sanchez, 1995). The number of such hypotheses and the efforts to test them under a variety of situations keep growing, which, indeed, is a very satisfying trend in the development of deductive, process-oriented research in agroforestry.

A related development is the formulation of quantitative models to describe specific aspects of agroforestry research. Ong's tree-crop interactions equation (Sanchez, 1995), which has undergone several modifications (Kass, 1996; Sanchez, 1996; Cannell et al., 1996), is an example. Others include van Noordwijk and Purnomosidhi (1995), Lawson et al. (1995), Gregory (1996),

Table 1. Nature[a] of agroforestry articles published in four international journals during the period from 1990 to 1996.

Journals	Number (and % of total) of articles in different subjects						
	Soil related	Plant related	Animal related	Interaction	Socioecon. related	Others[b]	Total
Agroforestry System	63 (10.17)	93 (15.02)	26 (4.20)	75 (12.11)	88 (14.21)	274 (44.26)	619
Forest Ecology and Management	28 (12.44)	141 (62.66)	8 (3.55)	5 (2.22)	13 (5.77)	30 (13.33)	225
Plant and Soil	28 (29.47)	54 (56.84)		10 (10.52)		3 (3.15)	95
Soil Biology and Biochemistry	12 (42.85)	12 (42.85)		2 (7.14)		2 (7.14)	28
Experimental Agriculture	8 (19.04)	17 (40.47)	5 (11.90)	4 (9.52)	2 (4.76)	6 (14.28)	42
Total	139 (13.77)	317 (31.41)	39 (3.86)	96 (9.51)	103 (10.21)	315 (31.22)	1009

[a] Categorization into different groups is based on subject-matter contents of papers according to the author's judgement.
[b] 'Others' include subject matter areas that are not listed here (e.g., concepts, modeling, and system/component descriptions), articles of a general nature that cover a number of topics, and special issues dealing with different aspects of a specific topic (e.g., agroforestry education).

and Ong et al. (1996). Also, several new quantitative models and concepts in computer-based modeling in agroforestry systems, mostly in temperate situations, were presented at an international workshop on modeling and fundamental research in agroforestry held at Montpellier, France, in June 1997 (selected papers from this workshop will be published as a separate special issue of *Agroforestry Systems* in 1998).

The status of research in several areas of agroforestry is reflected in the other nine papers included in the present volume. Other recent research publications on specialized topics – e.g., tree-crop interactions (Ong and Huxley, 1996), plant litter quality and decomposition (Cadish and Giller, 1997) and soil conservation (Young, 1997) – reflect the advances in those specific areas. Efforts to address specific land-management problems have resulted in some application-oriented agroforestry publications on subjects such as the management of imperata-infested grasslands in southeast Asia (Garrity, 1997), and improved fallow systems (Cooper and Buresh, in press). Collectively these efforts have helped to prove or disprove some notions and hypotheses about agroforestry, especially in the biophysical area. Some examples of such recent advances are given in Box 2.

Although there may not be a unanimity of views on the nature, pace, foci (geographical and thematic), and extent of accomplishments in agroforestry research during the past two decades, it is indisputable that agroforestry today is no longer a mysterious enigma that defies science and scientific principles, as it was perceived to be two decades ago. Today we do not just speculate about the advantages of trees and crop and/or animal associations, nor do we fantasize about their magical or imaginary virtues; we now have scientifically rigorous data to prove/disprove several of the assumptions and speculations that used to riddle discussions on agroforestry. Agroforestry is now on a firm scientific footing. Equally important, agroforestry is no longer simply a hand maiden of agriculture or forestry; it has a niche of its own. To reiterate Sanchez (1995), agroforestry is well on its way to becoming a specialized science at a level similar to those of crop science and forestry science.

The future

The gains and developments of two decades of agroforestry research are certainly impressive. The question, then, is what next? What direction should agroforestry research take? In order to answer this, we need to critically evaluate: (1) the extent to which the original research/development expectations and goals – whether stated directly or implicitly two decades ago – have been fulfilled, and (2) the current and emerging problems and expectations.

The charter of ICRAF's 1977 mandate states that research in agroforestry would and should lead to alleviation of hunger and poverty, reversal of environmental degradation, reduction of deforestation, and enhancement of

TROPICAL AGROFORESTRY RESEARCH

"THE PRESENT" (1990s)

Research Objectives

- Science of Agroforestry
  - Hypotheses Testing
  - Quantitative Models
- Technology Development

Research Areas

High Priority

Soil fertility/improvement
Biophysical interactions
Systems design

Low Priority

MPT improvement
Social/anthropological aspects
Animal component

Least/Little Priority

Economic evaluation
Pest and diseases
Nontimber tree products
Larger spatial-scale issues
- C sequestration
- Water quality
- Biodiversity conservation

fuelwood and fodder supply. To what extent have these expectations been fulfilled? The scenarios of land-use problems and trends in developing countries during the past 20 years could be good indicators, although these are certainly not related exclusively to the influence of agroforestry, and the effect of agroforestry practices will vary considerably among them. Some of these indicators during (or as close as possible to) this period are presented in Figures 1b–1d and Table 2, along with the overall demographic changes that occurred during that period (Figure 1a). Deforestation, shortage of fuelwood

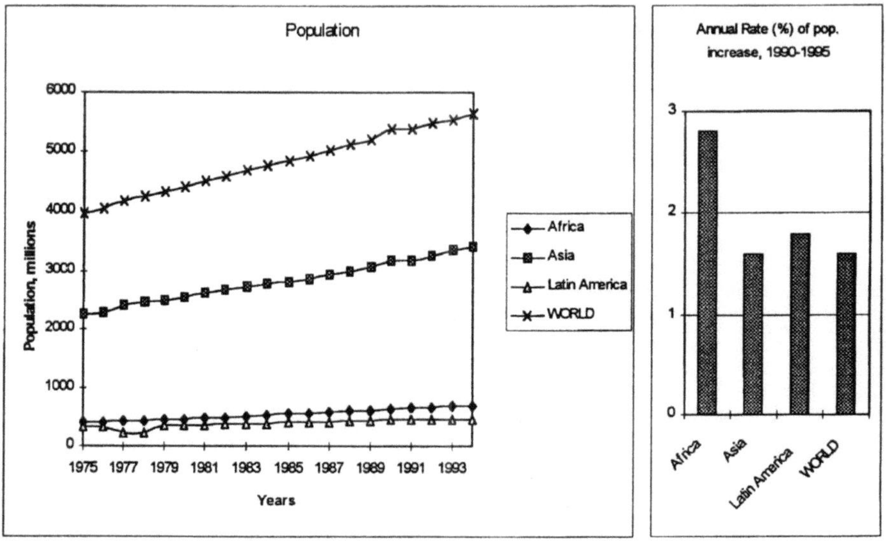

Fig. 1a. Population increase in the developing continents and the world, 1975–1994.
Source: United Nations (1996).

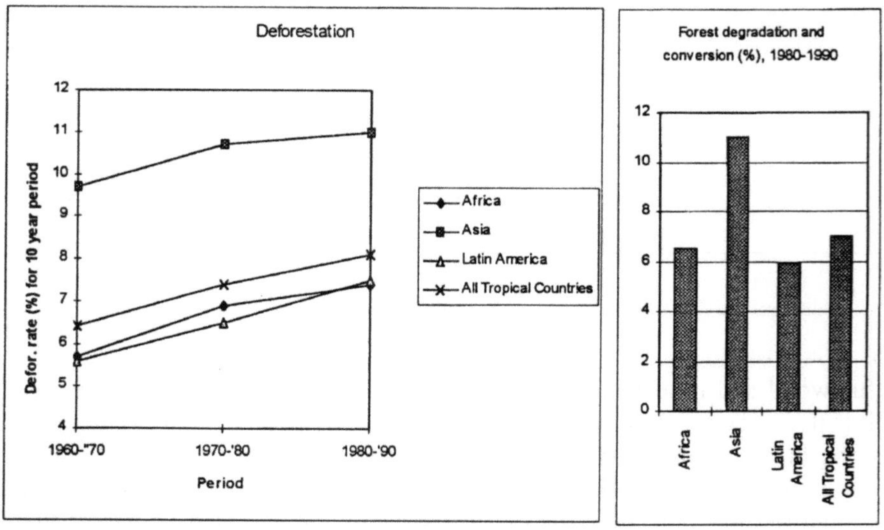

Fig. 1b. Rates of deforestation and forest degration in the developing continents.
Source: Singh and Marzoli (1995), WRI (1996).

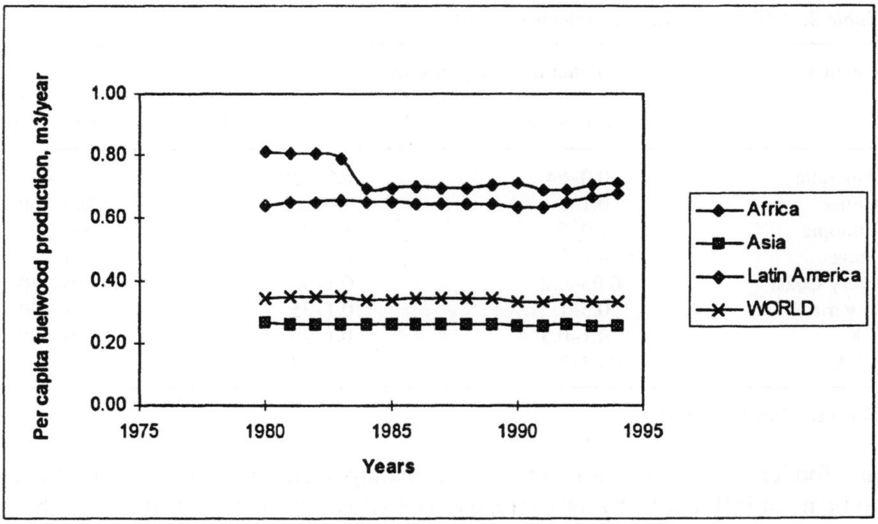

Fig. 1c. Fuelwood production in the developing continents and the world, 1980–1994.
Source: FAO (1995), United Nations (1996).

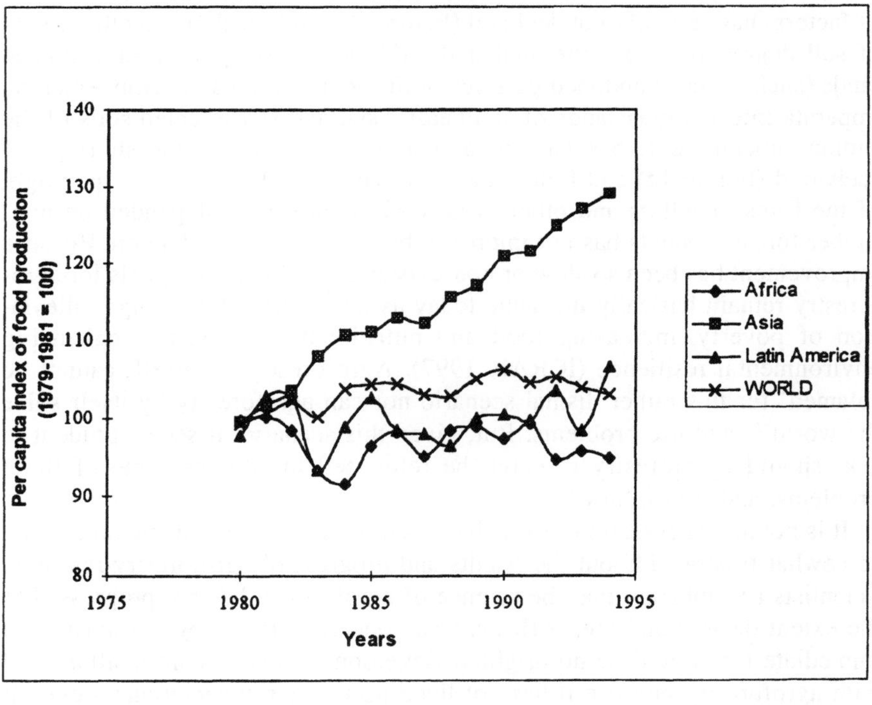

Figure 1d. Food production in the developing continents and the world, 1980–1994.
Source: FAO (1995), WRI (1996).

Table 2. Soil erosion rates in selected countries.

Country	Annual rates of soil erosion (Mg/ha)		
	Natural soil	Cultivated soil	Bare soil
Australia	0.0–64	0.1–150	44–87
China	0.1–2	150–200	280–360
Ethiopia	1.0–5	8.0–42	5.0–70
India	0.5–5	0.3–40	10–185
Ivory Coast	0.03–0.2	0.1–90	10–750
Nigeria	0.5–1	0.1–35	3–150
UK	0.1–0.5	0.1–20	10–200
USA	0.03–3	5–170	4.0–9

Source: Morgan (1995).

and fodder, and soil erosion are the land-management indicators that are likely to be most influenced by agroforestry. The situation in respect of any of these is not fully satisfying by any standards. While food production in Asia and Latin America has shown an upward trend, it has shown a disappointingly negative trend in Africa (Figure 1d), where agroforestry research has been most concentrated. Deforestation (which is, admittedly, influenced by a myriad of factors) has certainly not declined (Figure 1b); soil erosion and other forms of soil degradation continue unabatedly (Table 2); reclamation of degraded lands (such as the abandoned pastures of the acid lowlands of Latin America, imperata-infested grasslands of southeast Asia, and salt-affected soils of the Indian subcontinent) has lagged far behind expectations; the shortage of fuelwood (Figure 1c) and fodder has, if anything, only worsened; the plight of the forest dwellers and others who traditionally have depended on non-timber forest products has not improved by any account; and so on. Because improvement has been so slow or non-existent, ICRAF's main goals for agroforestry remain basically the same today as they were 20 years ago: alleviation of poverty, increasing food and nutritional security, and enhancing environmental resilience (ICRAF, 1997). Agroforestry, by itself, cannot be 'blamed' for this rather dismal scenario nor can agroforestry by itself solve the world's land-use problems. But, given this scenario, it seems prudent to ask: should agroforestry research be refocused to address some of these problems, and if yes, how?

It is not an exaggeration to say that scientists and development experts are somewhat frustrated about the results and progress of agroforestry research. Scientists are unhappy that the science of agroforestry has not progressed to the extent desired and are, perhaps, also concerned that progress during the immediate future will be no brighter. Development experts are disillusioned with agroforestry because it has not lived up to their expectations – even if some of those expectations were unrealistic. Perhaps these frustrations are also based on biased perceptions about research and science. A fundamental review of the concept of research is essential in this context.

Research, science, and technology

Although the terms 'research' and 'science' are often used together, they are neither inseparable nor not synonymous. It is true that research is done according to scientific procedures, but research does not, and need not, always result in advancement of basic science. According to dictionary definitions, science is systematized knowledge derived from observation, study, and experimentation carried on in order to determine the nature or principle of what is being studied; it refers to the state of fact or knowledge, as opposed to intuition or belief. Research is careful, patient, systematic, diligent inquiry or examination in some field of knowledge, undertaken to establish facts or principles, and involves laborious or continued search for truth (Webster's Dictionary, 1997). Thus, while all research must be carried out scientifically, the objective of the research could be the advancement of knowledge (science) or the application of knowledge; the latter aspect, which involves the study of the practical or industrial application, is usually referred to as technology.

The distinction between science and technology is important in this discussion. As summarized in Table 3, while the goal of science is pursuit of knowledge and its emphasis is on 'knowing,' technology relies on applica-

Table 3. Directions in agroforestry research: a conceptual approach to science and technology.

A. Basic differences

Descriptors	Science	Technology
Goal	Pursuit of knowledge	Application of knowledge
Action/Value	Knowing	Doing
Question	What is	What is to be
Motivation	Curiosity	Need
Procedure	Scientific method	Skillful synthesis

B. Directions in agroforestry research

Science		Technology
Inductive	Deductive	
Observation	Problem identification	Societal need
↓	↓	↓
Classification	Theory/Hypothesis	Analysis
↓	↓	↓
Generalization	Experimentation	Design
↓	↓	↓
Theory	Observation	Application (Advice to farmers, predictions to policy makers)
	↓	
	Verification	
"PAST"	"PRESENT"	"FUTURE"?

tion of knowledge and 'doing.' The motivation for science is curiosity, whereas technology is need-driven. Furthermore, although appropriate scientific methods are important in both science and technology, the latter involves more of a skillful synthesis and integration of techniques and procedures from a cross section of disciplines rather than following a specific set of methods of any specific discipline or its subset. If the objective of agroforestry research is to solve land-management problems, it seems that our efforts should be aimed at generating appropriate technologies. However, because technologies cannot be generated without application of science, scientific understanding of the principles that can be applied is the essential first step in the process of technology generation. Without pushing this discussion too far, suffice it to say that a judicious balance between (fundamental) science and (application-oriented) technology should be the strategy for agroforestry research in the future (Box 3).

TROPICAL AGROFORESTRY RESEARCH

"THE FUTURE"

- Manage

  Competition and
  Complexity

  for

  Profitability and
  Sustainability

- Generate Technologies based on scientific principles

- Special emphasis on

  Indigenous trees
  The "agro" part of agroforestry
  Impact assessment
  Policy environment
  Large spatial scale issues

Box 3.

Research areas to support the development of agroforestry technologies for the 21st century

If it is accepted that the main focus of agroforestry research should be to generate technologies for solving the land- management problems that we set out to address 20 years ago, and that research should be a means rather than an end in itself in the era of dwindling research support, it is imperative to properly prioritize the research agenda of agroforestry for the 21st century. Long lists of research topics that appear in many publications (for example, most other papers in this volume) need to be combined, coordinated, and prioritized according to a well-thought-out direction. Some general areas and topics, which ought to be considered while setting the agroforestry research agenda for technology generation in the future, are highlighted here.

Indigenous and underexploited trees

The importance of indigenous trees in rural households and agroforestry systems needs no special emphasis. Traditionally people throughout the tropics have depended on indigenous plants for a variety of their basic household needs, from food (mainly fruits) to medicines to fibers; these products have also generated the much-needed cash for purchased products. Examples of the use of a large number of such tree species in homegardens and other agroforestry associations can be found in almost all descriptions of traditional agroforestry systems (e.g., Nair, 1989). Exploitation of these indigenous trees through domestication and improvement was recognized as one of the most promising opportunities in agroforestry even during the early stages of organized agroforestry research (Nair, 1990). Systematic efforts in this direction, however, have not been undertaken until recently. Most tree selection and improvement efforts in agroforestry have been on the so-called multipurpose trees, and such efforts have often involved popularization of exotic species (Turnbull, 1987; MacDicken, 1994). Lately, interest and attention are being focused on the so-called 'Cinderella' species, i.e., indigenous fruit trees and other woody species that provide medicinal products, ornamentals, or high-grade timber, but whose value though appreciated by local people has been largely overlooked by experts (Leakey and Newton, 1994; Leakey et al., 1996). In a companion paper in this volume, Leakey and Simons (1997) outline the strategy, challenges, and ICRAF's programs for domestication and improvement of these 'Cinderella' trees in agroforestry systems.

Domestication, improvement, and exploitation of indigenous trees should be a major focus of agroforestry research in the 21st century. Domestication of these tree crops could increase yields, provide higher quality products, enhance commercial potential, and above all, contribute to the food security of the local population. Several reports that list a vast variety of such trees under different situations are available (e.g., Okafor and Fernandes, 1987; Smith et al., 1992; Leakey et al., 1996) but organized initiative in moving forward with the task of domestication and improvement is only just emerging (Leakey and Simons, 1997).

The 'agro' part of agroforestry

In spite of any semantic shortcomings of the word agroforestry (Nair, 1993), it is obvious that it literally entails two parts: 'agro' and 'forestry.' In most discussions, development efforts, and research initiatives, agroforestry is perceived as more a part of forestry than of agriculture. In reality, agroforestry is an aspect of neither agriculture nor forestry. Agroforestry has (or, should have) an identity of its own.

Putting aside semantic and 'niche' problems, it is clear that in most cases, the agronomic component of agroforestry systems is equally as important as the forestry (tree) component. However, all discussions on component improvement or research tend to center around trees – multipurpose trees, nitrogen-fixing trees, indigenous fruit trees, and so on – to the total exclusion of agronomic components. The reasons are several: trees are the 'central' components of all agroforestry systems; the greatest opportunities and needs in agroforestry are, or at least were, in the area of trees; trees provide production and service functions; the emphasis in agroforestry is on exploiting the trees and their attributes for enhancing productivity and sustainability of farming systems, and so on. However, the 'agro' components of agroforestry are important too, at least from the production point of view. Usually, in agroforestry experiments, the crop component is considered a 'given.' In most cases, the varieties used are improved genotypes that have been developed for high performance under conditions of optimum supply of light, nutrients, and water, and freedom from pests and diseases. Paradoxically, one of the compelling reasons for the development of agroforestry is the lack of such optimum conditions in many cases; agroforestry is expected to do well under conditions that are far from optimal for the performance of agronomic crops. Suffice it to say, crop varieties and cultivars that are adapted to less-than-optimal growth conditions are essential in agroforestry. Varieties of crop plants have been developed for some special conditions (water shortage, lower and higher pH ranges, and even nutrient deficiency), but breeding/selection of varieties adapted to low light conditions has perhaps never been attempted. There is a wide range of other issues related to crop selection and management in agroforestry systems. Agroforestry research in the 21st century should embrace agronomic components as enthusiastically as the tree components.

Impact assessment and evaluation procedures

A major problem facing agroforestry research is the lack of a comprehensive, robust, and widely applicable methodology for realistically assessing the benefits and impacts of agroforestry practices. Traditional agroforestry systems have evolved under site-specific conditions for meeting the land users' basic needs. Our information about them is mostly experiential rather than experimental. In experimental systems, the objectives are specific, and the results are therefore quantifiable according to the accepted criteria for assessing those objectives. Thus, agronomic or silvicultural productivity, soil improvement and protection, economic returns, and even social benefits of an agroforestry system can be measured according to the respective norms and procedures for each parameter. Indeed, methodologies for some specific subsets of the intricate puzzle of agroforestry have been developed – or at least attempted – by some researchers. Examples include: experimental designs and statistical analyses of agronomic-type field experiments (Rao and Coe, 1992; Coe, 1994), measurement of competition for light and water (Ong et al., 1996), and root interactions (van Noordwijk et al., 1996). Studies of this nature have been much less common in the socioeconomic arena. Although some notable initiatives have been made – calculation of a whole farm efficiency index (Johnson, 1994), bioeconomic modeling (Menz and Grist, 1996), and socioeconomic impact assessment of some agroforestry interventions (Current et al., 1995; Mercer and Miller, 1997) are among the few such examples – these, unfortunately, have not been vigorously pursued.

These analyses of single (or a small cluster of) factors cannot be used for realistic assessment of the impact of agroforestry. Agroforestry systems have multiple objectives, products, and services; the success of a system is not judged by the performance of just one factor, but by the performance of a combination of several factors, products, and services, each of which is a complex issue in itself. For example, the extent of complexity involved in biophysical interactions is evident from the discussion on that topic included in this volume (Rao et al., 1997). Assessments that focus on only one or a few of the factors that affect agroforestry systems will be partial and inadequate. The problem is further compounded by the temporal (short- vs. long-term) nature of the benefits. We need a measure of the combined value of all products and services at various time scales. It is, indeed, a formidable task.

In short, it is essential that the development of appropriate methodologies that embody economic, social, and environmental costs and benefits of agroforestry should be a priority item in the agroforestry research agenda for the 21st century.

Policy environment

Development of appropriate biological technologies forms only one, perhaps the first, step in the promotion of agroforestry; provision of public policies and institutions that support program implementation is an equally important requirement. Analysis of the experience of several countries in Latin America has shown that rapid and dramatic expansion of agriculture and plantation forestry occurred when supportive public policy was put in place through appropriate subsidies, national investment, and adequate extension programs (Southgate, 1992). The Green Revolution in Asia was similar: it too occurred in places where national policies supported and facilitated the availability and adoption of the technologies.

Policy here refers to the rules and regulations of government (as opposed to societal and cultural norms) that bind the whole citizenry of a political system. Passing land-utilization laws, collecting levies and other forms of commodity-specific taxes, providing (or denying) access to reserved forests, and regulating import or export of agricultural commodities are all examples of state policies or interventions that affect the lives of citizens (Blair, 1991), as well as affect land-use patterns and sectoral development. The lack of institutional support for agroforestry is a major policy-related constraint to agroforestry development. A number of factors can be cited to explain this: the absence of multidisciplinary institutions capable of addressing integrated natural-resource and land-use problems, little institutional recognition that small farmers must allocate their few resources toward the production of a variety of goods and services, a shortage of institutions with the mandate or capability of allocating resources across institutional boundaries, and institutional interest in minimizing the influence of competing organizations (Follis and Nair, 1994).

Sanchez (1995) identified two broad kinds of policy research: policies to favor agroforestry's contribution to sustainable land-use, and those that aim to reverse existing perverse policies. He went on to identify the three major policy research issues in agroforestry, each encompassing the two broad kinds of policy research: land and tree tenure, assistance to farmers when trees in agroforestry systems are not productive, and valuation of environmental externalities. In addition to those, particular attention should be paid to improving the institutional structures governing marketing of agroforestry products and providing information and extension needs, as has strongly been indicated in a study on agroforestry adoption in Central America (Current et al., 1995).

An enabling policy environment that favors smallholder rural development is an essential condition for the success of agroforestry. This issue, however, has not been addressed with deserving seriousness in agroforestry research.

Spatial issues: the need for larger-scale perspectives

Agroforestry field research has so far focused on smaller spatial scales or research plots in experiment stations or farmers' fields. Less than 5% of the publications included in Table 1 are based on global, regional, or even watershed scales. Publications that appear to take a broader perspective usually address the issues in a conceptual generalized manner, with little data from empirical investigations. As a consequence, the importance of agroforestry in issues that need to be assessed on a much larger scale than individual fields or plots – for example, carbon sequestration, water-quality improvement, biological conservation and biodiversity, and extraction of nontimber forest products – has been neglected.

Carbon sequestration in terrestrial ecosystems is a good example of such larger-scale spatial issues. Agroforestry systems such as fuelwood plantations, shelterbelt/windbreak systems and woodlots may have the potential to sequester C, or offset fossil fuel emissions by substituting sustainably produced fuelwood and fodder (Brown et al., 1993; Shroeder et al., 1993). Based on a preliminary assessment of national and global terrestrial C sinks, Dixon (1995) argues that there are two primary beneficial attributes of agroforestry systems in this regard: (1) direct near-term C storage (decades to centuries) in trees and soils and (2) potential to offset immediate greenhouse gas emissions associated with deforestation and subsequent shifting cultivation. Shroeder (1994) estimated that median carbon storage by agroforestry practices was 9, 21, and 50 Mg C/ha in semiarid, subhumid, and humid ecozones, respectively. Dixon (1995) estimated that within tropical latitudes, one hectare of sustainable agroforestry can provide goods and services with the potential to offset 5 to 20 ha of deforestation. The impact of agroforestry on terrestrial C budget is an area that needs to be researched.

Other environmental benefits such as water-quality improvement through agroforestry have received no attention in agroforestry research in the tropics. Results from temperate-zone agroforestry point to the beneficial effects of riparian buffer strips in reducing sedimentation load and nutrient pollution in agricultural landscapes (Schultz et al., 1995). These may not be relevant to most of the tropics, at least for now. However, other potential benefits such as amelioration of salt-affected soils through agroforestry have been amply demonstrated (Tejwani, 1994) but not adequately promoted.

Conservation of biodiversity is another issue in which agroforestry may have a major role to play but which has received scant attention. It is obvious that agroforestry systems are far more diverse than crop or tree monocultures. Furthermore, it is often argued that the most important way in which agroforestry can help conserve plant and animal biodiversity lies in reducing tropical deforestation (Grainger, 1993; Vosti et al., 1997). It may not be prudent, however, to expect that having a high number of economically useful plants per hectare would lead to biodiversity conservation. Similarly, in commercial prospecting among natural sources for new products (such as the biodiversity prospecting, which is usually promoted not because of the realization of the value of nature's diversity for ecological, aesthetic, or spiritual reasons, but by the lure for the discovery of the miracle plant or animal that contains the cure for such human diseases as cancer or AIDS), the value of biodiversity is not as high as some conservationists might suppose (Simpson, 1997). Therefore, Apart from a few attempts to list the number of economically useful plants that are found in some indigenous agroforestry systems (for example, Nair, 1989; Tejwani, 1994; De Jong, 1996; Lawrence, 1996), or isolated studies such as mammalian diversity in coffee agroforestry systems (Gallina et al., 1996), agroforestry research has seldom attempted to establish the importance of agroforestry as a viable option for biodiversity conservation.

Conclusions

Agroforestry research should involve building bridges *from* the past *through* the present *to* the future (Figure 2). During the past two decades, we have progressed through the stages of inductive reasoning and formulation of hypotheses to deductive experimentation. We should now place greater emphasis on harnessing the *science* of agroforestry to develop technologies of wide applicability. It is not a question of choosing either the advancement of science or the development of technologies; one cannot proceed without the other. However, we should not lose sight of the fact that agroforestry research was initiated to help solve problems. Therefore, the question should be how to blend science and technology in an optimum balance in order to develop science-based technologies that will address the problems. Future generations will judge the merits of our efforts based on our success in accomplishing this important task of balancing science and technology. It is heartening to note that ICRAF, the world's premier tropical agroforestry research institution, envisions (ICRAF, 1997) that agroforestry research by the year 2020 will have accomplished the goals that we established when we started agroforestry research 20 years ago, and helped evolve agroforestry as a dynamic, development-focused, applied ecological science.

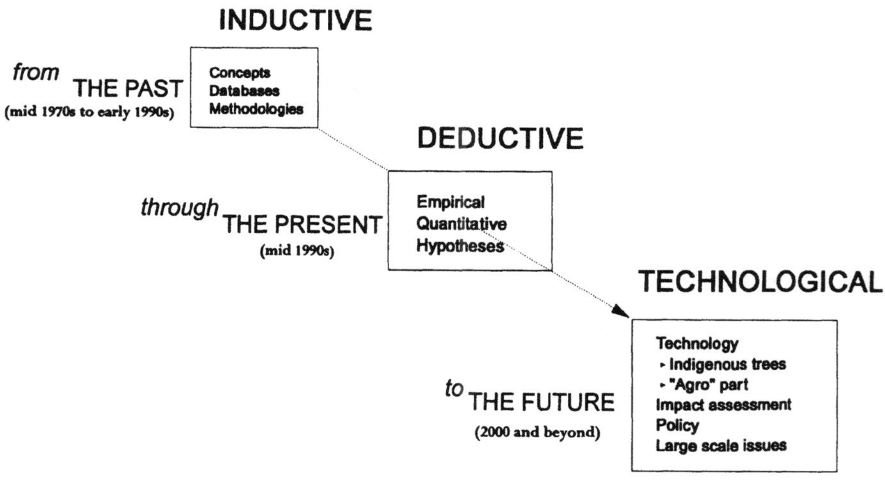

Fig. 2. Building bridges in tropical agroforestry research.

References

Avery ME, Cannell MGR and Ong CK (eds) (1990) Biophysical Research for Asian Agroforestry. Oxford Press, New Delhi, India

Beer JW, Muschler RG, Kass DCL and Somarriba EJ (1997) Shade management in coffee and cacao plantations. Agroforestry Systems 38: 139–164

Blair HW (1991) The uses of political science in agroforestry. In: Burch RB Jr and Parker JK (eds) Social Science Applications in Asian Agroforestry, pp. 85–109. Winrock International, Arlington, VA, USA

Brown S, Hall CAS, Knabe W, Raich J, Trexler MC and Woomer P (1993) Tropical forests: their past, present, and potential future role in the terrestrial carbon budget. Water, Air, and Soil Pollution 70: 71–94.

Cadish G and Giller KE (eds) (1997) Driven by Nature: Plant Litter Quality and Decomposition. CAB International, Wallingford, UK

Cannell MGR, van Noordwijk M and Ong CK (1996) The central agroforestry hypothesis: the trees must acquire resources that the crop would not otherwise acquire. Agroforestry Systems. 34: 27–31

Coe R (1994) Through the looking glass: 10 common problems in alley-cropping research. Agroforestry Today 6: 9–11

Cooper P and Buresh RJ (eds) Proceedings of the Improved Fallows Workshop. Agroforestry Systems (Special Issue) (In press)

Current D, Lutz E and Scherr S (eds) (1995) Costs, Benefits, and Farmer Adoption of Agroforestry. The World Bank, Washington DC, USA

De Jong W (1996) Swidden-fallow agroforestry in Amazonia: diversity at close distance. Agroforestry Systems 34: 277–290

Dixon RK (1995) Agroforestry Systems: sources or sinks of greenhouse gases? Agroforestry Systems. 31: 99–116

FAO (1995) FAOSTAT-PC, on diskette. Food and Agriculture Organization of the United Nations, Rome, Italy

FAO (1996) FAO Production Yearbook, 1995. FAO, Rome, Italy

Follis MB and Nair PKR (1994) Policy and institutional support for agroforestry: an analysis of two Ecuadorian case studies. Agroforestry Systems 27: 223–240

Gallina S, Mandujano S and Gonzales-Romero A (1996) Conservation of mammalian biodiversity in coffee plantations of Central Veracruz, Mexico. Agroforestry Systems 33: 13–27

Garrity DP (ed) (1997) Agroforestry innovations for imperata grassland rehabilitation. Agroforestry Systems 36 (Special Issue)

Grainger A (1993) Controlling Tropical Deforestation. Earthscan, London, UK

Gregory PJ (1996) Approaches to modelling the uptake of water and nutrients in agroforestry systems. Agroforestry Systems. 34: 51–65

ICRAF (1997) ICRAF Medium-Term Plan, 1998–2000. International Centre for Research in Agroforestry, Nairobi, Kenya

Johnson TR (1994) Agroforestry Performance Evaluation. MS Thesis, University of Florida, Gainesville, FL, USA

Kang BT (1993) Alley cropping: past achievements and future directions. Agroforestry Systems 23: 141–155

Kang BT, Reynolds L and Atta-Krah AN (1990) Alley farming. Advances in Agronomy 43: 315–359

Kass DCL (1996) Letter to the editor. Agroforestry Systems 33: 101–106

Kiepe P and Rao MR (1994) Management of agroforestry for the conservation and utilization of land and water resources. Outlook on Agriculture 23: 17–25

Lal R (1989) Agroforestry systems and soil surface management of a tropical alfisol. Parts I–VI. Agroforestry Systems 8(1): 1–6, 7–29; 8(2): 97–111, 113–132; 8(3): 197–215, 217–238, 239–242

Lawrence DC (1996) Trade-offs between rubber production and maintenance of diversity: the structure of rubber gardens in West Kalimantan, Indonesia. Agroforestry Systems 34: 83–100

Lawson GT, Crout NMJ, Levy PE, Mobbs DC, Wallace JS, Cannell MGR and Bradley GR (1995) The tree-crop interface: representation by coupling of forest and crop processmodels. Agroforestry Systems 30: 199–221

Leakey RRB and Newton AC (eds) (1994) Tropical Trees: Potential for Domestication and the Rebuilding of Forest Resources. HMSO, London, UK

Leakey RRB, Temu AB, Melnyk M and Vantomme P (eds) (1996) Domestication and Commercialization of Non-timber Forest Products in Agroofrestry Systems. Non-Wood Forest Products No. 9. FAO, Rome, Italy

Leakey RRB and Simons AJ (1997) The domestication and commercialization of indigenous trees in agroforestry for the alleviation of poverty. Agroforestry Systems 38: 165–176

Leakey RRB, Temu AB and Melnyk M. (eds) (1996) Domestication and Commercialization of Non-Timber Forest Products for Agroforestry. FAO (Non Wood Forest Products 9), Rome, Italy

MacDicken KG (1994) Selection and Management of Nitrogen-Fixing Trees. Winrock International, AK, USA

Menz KM and Grist P (1996) Increasing rubber planting density to shade Imperata: a bio-economic modelling approach. Agroforestry Systems 34: 291–303

Mercer DE and Miller RP (1997) Socioeconomic research in agroforestry: progress, prospects, priorities. Agroforestry Systems 38: 177–193

Morgan RPC (1995) Soil Erosion and Conservation, 2nd ed. Longman, London, UK

Nair PKR (ed) (1989) Agroforestry Systems in the Tropics. Kluwer, Dordrecht, The Netherlands

Nair PKR (1990) The Prospects for Agroforestry in the Tropics. Technical Paper No. 131, The World Bank, Washington, DC, USA

Nair PKR (1991) State-of-the-art of agroforestry systems. In: Jarvis PG (ed) Agroforestry: Principles and Practice, pp 5–29. Elsevier, Amsterdam, The Netherlands

Nair PKR (1993) An Introduction to Agroforestry. Kluwer, Dordrecht, The Netherlands

Nair PKR (1994) Agroforestry. Encyclopedia of Agricultural Sciences 1: 13–25. Academic Press, New York, NY, USA

Nair PKR (1996) Agroforestry directions and literature trends. In: McDonald P and Lassoie J (eds) The Literature of Forestry and Agroforestry, pp. 74–95. Cornell University Press, Ithaca, NY, USA

Okafor JC and Fernandes ECM (1987) Compound farms of southeastern Nigeria: a predominant agroforestry homegarden system with crops and livestock. Agroforestry Systems 5: 153–168

Ong CK (1994) Alley cropping – ecological pie in the sky? Agroforestry Today 6(3): 8–10

Ong CK, Black CR, Marshall FM and Corlett JE (1996) Principles of resource capture and utilization of light and water. In: Ong CK and Huxley PA (eds) Tree-Crop Interactions: A Physiological Approach, pp 73–158. CAB International, Wallingford, UK

Ong CK and Huxley PA (eds) (1996) Tree-Crop Interactions: A Physiological Approach. CAB International, Wallingford, UK

Raintree JB (1987) The state-of-the-art of agroforestry diagnosis and design. Agroforestry Systems. 5: 219–250

Rao MR and Coe R (1992) Evaluating the results of agroforestry research. Agroforestry Today 4(1): 4–9

Rao MR, Nair PKR and Ong CK (1997) Biophysical interactions in tropical agroforestry systems. Agroforestry Systems 38: 3–50

Sanchez PA (1995) Science in agroforestry. Agroforestry Systems 30: 5–55

Sanchez PA (1996) Letter to the editor. Agroforestry Systems 33: 106–108

Schroeder P (1993) Agroforestry systems: integrated land use to store and conserve carbon. Climate Research 3: 53–60

Schroeder P (1994) Carbon storage benefits of agroforestry systems. Agroforestry Systems 27: 89–97

Schultz RC, Colletti JP, Isenhart TM, Simpkins WW, Mize CW and Thompson ML (1995) Design and placement of a multi-species riparian buffer strip system. Agroforestry Systems 29: 201–226

Simpson RD (1997) Biodiversity prospecting: shopping the wilds is not the key to conservation. Resources for the Future 126 (Winter 1997): 12–15

Singh KD and Marzoli A (1995) Deforestation trends in the tropics: a time-series analysis. In: Conference on Potential Impact of Climate Change, San Juan, Puerto Rico, pp 8–9. FAO Rome, Italy

Smith NJH, Williams JT, Plucknett DL and Talbot JP (1992) Tropical Forests and Their Crops. Cornell University Press, Ithaca, USA

Southgate D (1992) Policies contributing to agricultural colonization of Latin America's tropical forests. In: Sharma NP (ed) Managing the World's Forests; Looking for Balance Between Conservation and Development, pp 215–235. Kendall/Hunt Pub, Dubuque, Iowa for the World Bank, Washington DC, USA

Tejwani KG (1994) Agroforestry in India. Oxford & IBH, New Delhi, India

Tejwani KG and Lai CK (1992) Asia-Pacific Agroforestry Profiles. FAO, Asia-Pacific Agroforestry Network Field Doc. 1. FAO, Bogor, Indonesia

Turnbull J (ed) (1987) Australian Acacias in Developing Countries, ACIAR Proceedings 16. Austr Cent for Int Agri Res, Canberra, Australia

United Nations (1996) 1994 Demographic Yearbook. UN, New York, USA

Van Noordwijk M, Lawson G, Soumare A, Groot JJR and Hairiah K (1996) Root distribution of trees and crops: competition and/or complementarity. In: Ong CK and Huxley PA (eds) Tree-Crop Interactions: A Physiological Approach, pp 319–364. CAB International, Wallingford, UK

Van Noordwijk M and Purnomosidhi P (1995) Root architecture in relation to tree-soil-crop interactions and shoot pruning in agroforestry. Agroforestry Systems 30: 161–173

von Carlowitz PG, Wolf GV and Kemperman REM (1991) Multipurpose Tree and Shrub Database: An Information and Decision Support System. ICRAF, Nairobi, Kenya & GTZ, Eschborn, Germany

Vosti SA, Witcover J, Oliveira S and Faminow M (1997) Policy issues in agroforestry: technology adoption and regional integration in the western Brazilian Amazon. Agroforestry Systems 38: 195–222

Webster's Dictionary (1997) Merriam Webster Inc., Springfield, MA, USA

WRI (1996) World Resources 1996–1997. World Resources Inst, Washington DC, USA

Young A (1989) Agroforestry for Soil Conservation. CAB International, Wallingford, UK

Young A (1997) Agroforestry for Soil Management. CAB International, Wallingford, UK

FORESTRY SCIENCES

1. P. Baas (ed.): *New Perspectives in Wood Anatomy.* Published on the Occasion of the 50th Anniversary of the International Association of Wood Anatomists. 1982 ISBN 90-247-2526-7
2. C.F.L. Prins (ed.): *Production, Marketing and Use of Finger-Jointed Sawnwood.* Proceedings of an International Seminar Organized by the Timber Committee of the UNECE (Halmar, Norway, 1980). 1982 ISBN 90-247-2569-0
3. R.A.A. Oldeman (ed.): *Tropical Hardwood Utilization.* Practice and Prospects. 1982 ISBN 90-247-2581-X
4. P. den Ouden (in collaboration with B.K. Boom): *Manual of Cultivated Conifers.* Hardy in the Cold- and Warm-Temperate Zone. 3rd ed., 1982 ISBN Hb 90-247-2148-2; Pb 90-247-2644-1
5. J.M. Bonga and D.J. Durzan (eds.): *Tissue Culture in Forestry.* 1982 ISBN 90-247-2660-3
6. T. Satoo: *Forest Biomass.* Rev. ed. by H.A.I. Madgwick. 1982 ISBN 90-247-2710-3
7. Tran Van Nao (ed.): *Forest Fire Prevention and Control.* Proceedings of an International Seminar Organized by the Timber Committee of the UNECE (Warsaw, Poland, 1981). 1982 ISBN 90-247-3050-3
8. J.J. Douglas: *A Re-Appraisal of Forestry Development in Developing Countries.* 1983 ISBN 90-247-2830-4
9. J.C. Gordon and C.T. Wheeler (eds.): *Biological Nitrogen Fixation in Forest Ecosystems.* Foundations and Applications. 1983 ISBN 90-247-2849-5
10. M. Németh: *Virus, Mycoplasma and Rickettsia Diseases of Fruit Trees.* Rev. (English) ed., 1986 ISBN 90-247-2868-1
11. M.L. Duryea and T.D. Landis (eds.): *Forest Nursery Manual.* Production of Bareroot Seedlings. 1984; 2nd printing 1987 ISBN Hb 90-247-2913-0; Pb 90-247-2914-9
12. F.C. Hummel: *Forest Policy.* A Contribution to Resource Development. 1984 ISBN 90-247-2883-5
13. P.D. Manion (ed.): *Scleroderris Canker of Conifers.* Proceedings of an International Symposium on Scleroderris Canker of Conifers (Syracuse, USA, 1983). 1984 ISBN 90-247-2912-2
14. M.L. Duryea and G.N. Brown (eds.): *Seedling Physiology and Reforestation Success.* Proceedings of the Physiology Working Group, Technical Session, Society of American Foresters National Convention (Portland, Oregon, USA, 1983). 1984 ISBN 90-247-2949-1
15. K.A.G. Staaf and N.A. Wiksten (eds.): *Tree Harvesting Techniques.* 1984 ISBN 90-247-2994-7
16. J.D. Boyd: *Biophysical Control of Microfibril Orientation in Plant Cell Walls.* Aquatic and Terrestrial Plants Including Trees. 1985 ISBN 90-247-3101-1
17. W.P.K. Findlay (ed.): *Preservation of Timber in the Tropics.* 1985 ISBN 90-247-3112-7
18. I. Samset: *Winch and Cable Systems.* 1985 ISBN 90-247-3205-0
19. R.A. Leary: *Interaction Theory in Forest Ecology and Management.* 1985 ISBN 90-247-3220-4

FORESTRY SCIENCES

20. S.P. Gessel (ed.): *Forest Site and Productivity.* 1986 ISBN 90-247-3284-0
21. T.C. Hennessey, P.M. Dougherty, S.V. Kossuth and J.D. Johnson (eds.): *Stress Physiology and Forest Productivity.* Proceedings of the Physiology Working Group, Technical Session, Society of American Foresters National Convention (Fort Collins, Colorado, USA, 1985). 1986 ISBN 90-247-3359-6
22. K.R. Shepherd: *Plantation Silviculture.* 1986 ISBN 90-247-3379-0
23. S. Sohlberg and V.E. Sokolov (eds.): *Practical Application of Remote Sensing in Forestry.* Proceedings of a Seminar on the Practical Application of Remote Sensing in Forestry (Jönköping, Sweden, 1985). 1986 ISBN 90-247-3392-8
24. J.M. Bonga and D.J. Durzan (eds.): *Cell and Tissue Culure in Forestry.* Volume 1: General Principles and Biotechnology. 1987 ISBN 90-247-3430-4
25. J.M. Bonga and D.J. Durzan (eds.): *Cell and Tissue Culure in Forestry.* Volume 2: Specific Principles and Methods: Growth and Development. 1987 ISBN 90-247-3431-2
26. J.M. Bonga and D.J. Durzan (eds.): *Cell and Tissue Culure in Forestry.* Volume 3: Case Histories: Gymnosperms, Angiosperms and Palms. 1987 ISBN 90-247-3432-0
    Set ISBN (Volumes 24-26) 90-247-3433-9
27. E.G. Richards (ed.): *Forestry and the Forest Industries: Past and Future.* Major Developments in the Forest and Forest Industries Sector Since 1947 in Europe, the USSR and North America. In Commemoration of the 40th Anniversary of the Timber Committee of the UNECE. 1987 ISBN 90-247-3592-0
28. S.V. Kossuth and S.D. Ross (eds.): *Hormonal Control of Tree Growth.* Proceedings of the Physiology Working Group, Technical Session, Society of American Foresters National Convention (Birmingham, Alabama, USA, 1986). 1987 ISBN 90-247-3621-8
29. U. Sundberg and C.R. Silversides: *Operational Efficiency in Forestry.* Vol. 1: Analysis. 1988 ISBN 90-247-3683-8
30. M.R. Ahuja (ed.): *Somatic Cell Genetics of Woody Plants.* Proceedings of the IUFRO Working Party S2.04-07 Somatic Cell Genetics (Grosshansdorf, Germany, 1987). 1988. ISBN 90-247-3728-1
31. P.K.R. Nair (ed.): *Agroforestry Systems in the Tropics.* 1989 ISBN 90-247-3790-7
32. C.R. Silversides and U. Sundberg: *Operational Efficiency in Forestry.* Vol. 2: Practice. 1989 ISBN 0-7923-0063-7
    Set ISBN (Volumes 29 and 32) 90-247-3684-6
33. T.L. White and G.R. Hodge (eds.): *Predicting Breeding Values with Applications in Forest Tree Improvement.* 1989 ISBN 0-7923-0460-8
34. H.J. Welch: *The Conifer Manual.* Volume 1. 1991 ISBN 0-7923-0616-3
35. P.K.R. Nair, H.L. Gholz, M.L. Duryea (eds.): *Agroforestry Education and Training.* Present and Future. 1990 ISBN 0-7923-0864-6
36. M.L. Duryea and P.M. Dougherty (eds.): *Forest Regeneration Manual.* 1991 ISBN 0-7923-0960-X
37. J.J.A. Janssen: *Mechanical Properties of Bamboo.* 1991 ISBN 0-7923-1260-0
38. J.M. Bonga and P. Von Aderkas: *In Vitro Culture of Trees.* 1992 ISBN 0-7923-1540-5

FORESTRY SCIENCES

39. L. Fins, S.T. Friedman and J.V. Brotschol (eds.): *Handbook of Quantitative Forest Genetics.* 1992 ISBN 0-7923-1568-5
40. M.J. Kelty, B.C. Larson and C.D. Oliver (eds.): *The Ecology and Silviculture of Mixed-Species Forests.* A Festschrift for David M. Smith. 1992 ISBN 0-7923-1643-6
41. M.R. Ahuja (ed.): *Micropropagation of Woody Plants.* 1992 ISBN 0-7923-1807-2
42. W.T. Adams, S.H. Strauss, D.L. Copes and A.R. Griffin (eds.): *Population Genetics of Forest Trees.* Proceedings of an International Symposium (Corvallis, Oregon, USA, 1990). 1992 ISBN 0-7923-1857-9
43. R.T. Prinsley (ed.): *The Role of Trees in Sustainable Agriculture.* 1993 ISBN 0-7923-2030-1
44. S.M. Jain, P.K. Gupta and R.J. Newton (eds.): *Somatic Embryogenesis in Woody Plants,* Vol. 3: Gymnosperms. 1995 ISBN 0-7923-2938-4
45. S.M. Jain, P.K. Gupta and R.J. Newton (eds.): *Somatic Embryogenesis in Woody Plants,* Vol. 1: History, Molecular and Biochemical Aspects, and Applications. 1995 ISBN 0-7923-3035-8
46. S.M. Jain, P.K Gupta and R.J. Newton (eds.): *Somatic Embryogenesis in Woody Plants,* Vol. 2: Angiosperms. 1995 ISBN 0-7923-3070-6
    Set ISBN (Volumes 44-46) 0-7923-2939-2
47. F.L. Sinclair (ed.): *Agroforestry: Science, Policy and Practice.* Selected Papers from the Agroforestry Sessions of the IUFRO 20th World Congress (Tampere, Finland, 6–12 August 1995). 1995 ISBN 0-7923-3696-8
48. J.H. Goldammer and V.V. Furyaev (eds.): *Fire in Ecosystems of Boreal Eurasia.* 1996 ISBN 0-7923-4137-6
49. M.R. Ahuja, W. Boerjan and D.B. Neale (eds.): *Somatic Cell Genetics and Molecular Genetics of Trees.* 1996 ISBN 0-7923-4179-1
50. H.L. Gholz, K. Nakane and H. Shimoda (eds.): *The Use of Remote Sensing in the Modeling of Forest Productivity.* 1996 ISBN 0-7923-4278-X
51. P. Bachmann, M. Köhl and R. Päivinen (eds.): *Assessment of Biodiversity for Improved Forest Planning.* Proceedings of the Conference on Assessment of Biodiversity for Improved Planning (Monte Verità, Switzerland, 1996). 1998 ISBN 0-7923-4872-9
52. G.M.J. Mohren, K. Kramer and S. Sabaté (eds.): *Impacts of Global Change on Tree Physiology and Forest Ecosystems.* Proceedings of the International Conference on Impacts of Global Change on Tree Physiology and Forest Ecosystems (Wageningen, The Netherlands, 26–29 November 1996). 1997 ISBN 0-7923-4921-0
53. P.K.R. Nair and C.R. Latt (eds.): *Directions in Tropical Agroforestry Research.* Selected Papers from the Symposium on Tropical Agroforestry organized in connection with the annual meeting of ASA (Indianapolis, U.S.A., 5 November 1996). 1998 ISBN 0-7923-5035-9

KLUWER ACADEMIC PUBLISHERS – DORDRECHT / BOSTON / LONDON

MIX
Papier aus verantwortungsvollen Quellen
Paper from responsible sources
FSC® C105338

If you have any concerns about our products,
you can contact us on
ProductSafety@springernature.com

In case Publisher is established outside the EU,
the EU authorized representative is:
Springer Nature Customer Service Center GmbH
Europaplatz 3, 69115 Heidelberg, Germany

Printed by Libri Plureos GmbH
in Hamburg, Germany